Insecticide and Fungicide Handbook

Insecticide and Fungicide Handbook

for crop protection

Issued by the
British Crop Protection
Council
and edited by
Hubert Martin DSc, ARCS, FRIC
and
Charles R. Worthing BSc, MA, DPhil

Fifth edition

Blackwell Scientific Publications

OXFORD LONDON EDINBURGH MELBOURNE

© 1976 Blackwell Scientific Publications
Osney Mead, Oxford
8 John Street, London, WC 1
9 Forrest Road, Edinburgh
P.O. Box 9, North Balwyn, Victoria, Australia

FIRST PUBLISHED 1963
SECOND EDITION 1965
THIRD EDITION 1969
FOURTH EDITION 1972
FIFTH EDITION 1976

Insecticide and fungicide handbook
for crop protection.–5th ed.
Bibl.–Index.
ISBN 0-632-00109-7
1. Title 2. British Crop Protection Council
3. Martin, Hubert 4. Worthing, Charles
Ronald
632'.951
632'.952   SB951.5
Insecticides

Distributed in the
United States of America by
J. B. Lippincott Company, Philadelphia
and in Canada by
J. B. Lippincott Company of
Canada Ltd, Toronto

Printed in Great Britain by
The Whitefriars Press Ltd
London and Tonbridge
and bound by
Mansell (Bookbinders) Ltd
Witham, Essex

Special note

Every effort has been made to ensure that the recommendations and statements made in this Handbook are correct but the British Crop Protection Council cannot accept responsibility for any loss, damage or other accident arising from carrying out the methods advocated in the Handbook.

Nothing in this Handbook shall be taken as a warranty that any substance or mixture of substances mentioned herein is not the subject of patent rights and the Council does not hold itself responsible for any infringement of said rights.

Member Organisations

BCPC BRITISH CROP PROTECTION COUNCIL

Association of Applied Biologists .

Agricultural Research Council

British Agrochemicals Association

British Association of Grain, Seed, Feed & Agricultural Merchants

Department of Agriculture & Fisheries for Scotland

Department of Agriculture, Northern Ireland

Department of the Environment

Ministry of Agriculture, Fisheries and Food, Pesticides Branch

Ministry of Agriculture, Fisheries and Food, Agricultural Development and Advisory Service

Ministry of Agriculture, Fisheries and Food, Agricultural Development and Advisory Service, Plant Pathology Laboratory

Ministry of Overseas Development

National Association of Agricultural Contractors

National Farmers' Union

Natural Environment Research Council

Society of Chemical Industry, Pesticides Group

Further information may be obtained from

The Secretary,
British Crop Protection Council,
160 Great Portland Street,
London W 1N 6DT

Contents

11   Pests and Diseases of Vegetable Crops

12   Pest and Disease Control in Outdoor Ornamentals and in Turf

13   Pests and Diseases of Forest Crops

Contributors

Foreword to first edition

It is now generally accepted that, in the present state of knowledge, pesticides are indispensable for food production. By definition, their use is detrimental to certain forms of life—the pest and pathogen—and the danger is that this toxicity may be exerted in unwanted directions, to the detriment of the user, the consumer and to the biological environment in which the materials are used. These hazards are scrutinized by appropriate authorities who report to the Ministers concerned. If necessary, safeguards are put into effect, either by Statutory Regulations or by advisory means. The adequacy of these safeguards is obviously dependent on a strict adherence, by the user, to the rules and recommendations proposed. The main purpose of this Handbook is to place before the grower and adviser the information necessary for the correct use of pesticides in crop protection.

The chemist has been successful, in recent years, in finding new compounds of more selective toxicity and, yearly, the list of hazardous pesticides is being reduced by the introduction of less noxious alternatives. A second purpose of this Handbook is to record this progress and to advise the grower on better methods of crop protection. In some instances, the new methods have been described in advance of their inclusion in official recommendations; if so, they must be regarded as tentative and for the information of the keen grower.

The material of the Handbook has been assembled by the Recommendations Committee of the British Insecticide and Fungicide Council, with the unstinted help of colleagues within the National Agricultural Advisory Service or at the state-aided Research Stations. In this first edition it has been found necessary, for reasons of space, to limit the text to the subject of crop protection, but the inclusion, in future editions, of farm storage may become possible. Suggestions for improvement, and correction of any errors or omissions, would be most welcome and should be sent to the Secretary of the Recommendations Committee, Lenton Experimental Station, Lenton House, Nottingham.

Every effort has been made to ensure that the recommendations are correct, but the British Insecticide and Fungicide Council cannot accept responsibility for any loss, damage or accident arising from carrying out the methods advocated in this Handbook.

Ministry of Agriculture, Fisheries　　　　　　　　H.G. Sanders MA, PhD
and Food
July 1963

Notes for readers

In addition to commoner abbreviations, the following have been used:

A.D.A.S.	M.A.F.F. Agricultural Development and Advisory Service
a.i.	active ingredient
A.L.	Advisory Leaflet (A.D.A.S.)
BOV	brown oil of vitriol
BSI	British Standards Institution
ct	concentration × time product
D.A.F.S.	Department of Agriculture and Fisheries for Scotland
e.c.	emulsifiable concentrate
F.C.L.	Forestry Commission Leaflet
HV	High Volume (the use of over 1000 litres/ha of spray on bushes and trees; of over 700 litres/ha on ground crops)
LV	Low Volume (the use of 200–500 litres/ha of spray on bushes and trees; of 50–200 litres/ha on ground crops
M.A.F.F.	Ministry of Agriculture, Fisheries and Food
MV	Medium Volume (see HV and LV)
N.I.A.B.	National Institute of Agricultural Botany
rev/min.	revolutions per minute
s.c.	soluble concentrate
s.e.	stock emulsion
STL	Short-term Leaflet (M.A.F.F.)
v/v	volume per volume
w.g.	water gauge
W.H.O.	World Health Organization
w.p.	wettable powder
w/v	weight per volume
w/w	weight per weight

In the description of the diseases and pests of crop plants, the significance of each is indicated by asterisks thus:

*** of major importance
** of local importance
* of minor importance

Chapter 1
The biological background

Agricultural and horticultural practices aim to establish reasonably uniform crops over relatively restricted areas. These conditions provide a potentially favourable environment for the development of pest and disease epidemics. But many other factors play their part both in determining the likelihood of an attack and the rate at which the infestation or infection may increase and become of economic significance. The influence of these factors depends to a large extent on the biology of the insects and disease organisms concerned, and varies greatly from one to another. An appreciation of the interreactions of these factors is essential if control measures are to be effective. The various recommendations given in later chapters of this book have been determined with the biological background of the relevant pest or disease constantly in mind.

The approach to pest and disease problems is essentially different, hence this chapter has been divided into two parts. When insects become pests on crop plants, they can be seen and destroyed: if an attack cannot be entirely prevented it can often be cured. On the other hand a disease only becomes apparent when the symptoms are seen on the plant, and only much later does any sign of the pathogen itself appear: prophylactic methods of control are therefore necessary and prevention rather than cure is a more practical possibility.

Pests

1.1 Insect life histories

Most insect species lay eggs although, in certain groups, living young may be produced at some stage in the life cycle. The newly hatched insect is outwardly different from the adult and develops into the adult form by a series of moults during each of which the old, tight skin is cast off, leaving the insect with a new skin previously secreted below the old one.

There are two major divisions in insect classification depending on the degree of transition which occurs between the immature and mature forms. In the first, the newly hatched insect is similar to the adult but has undeveloped reproductive organs and wings. Immature stages are known as nymphs and increase in size by a number of moults until they become adults. The new skin is soft but soon hardens. Growth in size therefore occurs only on moulting. During this period of growth the genitalia and wings develop, becoming functional when the insect reaches maturity. Aphids typify this form of development and adults and nymphs at all stages can often be seen on an infested plant. The other type of transition occurs in insects such as moths and butterflies, beetles, sawflies and two-winged flies in which the juvenile form is a larva that differs fundamentally from the adult, usually having different mouthparts and a totally different feeding habit. When the larva is fully grown it changes to the adult form by passing through an outwardly quiescent but internally active stage—the pupa.

In the first group, the insects which feed on plants do so in both the adult and nymphal stages. For example, aphids at all stages of growth may be found feeding on apple, plum, soft fruit, beans, brassicas, lettuce and cereals. Capsids feeding on fruit trees, thrips on peas and whitefly on glasshouse crops are also included in this category.

In the second group, most plant destruction and damage occurs as a result of feeding by the larval stages. Caterpillars of moths, butterflies and sawflies are voracious feeders: larvae of codling moth and apple sawfly feed inside apples, pea moth caterpillars eat peas within the pods, cabbage caterpillars consume brassica leaves, and cutworms (larvae of certain noctuid moths) feed on the roots and shoot bases of various plants. Among the beetles, raspberry beetle larvae feed on the developing fruits of raspberries and loganberries, stem weevil larvae tunnel in the stems of brassicas, wireworms (larval click beetles) attack the underground parts of such plants as cereals, potatoes and sugar beet, chafer grubs feed on plant roots particularly those of grass, and grain weevil larvae destroy individual cereal grains by feeding within them. Among the two-winged flies, pear midge larvae infest pear fruitlets, larvae of mangold fly mine the leaves of sugar beet plants, wheat bulb fly larvae and frit fly larvae attack the shoots of cereals, larvae of onion fly and narcissus flies feed inside the bulbs of their host plants, larvae of cabbage root fly and carrot fly feed on the roots of appropriate vegetables, and leatherjackets (larval crane flies) are general root feeders. Apart from the plant-damaging larvae, the second group of insects also includes adults which feed on plant tissue. Adult weevils of different species feed on pea, bean and clover leaves and the bark of young trees; flea beetle adults eat holes in the leaves and stems of seedling brassica and other crops; pygmy beetles attack the leaves, stem and roots of beet and mangolds; and adults of the strawberry seed beetle eat the seeds from strawberry fruits. Worker wasps can cause damage to apples, pears, plums and autumn strawberries by feeding on the ripening fruits.

Of the other creatures that are allied to insects, those which damage plant growth include millepedes, symphylids and spider mites; slug and eelworm damage also falls within the province of the entomologist. Millepedes at all stages in the life cycle feed on vegetable matter such as roots, bulbs and tubers and they are often abundant at the sites of primary damage caused by other pests such as slugs. Symphylids most commonly attack glasshouse crops such as tomatoes, cucumbers and lettuce, feeding on the root systems until the plants wilt and die. Red spider mites are most destructive, feeding on glasshouse crops and also attacking outdoor crops especially when the weather is hot and dry; the mites suck sap from the leaves causing them to wither and die and sometimes spin webs over the foliage.

1.2   Destructive insects

The foregoing paragraphs show that insect damage to crops is mainly associated with feeding; reference must now be made to various types of feeding. Some insects feed by chewing plant tissue and others by sucking sap from the plant. Secondary damage occurs when waste products such as excreta and honeydew spoil the produce.

(a) **Chewing insects** damage crops in many ways according to the form of the mouthparts and the part of the plant attacked. *Leaf eaters* include various caterpillars such as those of cabbage white butterflies and gooseberry sawfly which cut away and eat sections of leaf tissue. Adult weevils make U-shaped notches in the margins of pea, bean and clover leaves, often considerably reducing the leaf area. *Miners* such as larvae of celery fly and mangold fly tunnel and feed between the upper and lower surfaces of the leaf; damage is most serious when the plant is attacked in the seedling stage. On a crop of chrysanthemums, leaves may be so badly disfigured by the mining of the chrysanthemum leaf miner that they have to be removed before the flowers are sold. *Stem borers* such as the larvae of frit fly and wheat bulb fly, which have mouthparts reduced to a pair of hook-like organs, rasp away the internal conducting tissue of cereal plants. Shoots are frequently killed in this way and the crop loss is extensive. *Root eaters* diminish the vigour of the plants by eating and severing the roots. Wireworms, cutworms and leatherjackets occur generally on farm and market garden crops; weevil larvae also attack the roots of a wide range of crops, while pea and bean weevil larvae destroy root nodules. Some insects such as swede midge lay their eggs near the growing point of their host plants and subsequent feeding by the larvae may destroy the shoot, causing such damage as blindness in cauliflowers. Several insects attack flower buds: blossom weevils lay their eggs in the flower buds of apples and strawberries causing the buds to wither and die, while blossom beetles destroy the flower buds of brassicas. Fruits are infested and rendered unmarketable by insects such as codling moth and sawfly caterpillars, whereas seeds are eaten by others such as strawberry seed beetle and cabbage seed weevil. Although insect damage to a plant may be relatively insignificant in itself, the damaged part of the plant often allows entry of disease organisms, which cause the plant to decay or rot.

(b) **Sucking insects** have mouthparts adapted for piercing plant tissue and sucking the sap. Within the needle-like proboscis, or beak, are two fine tubes through one of which saliva from the insect passes into the plant and, through the other, sap from the plant passes into the insect. Examples of sucking insects are capsids, aphids, suckers, scale insects and whiteflies.

Aphids attack a wide range of plants and in temperate areas of the world they cause great damage. Their capacity for rapid build-up on crops is brought about by the fact that they produce living young, which results in a telescoping of generations and enables them to produce more generations in a year than exclusively egg-laying insects. Apart from the debilitating effect on the plant of large numbers of aphids feeding on the sap, otherwise profitable parts of the plant are often rendered unsaleable because of physical deformation or fouling by cast skins and honeydew. The latter is made up of the unused part of the sap which has passed through the aphid body and been ejected. Moulds grow freely on these sugary excreta which then become even more objectionable.

Although the entry of fungi and bacteria into the punctures made by sucking insects can cause secondary damage, the greatest potential danger lies in the transmission of viruses by aphids to the plants. Viruses such as raspberry mosaic, sugar beet yellows, barley yellow dwarf and those which attack brassicas and potatoes are spread within and between crops by winged aphids. Virologists recognize two main types of virus— the persistent and the non-persistent. To acquire and transmit a persistent virus, an aphid must feed on an infected plant. The sap containing the virus passes from the plant into the insect gut. From here the virus migrates to the salivary glands from where it can be injected with the saliva into a healthy plant when the insect is feeding. The migration of the virus from the gut to the salivary glands of the insect may take hours or even days. On the other hand, non-persistent viruses are easily acquired by the aphid while merely probing within the epidermal cells of the plant, a common procedure before feeding commences. The virus is then carried on the mouthparts and is thus readily transmitted from one site to another in a few minutes.

The general effect of virus infection is to reduce the growth and yield of the plant and to reduce the vigour of the vegetative offspring.

1.3   Useful insects

(a) **Pollinators.** Flower-visiting insects, including hive bees, are of great importance in the pollination of seed-producing crops and fruit. The use of insecticides when bees are working or overflying the crop can seriously decrease their numbers and possibly the yield of the crop. Hazards to bees can be greatly reduced if insecticides are applied at times when bees are not foraging or when the crop and the plants within it are not attractive to them. Although it is usually safe for bees to forage in an area 48 hours after application of an insecticide, some materials which are labelled as dangerous to bees may be residually toxic for longer periods. Different formulations of the same insecticide may vary in their toxicity to bees: for example, granular formulations of systemic insecticides are generally very much safer than sprays where bees are working the crop.

(b) **Parasites and predators.** Insects become pests usually because they grow, feed and breed quickly on their host plants. Numbers usually fluctuate, a decline being generally associated with unfavourable weather such as low temperatures or heavy rain for a prolonged period, with a dwindling supply of food as the crop is progressively destroyed or becomes too small to support the rapidly increasing numbers of insects, or with the activities of insect enemies. The latter may be predators of the pest, using it as a source of food, or parasites which use the pest insect as food and shelter for their developing young. The interrelationships between pests and their enemies are complex and vary from one example to another. In general, however, as the pests increase in density, the available enemies are able to find them more quickly, with the result that the populations of parasites and predators increase. Eventually there are so many individuals searching for prey that the pest numbers decline and shortly afterwards the parasites and predators are themselves restricted for want of food. Thus, many pest fluctuations may be due to the effect of their enemies, although it must be noted that a pest insect can often reach epidemic proportions before its enemies are sufficiently numerous to effect any worthwhile degree of control.

A good example of the value of predatory and parasitic insects can be seen in almost any large colony of aphids. Observation will show that ladybirds of various species eat these pests and also lay their eggs on the infested plants; when the larvae hatch they thrive on the ample food supply with which they are surrounded. Hover fly adults feed on nectar but lay their eggs on aphid-infested plants; the larvae move over the plant searching for aphids and may eat 400–500 before they pupate. Larvae of certain midges and lacewings are also commonly found feeding on aphid colonies. In addition, minute wasps oviposit into aphids and the larvae feed on, and then pupate within, the host body. Parasitized aphids are easily recognized for they are swollen, shiny brown and attached to the plant by silk produced by the developing parasite. This is not the whole story for there are 'lesser fleas'—hyperparasites which lay their eggs on the developing parasite larvae within dead aphids. These are obviously not useful insects as they destroy the potentially valuable parasites.

1.4   Control measures

Various types of control will now be mentioned. Any animal which destroys part of a pest population is exerting some measure of *biological control*; an animal which is a pest in one region may be prevented from establishing itself in another by *legislative action*; pests can be influenced by the *cultural operations* practised by the grower; and, finally, if a population does increase to the level at which it is causing economic damage, it can usually be controlled by *chemical means*.

(a) **Biological control.** There are examples of serious pest problems having been solved by the introduction of parasites and predators, but these are relatively rare (about 20 cases in the last 50 years). In Britain, considerable attention is being given to the biological control of pests on crops which are grown in glasshouses. For example, the glasshouse red spider mite, which is common on cucumbers, tomatoes and chrysanthemums, can be successfully controlled by the imported predacious mite *Phytoseiulus persimilis*. This predator has a shorter life cycle than that of the spider mite and is very efficient at finding its prey. However, its survival is dependent on the maintenance of a low level spider mite population; this means that it is sometimes necessary to infest the plants artificially with the pest as well as to introduce the predator. Another pest, the whitefly found on plants such as cucumbers and tomatoes, may be controlled by the introduction of the parasite *Encarsia formosa*, which attacks the immature stages of the whitefly. The parasite completes its life cycle in about 5 weeks and each female may lay up to 50 eggs, so that a severe whitefly infestation can be controlled in a few months. To ensure that the parasite does not die out, parasitized whiteflies are periodically introduced into the glasshouse.

More details of these and other examples of biological control can be found in MAFF Bulletin 20, 'Beneficial Insects and Mites'.

(b) **Legislative control.** The Plant Health Act, 1967, which replaced the Destructive Insects and Pests Acts, 1877–1927, forms the basis of plant health legislation in Britain, the object being to prevent the introduction or spread of the more important foreign pests and diseases and to minimize the planting of diseased material. Two examples will suffice to illustrate this aspect of control.

Colorado beetle, which spread from the USA into Canada and thence to Europe in 1921, is not established in Britain. Despite stringent precautions at all sea- and airports, there is a potential danger of the importation of plants carrying larvae, pupae or adult beetles. Anyone finding this pest is required to inform the Ministry of Agriculture; if the presence of the insect is confirmed, control measures are undertaken at public expense, the Ministry's policy being to eradicate the infestation. During the years 1941–52, a number of breeding colonies of beetles were found and subsequently exterminated. Since then, there have been no cases of infestations developing in the field, although each year live beetles are intercepted on imported produce and products.

Beet cyst eelworm attacks sugar beet and can cause severe crop losses. To prevent this happening on a large scale, the growing of susceptible crops on land known to be infested is regulated under the Beet Eelworm Orders, 1960 and 1962, and the sale of plants grown on infested land is forbidden. A proper crop rotation does much to prevent a rapid build-up of this pest and is the best possible insurance against trouble on farms where sugar beet is grown extensively.

(c) **Cultural control.** Crop rotations, normal farm management, hygiene and the use of healthy planting material do much to prevent unnecessary losses due to insect and eelworm pests. The time of sowing the crop can sometimes be adjusted to avoid or reduce attack by certain pests. Practices such as manuring and irrigation, which assist plants to grow strongly and healthily, also help to prevent undue damage by pests.

The rotation of crops is a normal practice and is essential to prevent the build-up of eelworms which attack crops such as potatoes, cereals and bulbs. Changing an accepted sequence of cropping is often limited by practical considerations, but where heavy infestations of wheat bulb fly occur regularly, wheat should not be grown after a crop which provides the uncovered earth necessary in July–September for egg laying by the adult flies.

Cultivation of the soil at certain times of the year exposes many soil-living insects to predation by birds. Thus, populations of leatherjackets and wireworms can be reduced by birds following the plough in autumn and spring. Using the roller on crops attacked by wheat bulb fly restricts the movement of larvae through the soil in seach of host plants, may crush larvae within the stems and encourages tillering.

Badly drained fields provide favourable habitats for slugs, especially when the humus content of the soil is high. Damage consists in the hollowing of cereal grains a short time after sowing, holing of plant stems, shredding of leaves, and the formation of irregular holes in tubers and roots. As these pests are difficult to control chemically, breeding sites should be reduced and conditions made less desirable, for example by drainage.

Attention to the sowing date of a crop will often ensure that plants are not in a susceptible stage when a particular pest is ready to lay eggs. In eastern England where carrot fly is common, risk of infestation can be lessened by delaying sowing until the end of May. The crop will largely escape damage by the first generation, and the second generation of flies will be correspondingly smaller.

At the other end of the season, crops should not be left in the ground when they have completed their growth if they are susceptible to pest attack. For example, wireworms attack mature potatoes causing a serious decline in quality. As this damage increases the longer the crop is left in the soil, lifting should begin as soon as possible after maturity is reached.

(d) **Chemical control.** There can be little doubt that the use of chemicals for the control of insect pests is of primary importance in agriculture; later chapters in this handbook give details of the wide variety of crop pests which can be controlled by chemical means.

A competent grower will minimize the risk of pest attack as far as possible by using cultural control measures. Resort to chemical control means that additional money must be spent on the crop, but this may mean the difference between an economic return and a loss.

Earliest records of insect control refer to the use of materials which were so noxious to humans that their effects on insects were unquestioned. By the second half of the nineteenth century, substances were being used which are still in use today: arsenical compounds were proving reliable against Colorado beetle and gypsy moth in the United States of America and nicotine, which had been prepared for years as a simple infusion of tobacco leaves, effectively controlled aphids, thrips and mites. In the 1920s pyrethrum, produced from the powdered flowers of the pyrethrum plant, entered the insecticide market for the first time; it was the principal insecticide of medical importance during World War II until 1942, when the use of DDT became widespread. However, pyrethrum, because of its rapid paralysing effect on insects and its relative harmlessness to mammals, is still extensively used, especially in fly sprays. The other principal insecticide available before 1939 was derris, the ground root of species of the genera *Derris* and *Lonchocarpus*.

Since 1945 the number of synthetic organic insecticides available for agricultural and horticultural use has increased enormously; the 3 main groups which we shall consider are the chlorinated hydrocarbons such as DDT. the organophosphorus insecticides such as parathion, and the carbamates such as carbaryl. DDT, probably the best-known of the chlorinated hydrocarbons, was developed during the early 1940s and used mainly against disease-carrying insects. It was immediately successful because of its effectiveness at very low concentrations and its usefulness against a wide range of insects. Its introduction was rapidly followed by the discovery of the high insecticidal activity of such compounds as gamma-HCH, aldrin, dieldrin and others of the chlorinated hydrocarbon group.

The general action of the chlorinated hydrocarbons depends on the insects eating treated vegetation or walking over a deposit of the insecticide applied to the plant or other surface. To be effective, careful application is required to ensure that the feeding site, or the area over which the insect will walk, is covered with insecticide.

A large number of phosphorus compounds are used as insecticides. They are inhibitors of enzymes which occur in insects, particularly in the tissues of the nervous system, and generally have a very rapid action. They are less persistent than chlorinated hydrocarbon insecticides and, therefore, do not create the same problem of residues which remain active for an unnecessarily long time in the plant or the soil.

Several organophosphorus insecticides are systemic, that is they are absorbed into the sap stream of the plant. The insecticide may be applied to the soil, from which it is taken up by the plant roots, or by spraying the plant, when it is absorbed through the leaves. Sap-feeding insects, for example aphids, are more readily killed by systemic insecticides than by those which have a contact action; their parasites and predators are not affected if the plants are not sprayed.

The first carbamate insecticide, carbaryl, was introduced in the mid-1950s. It is effective against many of the insects controlled by DDT but is less persistent. Several other carbamates have since come on to the market for the control of a variety of pests; these compounds include methiocarb which is used mainly against slugs and snails.

Insecticides belonging to many other chemical groups have now been synthesized with the result that the range of physical and chemical properties of insecticides is very wide. They also vary greatly in their toxicity to man. Suitable formulation of those insecticides having a high mammalian toxicity can often make them less dangerous for purposes of transport and application without reducing their effect on the pest.

One of the drawbacks of the chemical approach to pest control is the general effect on the whole fauna of the plant and soil. Although pest insects are destroyed, so too are many parasites and predators living within the area of application of the insecticide. This effect was known when tar oil washes for fruit trees were first introduced, but another unforeseen result soon produced a problem as great as the one which had been solved: various insects and the fruit tree red spider mite began to appear in large numbers on the trees. With the advent of DDT, the mite began to occur in numbers never seen before. The killing of predatory and parasitic insects which previously kept the mite at low levels was a disaster which serves to underline possible side-effects of spraying with non-selective insecticides. From its role as a relatively insignificant member of the fruit tree fauna, red spider mite is now one of the most troublesome pests with which the fruit grower has to contend. Increases in the numbers of conifer spinning mite on conifers are similarly attributable to the use of DDT or HCH to control other pests.

1.5 Economics of treatment

From the foregoing sections it is apparent that insects attack a large number of crop plants and that vast sums of money may be spent annually in attempts to control them. Although casual observation may show that damage is occurring, intensive work is necessary to show the extent of the damage, either in terms of loss of crop or lower profit obtained on selling the produce. The majority of the estimates of crop losses quoted early in this century were made as a result of intelligent observation by field entomologists. More recent work has aimed at determining these losses by critical field experimentation followed by systematic survey work. Crop losses may be expressed as the average loss of crop in a particular year, or the area of land equivalent to that from which the entire crop was lost, or in terms of financial returns. The last is the most difficult to evaluate because prices fluctuate with the supply/demand ratio, so that when a commodity is scarce it is more expensive than when it is freely available. Thus, a 15 per cent loss of crop might not lead to a proportionate drop in income from that crop.

A further difficulty which applies to all loss assessment is that although the individual effects of various pests and diseases of a crop may be determined, they may not be additive. Thus, an estimated loss of crop due to attacks by more than one pest may be much greater than the actual loss. In addition, the compensating factor which results in healthy plants adjacent to dead or dying ones giving more than the average yield can upset predictions of loss of yield. If a crop of uniform size is required for some particular market, the last factor can be of importance.

When the economic aspects of crop loss are being considered, the over-riding question is whether the value of the extra crop to be expected as a result of pest control is greater than the cost of chemicals and applications. On quality crops such as fruit, protection is nearly always worth while for the slightest blemish reduces the quality rating on the open market and the value is correspondingly decreased. On the other hand, there is little point in protecting fodder crops from insect attack unless this results in a substantial increase in yield. Between these extremes there is a host of problems associated with the limits to which pest control should be taken. One insecticidal application may give only 70 per cent control of a pest: is it worth while making a second application to control a further 20 per cent and another to obtain almost complete control? Definite answers to such questions relating to specific pests and crops are difficult to give, mainly because there are few results available of actual damage assessment. Obviously too, conditions from field to field and from season to season vary enormously. Usually, the grower relies on his experience, or that of his advisers, in deciding whether and when to spray. But often by the time an insect has reached pest status, the damage to the crop has been done. For example, by the time colonies of wingless aphids are noticed feeding on crops, their winged parents have possibly spread virus to several localities in the area. Extensive work on the relationship between cabbage aphid attack on Brussels sprouts and yield has indicated that early infestations of a few aphids per plant are responsible for as much damage as that caused by the large numbers of aphids seen on plants during August and September.

Alternatively, it has been shown experimentally that an insect attack which looks dangerous may be causing little damage; extensive defoliation and loss of stand in the seedling stages of some crops has very little subsequent effect on yield. In experiments, loss of 50 per cent of the leaf area of sugar beet seedlings at the 4–8 leaf stage, simulating the type of damage caused by such pests as mangold fly, reduced the yield of roots by only 5 per cent, while destruction of up to one-half of the initial plant population resulted in a 10 per cent loss of yield. In experiments on peas, loss of 12·5 per cent of the leaf area when the plants were at the 4 expanded leaflet stage led to a loss of 8 per cent of the yield of shelled green peas. It was suggested that damage of this level by the weevil *Sitona lineatus* was the exception rather than the rule and that, if done at all, spraying should be restricted to the headlands of fields.

Finally, spraying a crop against a particular pest may result in a profit margin that is so small that it is essential that the chemical be applied at the critical time. Work in East Anglia on dry harvesting peas has shown how vital is the proper timing of spray applications.

This brief account, although restricted to a few examples, is sufficient to show that the economics of chemical pest control is a complicated subject and the grower needs to be up to date in his approach to the problem, seeking unbiased advice whenever possible before committing himself to heavy expenses which may not be recovered.

1.6   Development of resistance to pesticides

A population of insects on a crop is largely made up of individuals which are essentially the same genetically, a few differing slightly from the rest. When repeated insecticidal application results in the appearance of resistance among the population, the process has been one of preferential selection. Those insects with a genetic make-up favouring survival under the conditions which destroy the main part of the original population become dominant and subsequent populations are composed of this strain. Such resistance may be due to the insect not being affected by the insecticide itself or to its behavioural pattern effectively isolating it from sites which have received the insecticide. Such resistance can be seen when applications of chemicals that once controlled a pest no longer do so, most frequently on crops on which routine control measures are the rule rather than the exception. Thus, 'insurance' spraying of orchards and the use of persistent insecticides as soil treatments may eventually produce resistant populations of the insect species they initially controlled.

There are many examples of resistance among insects in the United States of America and in the tropical regions of the world, where insecticides are used more freely than they are in Britain and where insect life cycles are shorter owing to the more favourable climatic conditions. Codling moth in America and Australia, and mosquitoes and flies in the tropics are but 3 examples of insects in which resistance has been found. In Britain, glasshouses provide the best conditions for the development of resistance, and resistant populations of red spider mite, whitefly and *Myzus persicae* exist under glass in many areas. Examples of resistance outdoors in this country include red spider mite on fruit trees, cabbage root fly on brassicas, carrot fly on carrots and damson-hop aphid on hops.

An important aspect of this problem is that pests resistant to one chemical may also show, or quickly develop, resistance to closely related compounds. Thus a whole group of pesticides may be rendered ineffective. When attempting to control a resistant population by chemical means it is, therefore, necessary to select a pesticide belonging to a different chemical group, for example substituting an organophosphate for an organochlorine compound. This approach depends on the development of resistance not outstripping the development of new groups of pesticides. In glasshouses where many generations of a pest occur during the course of a year and the development of resistance is correspondingly rapid, the danger of exhausting the supply of effective chemicals has led to the introduction of the biological control measures already mentioned.

1.7   Effects on soil

Some insecticides are applied directly to the soil to control insects attacking plants at, or below, the soil surface. Before 1964, aldrin and dieldrin were widely used both as seed dressings and as directly applied soil insecticides. In 1964, however, the Advisory Committee on Poisonous Substances used in Agriculture and Food Storage recommended, and the Government accepted, restrictions in the use of aldrin and dieldrin. The situation was reviewed again in 1969 when restrictions were placed on the use of DDT. The crops which may still be treated with these insecticides are listed in official recommendation sheets for the safe use of chemicals. Other chemicals, mainly in the organophosphorus group, are available as alternatives; for example, disulfoton and phorate for the control of carrot fly, and chlorfenvinphos for the control of carrot fly, cabbage root fly, frit fly and wheat bulb fly.

Persistent insecticides may drop to the soil as run-off from treated vegetation, substantial amounts being found beneath treated crops. These insecticides remain active in the soil for periods running into years and there are many problems associated with this persistence. It has been observed that the incidence of cabbage root fly on brassica crops grown on previously treated land has risen due to the adverse effects of aldrin on the beetle predators of this pest. In addition, it is apparent that there will be long-term effects on the soil fauna though these have not yet been critically evaluated. Preliminary results suggest that populations of beetle and fly larvae and some mites decline on insecticide-treated soil and that aldrin reduces springtail populations whereas DDT has the opposite effect. The mechanics of these population fluctuations and the interactions between predators, prey and insecticides are currently under investigation.

Another problem is the extent to which plant tissues become contaminated by insecticide passing from the soil into the crop; aspects include insecticide residues which may be harmful if ingested and the tainting of crops such as potatoes and carrots grown in soil previously treated with HCH. The theory of plant absorption and retention of insecticides is attracting much attention and research, but the measurement of residues is a time-consuming task. There are, however, recommendations concerning the minimum periods which must elapse between the treatment and harvesting of various crops.

1.8   Phytotoxicity

Chemicals applied to plants to control pests and diseases sometimes have a phytotoxic effect. This may amount to no more than a temporary check in growth, as may occur when seed treatments retard germination, but, in severe cases, extensive necrosis of leaf tissues and possible death of plants may result.

Knowledge of possible phytotoxic hazards is obviously greatest for those materials which have been widely used for many years. It is well known, for example, that some varieties of black and red currant are sulphur shy; that DDT is toxic to cucumbers and related plants and has a harmful effect on certain barley varieties; that dimethoate and formothion should not be used on chrysanthemums. Where appropriate, such information is given in the text of this book, but it should be noted that new cases of phytotoxicity are likely to occur, particularly where newer materials and formulations are used.

1.9   Conclusions

At present, control of many pests can only be achieved by chemicals, although the conditions which lead to their application need much more critical appraisal. Faced with secondary effects such as the development of resistance, residues in soil and plants, and the effects on insect parasites and predators, there needs to be more discrimination as to when insecticides should be applied. More detailed ecological investigations need to be made to determine the value of beneficial insects, and possibilities should be explored of changing accepted cultural techniques to produce habitats less acceptable to pests. At the same time, more selective insecticides are required so that pests may be destroyed without upsetting the attendant complex of non-damaging insects. The development and production of more specific insecticides implies higher costs; these might well be absorbed by the grower and cost/benefit ratios remain acceptable, if less spraying were done on an insurance basis or in an attempt to keep a crop completely clear of pests.

The application of pesticides should never be a hit or miss affair; this aspect of pest control should be critically timed and conscientiously carried out. For most pests it is necessary to obtain skilled entomological advice on the time at which the crop should be sprayed, but this is of little use if only a small proportion of the population is touched by the pesticide. When controlling virus vectors, the treatment must be as near 100 per cent effective as possible.

Plant diseases

Most plants are liable to be attacked by several and sometimes many different pathogens: usually only a few are of major significance and the others occasionally of importance under certain conditions. Over the years, changes, often associated with changes in agricultural or horticultural practices, may alter considerably the pattern of predominant troubles. Most pathogens are restricted to relatively few hosts and these are often fairly closely related botanically. There are, however, several pathogens with a wide host range: the ubiquitous grey mould can be found on many plants, the Verticillium wilt fungus and cucumber mosaic virus have a very wide host range and Armillaria root rot can attack most woody plants. Some pathogens can be recognized by the characteristic symptoms that they produce and the common name for the diseases is often descriptive of this symptom: red core of strawberries and eyespot of wheat are 2 good examples. But in general plants often react similarly to invasion by different pathogens. Attack on the roots results in the yellowing of leaves and gradual death of the plant and only laboratory examination will reveal the culprit. Sometimes the causal fungus can be seen on the roots: Armillaria root rot and take-all of cereals both produce characteristic mycelium on the surface of affected roots. The fungi attacking through the roots and causing wilts are often identifiable only after they have been isolated in the laboratory from the affected stems. For all the many fungi causing leaf spots, stem cankers and fruit rots, the presence of characteristic spores or other structures enables the causal pathogen to be identified. For both bacterial and virus diseases much reliance has to be placed on the symptoms produced: the identification of the pathogens often involves the inoculation of plants or parts of plants.

Most plant pathogens belong to one of the 3 groups: fungi, bacteria and viruses. A brief outline of the main characteristics of these groups and their sub-groups may help those who are unfamiliar with them to appreciate the salient characters.

The fungi causing plant diseases can be grouped into 4 sub-groups. These sub-groups are distinguished from one another by the microscopic characters of the mycelium, of the spores and the organs on or in which the spores are developed.

The first sub-group comprises the Phycomycetes: they typically produce both motile spores, which depend on water for their spread, and long-lived resistant resting spores. In this group are the soil-borne organisms causing club root of brassicae, wart disease of potatoes and crook root of watercress; also included are the downy mildews with their best-known representative potato blight and the damping-off fungi belonging to *Phytophthora* and *Pythium*.

The second sub-group consists of the Ascomycetes, again typically producing 2 different types of spore, one produced in vast numbers on the affected parts and the other developing either in a thick-walled perithecium, usually formed in or on dead host tissue, or a cup-shaped apothecium, normally arising from a resistant structure, the sclerotium. Included in this group are the perithecia-forming powdery mildews and the fungi causing apple scab, blackcurrant leaf spot and rose black spot: Sclerotinia disease and the closely allied clover rot belong to those producing apothecia.

The third sub-group contains the Basidiomycetes. They also often produce 2 or more kinds of spores but one is always formed on a characteristic structure known as a basidium. This group includes the smuts represented by loose smut of wheat and barley, and the rusts, species of which can be found on most plants. Also included are the so-called 'higher' fungi represented by the silver leaf fungus which produces small purple brackets on the dead branches and the Armillaria root rot fungus which develops clumps of tawny toadstools at the base of infected trees and shrubs.

The fourth sub-group consists of the Fungi Imperfecti; a miscellaneous group for most of which only one kind of spore is known and some have apparently none. Included here are leaf mould of tomato, leaf spot of celery and white rot of onion.

There are relatively few bacteria causing plant diseases in this country and they can conveniently be divided into 5 main groups corresponding to the genera of bacteria. These genera are differentiated largely on biochemical, physiological and to some extent also on pathological grounds. In the genus *Agrobacterium* are included crown gall and leafy gall; in *Corynebacterium*, silvering of beet; in *Pseudomonas*, bacterial canker of cherry and plum, halo blight of beans, and halo blight of oats; in *Erwinia*, fireblight and some of the important soft rotting organisms; and in *Xanthomonas*, hyacinth yellows and begonia bacterial blight.

The viruses, perhaps most aptly described as possibly the simplest type of organism, have so far defeated attempts to classify them satisfactorily. The common names given to them are usually descriptive of the symptoms on one of the host plants, often the one on which the virus was first found. Viruses are normally systemic throughout infected plants and consequently are readily transmitted in cuttings, layers, tubers, etc., of all vegetatively propagated plants. Unlike fungi and bacteria, viruses can only be spread from plant to plant by some external agency; insects, such as aphids, thrips and leaf hoppers, are the most common vectors, though some viruses are spread by nematodes.

1.10   Factors influencing disease incidence

There are three primary requirements for the development of a disease: firstly the pathogen must be present or be introduced; secondly the host plant, or certain parts of it, must be susceptible to the particular pathogen, thirdly the environment including soil and weather conditions must be favourable for infection to take place and for spread to occur.

Knowledge of the likely source of each pathogen is of fundamental importance for the successful application of any control measure. Most pathogens need to survive for a considerable period of the year or sometimes longer in the absence of either their host or the tissues which they attack. They may be soil-borne, they may survive in debris from the previous crop, they may be seed-borne or perennate in buds or rootstocks or other parts of perennial plants. They may find volunteer seedlings or other hosts and then spread into the new crop from nearby, or occasionally far distant, sources as airborne spores, or for the viruses with their insect vector. Each of these sources will be considered in greater detail in the section below on control measures.

As has already been mentioned, most pathogens are restricted in their ability to attack plants and even the rusts, powdery mildews and downy mildews, which as groups cause rather similar symptoms on very many plants, are often individually specific and varieties of the same plant may or may not be susceptible. The factors responsible for these differences in resistance to attack are varied and often obscure but such resistance can be controlled by the genetic make-up of the plant and plant breeders have been able to incorporate resistance into new varieties of crop plants. In this country most success has been obtained with potatoes resistant to wart disease, with wheat resistant to yellow-rust, with strawberries to red core and hops to Verticillium wilt. The appearance of races of fungi able to attack such resistant varieties is a constant source of concern to plant breeders.

Most fungal and bacterial pathogens attack only certain parts of plants: root pathogens rarely extend to above-ground parts and pathogens attacking the leaves, stems, flowers and fruits are often most damaging to one of these parts and only occasionally occur on the others. Many of these pathogens infect the tissues when they are at a certain stage of growth or maturity. For instance, the apple scab fungus infects the very young leaves: the typical greenish brown patches may take up to a month to become visible and the disease then appears on old leaves. It is thus understandable that the critical time for the establishment of scab on apple trees is in the spring when for some weeks, particularly the period from bud burst to green cluster, much of the leaf area on the tree consists of young leaves. For some storage diseases too, the age of the tissues they infect is of significance. The fungi causing dry rot and gangrene of potato can infect tubers at lifting time but they normally do not invade the tubers until later in the storage season when the tissues become more susceptible. A similar effect occurs with Gloeosporium rots of apple which become increasingly important causes of loss during storage: the commercial prolongation of the storage season can result in greater losses due to such storage diseases.

The nutrition of the host can influence the susceptibility of plants to disease. It is sometimes difficult to dissociate the effects of nutrition from the inevitable modifications in the microclimate and other environmental factors but there is no doubt that nutrients, particularly nitrogen, phosphorus and potassium, can affect the incidence of some diseases. In general excesses or deficiencies often lead to greater susceptibility. Nutritional effects can alter symptom expression of virus diseases.

The environment plays an all-important part in determining the incidence of soil- and air-borne pathogens. Many soil-borne pathogens are sensitive to soil reaction. For example, alkaline conditions are unfavourable for the club root fungus. It can survive and attack under these conditions but a narrower range of other factors needs to be satisfied if the disease is to occur; the temperature becomes more critical and the spore load in the soil needs to be higher than in a more acid soil. On the other hand an alkaline soil favours attacks of potato common scab and cereal take-all, which tend to be most troublesome in chalkland areas. The temperature of the soil can also affect the incidence of diseases. Under cold, wet conditions damping-off fungi can cause serious losses of germinating seedlings. Early sown peas are often severely attacked by soil-borne fungi unless protected by seed treatment with a fungicide. Potatoes planted into cold, dry soils can be attacked by the black scurf fungus often present on the surface of the tuber or in the soil: under good growing conditions the sprouts normally grow away unaffected.

For all fungi and bacteria the weather, both directly and indirectly in its influence on the microclimate around the plants, inevitably affects profoundly the occurrence and development of all the diseases that these organisms cause. Weather conditions affect their take-off and spread from the source, their transport to the host plant, whether they infect or not and, by affecting the subsequent local spread, determine the development of the epidemic. For most fungi and bacteria, temperature and humidity (or wetness) are the two most important factors. Temperature is often the factor limiting the geographical range of fungi and bacteria, whereas humidity determines their incidence in a particular crop in any one year. This close association of diseases with weather conditions is widely appreciated but the actual conditions that individual pathogens respond to vary greatly and are often critical. For most diseases the symptoms appear after the pathogen has become well established on its host and the conditions leading to the initial development took place weeks or months previously and subsequent favourable weather conditions will promote its rapid spread. Knowledge of the specific conditions enabling early infections to take place and subsequent spread to occur provide the basis for forecasting schemes such as those developed in this country to help growers combat apple scab and potato blight. Much more needs to be discovered about the life histories and weather relationships of our other important pathogens.

Cultural operations can also influence the occurrence of diseases. The date of sowing can be a significant factor in the development of barley powdery mildew: late sown barley crops are often growing rapidly in early summer when spores are available on overwintered plants and the weather often suitable for disease development and spread. On some other crops it is the early sown ones that are likely to be the most affected. Wounding by pruning operations or during harvesting and subsequent handling can also predispose plants to infection. The silver leaf fungus invades pruning wounds on fruit trees and many storage diseases and particularly soft-rotting bacteria follow damage to the tissues.

1.11   Control measures

With very few exceptions the pathogens causing plant diseases penetrate into the host tissue and once they have done so are very difficult to control by the application of chemicals: measures can be taken to prevent spread but even these are unlikely to be effective unless they are applied at an early stage of the development of the disease. As fungi are themselves plants, the control measure that will kill the pathogen and leave the host unaffected is not always easy to find and phytotoxicity is a constant problem. For a few diseases heat treatment will eradicate the pathogen: hot water treatment of celery seed for the control of celery leaf spot, and of brassica seed for the control of canker affords effective disinfection but the temperature of the bath needs to be regulated carefully to prevent serious loss in germination. Heat therapy of chrysanthemum plants for the elimination of a virus disease is now being used commercially.

The powdery mildews are exceptional in growing superficially on leaf surfaces with only small 'pegs' penetrating into the cells below. It is possible, though not easy, to eradicate such infections with appropriate spray materials. It is also possible to eradicate early infections on other plants but, with apple scab for instance, sprays must be applied within a few days of infection: only occasionally can the fungus in established lesions be killed.

For the vast majority of diseases prevention rather than cure is essential if losses are to be avoided. Whenever research is focused on to a disease detailed studies of the life history of the pathogen become necessary and they almost invariably reveal that the adoption of careful cultivation practices must form the basis of any general protective programme for its control: the judicious use of chemical treatments can then be expected to play a decisive part. Assessment of the likelihood of a disease appearing and developing severely enough to justify a control programme is one of the many difficult problems that a grower has to contend with. Frequently it is not until a severe attack has been experienced that some action becomes necessary and then drastic measures may need to be taken and followed up the next season. The most important protective measure is the prevention of the introduction of the pathogen into the crop: the most dangerous source of infection in any crop is a diseased plant within it. It is perhaps unnecessary to stress the importance of planting only healthy planting stock in land in which soil pathogens have not been allowed to increase by repeated successive cropping. The following very brief summary of the control measures available will show that variations on this general theme, depending upon the source of the particular pathogen, form the basis for the control of all diseases.

Contaminated soil harbouring soil-borne pathogens causing root rots and wilts poses many problems. Under field conditions the careful selection of cropping sequences with appropriate intervals between susceptible crops will keep, or reduce, the pathogen to a low level insufficient to cause serious losses to a subsequent host crop. The length of the interval varies from pathogen to pathogen and depends on the form in which it survives. It may persist on roots and stubble and die out as these disintegrate: hence for some cereal foot rots up to 3 years is needed. Other pathogens survive as resting spores or resistant sclerotia and for club root and clover rot an interval of up to 8 years may be necessary. No methods of eliminating such pathogens from large areas of soil are available but for some soil-borne diseases there are chemical control measures which effectively protect plants planted out in the soil: young brassica plants can be grown in club root contaminated soil after they have been dipped in a fungicide and onion seedlings can be protected from attack by white rot and smut by seed treatment. Under glasshouse conditions the build-up of soil-borne pathogens can be dangerously rapid and the economics of the industry are such that rotation of crops is impracticable. Routine soil disinfection has become an essential part of the intensive cropping systems associated with the production of crops of tomatoes, lettuces, carnations and other plants in glasshouses. But treated soil may be quickly invaded by reintroduced pathogens and even greater care is needed to ensure that planting stock is healthy.

Infected debris is another common source of infection and much can be done by general hygienic measures to eliminate this source. The effective disposal of tomato haulms, which can become a prolific source of Didymella stem rot infection, and of hop vines from gardens affected by Verticillium wilt should be routine procedures. Blighted potato tubers discarded at riddling near clamp sites can become the starting point for early local spread of the disease. For some diseases for which infected debris constitutes a major source of infection, no commercial methods of destruction have as yet been devised. Overwintering dead leaves containing the perithecia of apple scab and blackcurrant leaf spot, on the orchard and plantation floor, are not easy to destroy and spraying of the young susceptible growth is at the moment the only practicable control measure.

Infection within the tissues of planting material is much more common than is generally appreciated: seeds, runners, tubers or cuttings all frequently carry unseen pathogens. The importance of maintaining stocks of healthy material for planting cannot be over-emphasized. For virus diseases of those plants that are vegetatively propagated, this is of the utmost importance and the Ministry of Agriculture, Fisheries and Food operates certification schemes for potatoes, hops, strawberries, raspberries and fruit tree rootstocks to ensure that a supply of healthy planting material is available for growers. Methods for testing cuttings of carnations for freedom from wilt pathogens have been devised and are being used by some commercial raisers of cuttings. A number of pathogens are commonly seed-borne, either occurring as contaminants on the surface of the seed or penetrating into the internal tissues of the seed. Seed of cereals and sugar beet can be effectively disinfected by seed treatment with mercury compounds and this is normally recommended as a routine treatment: hot water treatment for the control of the internally-borne loose smut of cereals, leaf spot of celery and canker of broccoli has been referred to earlier; and a thiram soak treatment is very effective for the control of some internal fungal pathogens. Seed potato tubers may carry blight and the fungus growing up into shoots from these tubers can be responsible for early local outbreaks from which spread will occur. Tulip fire is carried on the bulbs and infection of the young shoots as they emerge from the soil provides a source of spores which in their turn infect the leaves and flowers of nearby plants. Many other bulb diseases are also carried over from season to season in this way and very careful sorting of the bulbs before they are planted to eliminate those that show signs of a rot or indeed any signs of infection will help to prevent an attack on the new crop. The hop downy mildew fungus overwinters in the rootstock and appears in the spring on the infected 'spikes'. Apple powdery mildew overwinters in buds which became infected the previous summer: the mildewed trusses which appear in the spring are prolific sources of spores which give rise to infections of the new leaves and buds to complete the cycle. Systematic cutting of the mildewed trusses in the spring together with a routine spray programme with a suitable fungicide will reduce incidence of this disease. The fungus causing peach leaf curl also overwinters in the buds or on the stem. Some fungi and bacteria establish themselves as cankers on fruit trees and sometimes cutting-out can be a very useful aid to their control.

All these sources of initial infection are often few and far between and difficult to detect but, when conditions are favourable, the speed with which they can build up, spread and cause widespread damage is remarkable. The application of an appropriate protective fungicide before infection is noticed and has made an appreciable headway can often prevent the development. But to ensure that this protective treatment is effective many other factors need to be considered. The timing of the application of the fungicide, both for the first application and for subsequent treatments to cover new growth and to renew deposits that have been washed off by weathering, can play a very significant part. Experiments and experience have established useful routine spray programmes for several diseases and new information is continually being obtained which enables some of the more empirical programmes to be replaced by more concise and rational ones. The method of application of the fungicide on the diverse types of plants involved in row crops, orchards, etc., provides another set of problems: these will be dealt with in a later chapter. The range of fungicides available also poses questions for the grower. The recent introduction of systemic compounds has widened the choice. These systemics when applied to the roots as a drench, to the seeds or to the leaves move within the plant and give protection to some other parts. The extent to which this occurs varies considerably; some compounds move only within the part to which they have been applied while others penetrate the whole plant and provide protection over a prolonged period. Many systemic and conventional protectant fungicides are relatively specific in their action. But this specificity can complicate matters particularly when more than one disease is, or may be, present on the same plant and two spray programmes with two different fungicides need to be integrated into one programme and into which insecticides also need to be incorporated. There are also considerations of cost, of possible phytotoxicity and almost certainly also other more local or personal aspects to be taken into account.

There can be no doubt that an appreciation of the biological background to each disease in relation to the cultivation of its host plant will provide the sound basis for the effective control measures to be described in the following chapters.

Chapter 2
The chemical background

Although the use of chemicals for the control of the pests and diseases of crop plants is of long history, the greatest development, the advent of the synthetic pesticide, is but 40 years old. The range and number of compounds now available as pesticides is of surprising extent and their classification has to be on the basis of their chemistry though the grower will probably have no inclination to delve deeply into this subject. Nor will it be necessary for him to do so, for the chemicals can be classified on the basis of the particular pests or diseases against which they are used, a classification helpful in its indications of the method of usage but which suffers the defect that a given chemical may appear in more than one category.

2.1   Insecticides

Compounds used mainly against insect pests are grouped as insecticides—an old term used in the days before it was found necessary to differentiate between the true insects and certain other groups of arthropods such as the red spider mites which have become troublesome pests. This wide, older definition is still of use and if a distinction is necessary, the 'insecticides' used against mites may be termed acaricides. Even among those compounds used against true insects, it is useful to classify according to the main group of insects against which they are effective, hence such terms as aphicide—effective against aphids—scalecide—effective against scale insects—etc.

Insecticides may also be sub-divided according to the way in which they penetrate into the insect body and thereby incapacitate it. Some, of necessity of greater or lesser volatility, enter the insect through its breathing pores and hence act as fumigants, though this term is seldom used outside of actual fumigation, a process not widely used by growers nowadays except for soil treatment. Nicotine, for example, the most effective of the older aphicides, acts mainly as a fumigant though applied in liquid or solid dilutions to its victims exposed on the leaf surface. In soil use, however, it is possible to use an insecticide of comparatively low volatility for the vapour is entrapped by the soil particles and retained for times long enough to be fatal to the soil pests against which it is applied.

Or the insecticide may be swallowed by the insect as it eats its way through the foliage, when it is called by the self-descriptive term of stomach poison. This term indicates that the poison must be applied to the foliage or plant tissue which the insect will eat and that, to be effective, it must pass from the gut into the body tissues of the insect, considerations to be borne in mind in deciding the time of application of the insecticide.

A third and the most important of the types of insecticide is the contact poison—a chemical which kills by contact with the exterior of the insect and which, therefore, must be capable of passing through the insect integument. Obviously such an insecticide must be applied at a time and in such a manner that the insect cannot avoid contact with it. With the older contact poisons such as nicotine and pyrethrum, insecticidal activity was soon lost after application—hence the term direct contact insecticide. But many modern contact insecticides retain their activity long after application and the insect may be killed even though only its feet come into contact with the insecticide deposit. The time of application is therefore less critical and the insecticide is able to protect the plant from infestation for some time after application; it is a protective contact insecticide.

The contact insecticides also include those compounds, such as the tar oils and DNOC applied when the tree is dormant. Such sprays were known as winter washes, though the term ovicide better indicates their purpose, for their main practical use is against the eggs of mites and insects.

Finally there are those insecticides which, when applied to the foliage or roots of growing plants, are taken up by the plant which thereby becomes systemically toxic to pests feeding on its sap or tissues. Such compounds are known as systemic insecticides. Their greatest role is the protection of young seedlings for which purpose they are applied in suitable form at the time of sowing or the emergence of new growth.

2.2   Fungicides

Fungicides applied to foliage fall into two groups; those used against fungi the mycelium of which is exposed on the surface of the leaf, for example, the powdery mildews; in general, these fungicides kill by direct contact. But a larger group of fungi causing foliar diseases live within the plant tissue inaccessible to a surface-applied chemical. Against such fungi the primary object is the protection of the foliage from infection—such fungicides are protective and clearly should be applied at a time prior to the arrival of the fungal spores which otherwise would infect the leaves. However if application is too late and the plant has become infected, it is still possible in certain cases to apply a fungicide which will kill the infected tissue, such a compound is usually called an eradicant fungicide.

Fungicides applied to 'seed'—using the word in its widest sense to include tubers, corms, etc.—are primarily direct in action against surface-borne spores but, in some cases, the fungicide is required to persist on the seed coat long enough to be effective against dormant mycelium contained within the seed. When applied to the 'seed' before sowing, the fungicide is called a seed disinfectant or seed dressing, though the latter term may include treatments not intended to counter seed-borne fungi or soil pests.

To these categories can now be added the systemic fungicides which, applied to seed, roots or foliage, render the plant resistant to fungal attack.

2.3   Formulation

At one time the grower intending to use pesticides would purchase the active component in more or less pure form, as with flowers of sulphur, or he would prepare the sprays he used from simple ingredients at the site of use. Bordeaux mixture, for instance, would be made in the spray tank from hydrated lime and a stock solution of copper sulphate; or he would dissolve soft soap in the spray tank and add to it the required amount of nicotine in the form of the 95 per cent alkaloid. But with the modern pesticide it is rarely that he would handle the separate components for they are marketed only as formulated products ready for use.

FORMULATION AS SPRAYS

If for spraying, such formulated products need only to be added to water in the spray tank. As many of the active pesticides are insoluble in water, they must be formulated to a product which rapidly disperses to give a uniform dilution. If solid, it may be ground with, or treated with, other components to give a powder which, on addition to water, disperses automatically to give a suspension of fine particles. Such a preparation is called a wettable powder, usually abbreviated to w.p., the special properties of which rest in the surface-active component (or wetter) used to render the particles easily wetted by water. The content of active ingredient (a.i.) is usually adjusted to a suitable round number by the addition of inert carrier.

Alternatively if the a.i. is liquid or if it is to be used in a liquid such as oil which is not miscible with water, the usual method of formulation is the so-called emulsifiable concentrate (e.c.), at one time also called a miscible oil. When added to water the emulsifiable concentrate spontaneously disperses as fine droplets to form an emulsion. In this case, the special property of self-dispersion is imparted by the use of surface-active components (surfactants) which serve to prevent the re-coalescence of the oil droplets thereby giving an oil-in-water emulsion. Because by far the greater bulk of the emulsion when diluted to spray strength is water and because of the powerful emulsifiers now available, the so-called inverted emulsion, in which the water droplets are dispersed in oil, is no longer likely to trouble the grower. The older stock emulsion (s.e.) a concentrated emulsion prepared by milling oil, water and emulsifier and containing around 70 per cent of oil, is often marketed these days.

Modern formulations have made it a simple matter for the grower to prepare his sprays. Rarely will he be required to do more than to add the requisite amount of wettable powder or emulsifiable concentrate to water in the spray tank. The need for the addition of supplementary materials to improve the wetting properties or the adherence of the spray deposit sometimes arises and care is still necessary when using mixtures of formulations to ensure that there is no unwanted interaction between the components of different formulations. In technical jargon the formulations must be compatible and the label will provide the necessary precautions. But a word on the reasons for care will be helpful. Wetting agents, emulsifying agents and other forms of surfactant fall into 3 broad groups according to their chemistry. Soaps and many of the modern surfactants used in wettable powders are the alkali salts of certain organic acids and, because the surface activity is due to the structure of these acids, they are called anionic surfactants. In a contrasted group, the special properties which confer surface activity reside in the basic part of the compound, they are cationic surfactants. If cationic and anionic wetters are mixed a reaction will occur resulting in the precipitation of the acidic and basic components as a water-insoluble grease with consequent loss of surface-active properties. Among the cationic surfactants are several compounds used as fungicides and care is therefore needed when their formulations are mixed with others.

In the third group of surfactants, these difficulties are avoided for, not being salts, they are not ionized in solution, hence they are called non-ionic surfactants. This particular type of incompatibility will not arise.

Wettable powders and emulsifiable concentrates are prepared for dilution with water to spray strength, the extent of dilution being determined by the amount of a.i. or of spray to be applied per hectare. It requires energy to break up the spray into droplets and to impart to these droplets the velocity needed to carry them to the surface to be sprayed. At one time this energy was supplied, via the spray pumps, to the spray itself and, in this so-called conventional or HV spraying, large amounts of spray were applied per hectare, naturally of a dilute spray. This method is still in wide use but machines are now available which use a strong current of air to convey the spray droplets to their target—a method which permits the use of smaller amounts per hectare—naturally of a more concentrated spray—hence the technique is called concentrate or LV spraying. These matters are discussed in Chapter 3 but a more recent development may be mentioned here though it is applicable only to the spraying of ground crops. In this method the spray is gravity fed and merely shaken on to the foliage by a to-and-fro movement of the nozzles.

The production of droplets by the use of a strong current of air is called atomization, though this term exaggerates the degree of reduction of droplet size. For this purpose, a solution of the a.i. in a volatile solvent, an 'atomizing solution', is often used. A more accurate description of the spray so produced is 'aerosol', though this term is better limited to the spray fog produced by a device in which the liquid is dispersed from a pressurized vessel containing a solvent of low boiling point which, in fact, boils when the pressure is released. The formulation used is here a solution of the pesticide in an organic solvent containing a suitable propellant liquid, such as chloromethane. The pressurized can has to be of small dimensions and the aerosol is a handy tool for the amateur when the target to be sprayed is small or for use in an enclosed space such as the frame or glasshouse.

More recently, ultra-low volume (ULV) spray applications produced by thermal fogging machines have been introduced for use on glasshouse and other protected crops. In this system, the pesticide is mixed with a suitable carrier/fogging agent and the fog is produced by injecting this mixture into the exhaust system of a petrol-driven motor. 'Wet' or 'dry' fogs (which are related to particle size) can be regulated by the size of the dosage jet selected. It is claimed that almost all the pesticides normally used as HV sprays can be applied in fog form from these machines, provided the correct carrier agent is used.

An alternative ULV system of spraying, producing larger droplets (60–110 microns), is provided by the 'Turbair' spraying machine. In this method, specially formulated oil-based pesticides are gravity fed through a finely calibrated metering device on to a rapidly rotating disc and the resulting droplets are carried to the target in a powerful air-stream created by the fan attached to the machine.

An alternative for use in an enclosed space is the 'smoke' in which the pesticide is dispensed by the use of heat usually generated from a pyrotechnic mixture, a method applicable only if the pesticide is heat-stable.

FORMULATION AS DUSTS

If the pesticide is to be used as a dust, formulation is a comparatively simple matter, the compounding of the active ingredient into a form suitable for the purpose for which the dust is to be used. In most cases this process is the intimate grinding of the a.i. with an inert carrier or filler but certain qualities are required, such as freedom from caking tendencies when the product is on the shelf or in the store. If intended for application through a dusting machine, it obviously must flow readily and not 'ball' in the hopper or clog the air-ducts. Ground sulphur, for example, is notoriously prone to ball in the hopper of the dusting machine but this tendency, due probably to static electricity, is easily cured by the admixture of a small proportion of some other dust.

If to be used for the treatment of 'seed', the dust should adhere well to the dressed seed yet not so readily disperse that a dust hazard to the operator is created. This fault can be corrected by the addition of anti-dust agents.

FORMULATION AS GRANULES

For certain uses, there are advantages in the application of the a.i. in the form of small granules, a method which has become widely used for the treatment of seedling crops with systemic pesticides (see p. 80). The ease of handling and application greatly reduce hazards to operators.

The latter fact is recognized in the Regulations of the Health and Safety at Work, etc., Act and the measures to be taken in the protection of personnel. Moreover, certain of the more hazardous pesticides have been cleared under the Pesticides Safety Precautions Scheme, for use only in granular form. For these purposes the expression 'specified substance in granular form' means a preparation which: (a) consists of absorbent mineral or synthetic solid particles some or all of which are impregnated with a specified substance, the size of particles being such that—(i) not more than 4 per cent (including the 1 per cent referred to below) by weight of the preparation is capable of passing a sieve with a mesh of 250 microns; and (ii) not more than 1 per cent by weight of the preparation is capable of passing a sieve with a mesh of 150 microns; and not more than 25 per cent of the aggregate weight of the particles capable of passing a sieve with a mesh of 150 microns shall consist of a specified substance; (b) has an apparent density of not less than 0·4 g/ml if compacted without pressure; (c) and of which not more than 12 per cent by weight consists of a specified substance.

FORMULATION AS SMOKES

For the treatment of enclosed spaces such as the glasshouse, it may be possible to disperse the pesticide as a 'smoke' by heat or by incorporation in a pyrotechnic mixture. The method, applicable only to thermostable pesticides, shares the defects and advantages of fumigation; excellent penetration to all corners of the treated space; uniform deposition; briefer persistence. The foliage should be dry but the soil moist; seedlings or newly-transplanted plants should not be treated nor should the treatment be carried out in bright sunlight.

2.4   The active components

It is now necessary to consider the chemicals employed as the active components of the formulated pesticide. As most of these chemicals now have common names agreed by the British Standards Institution, they may be simply arranged in alphabetical order. In this way, the reader may, at this stage, be spared the long, and to him probably unintelligible, chemical name though it is important that he should know something of the chemical affinities of the compound for, should he be unfortunate enough to encounter resistant strains of pest or fungus he will have to select an alternative pesticide outside the particular group to which resistance has arisen. This chemical information has, for convenience, been placed in the Tables of the Appendix.

GENERAL NOTES ON PRECAUTIONS

It is self-evident that a compound used because it is toxic to an insect or fungus, may also be toxic to other forms of life. Precautions have to be taken to ensure the pesticide is used in a manner which will not expose other organisms, whether man, stock or wildlife, to danger. This topic is discussed in detail in Chapter 4 but there are general precautions to be taken even if the pesticide is not specified in the Health and Safety (Agriculture) (Poisonous Substances) Regulations and certain of these precautions may, for emphasis, be dealt with here for they will bear repetition.

For the protection of operators handling formulations for dilution before use, that is to say w.p., s.c. and e.c. formulations, the instructions may include: Wear rubber gloves when handling the concentrate; wash off with soap and water any concentrate on the skin; avoid inhaling mist from diluted material during application; wash the hands and other exposed parts of the body with soap and water on the completion of any operation and before eating, drinking or smoking. For those handling dust formulations, the instructions may include: Do not handle the dust unnecessarily; avoid inhaling the dust; wash the hands and other exposed parts of the body on the completion of the operation and before eating, drinking or smoking.

Operators engaged in fumigation or the use of aerosols may take the following precautions: wear rubber gloves and face-shield and, if inhalation is unavoidable, wear an efficient respirator or suitable mask; if the skin is contaminated, wash thoroughly with soap and water and wash before eating, drinking or smoking; also wash rubber gloves inside and out after use and do not wear contaminated clothing longer than necessary.

For the protection of the consumer against possible contamination of his food, a time interval may be laid down under the Pesticides Safety Precautions Scheme between the last application of the pesticide and the harvesting of the treated crop. As this time is determined by the persistence of the pesticide deposit, it will be stated in the description of the compound concerned. Similarly for the protection of stock and poultry and for those engaged in the cultivation, thinning, pruning, etc., of the treated crop, access to the treated area should not be allowed until the lapse of a time interval specific for each chemical. Empty containers should be destroyed or disposed of out of reach of children see 'Code of practice for the disposal of unwanted pesticides and containers on farms and holdings', issued by MAFF. As many pesticides are powerful fish poisons, care is required to avoid the contamination of ponds, rivers or streams by the pesticide itself, by drainings from spray tanks, by wash water used to clean appliances or by empty containers.

Pesticides, and in particular insecticides, should not be applied to crops in flower, for bees and other pollinating insects would then be exposed; but these beneficial insects also visit flowering weeds, hence grass orchards should be mown before spraying and flowering weeds in plantations should be kept down.

Given effective pesticides, the responsibility for their safe use rests solely on the user. The discharge of this responsibility requires only common sense but, for the assistance of the user, his role has been defined in the 'Code of Conduct', a subject which belongs to Chapter 4, p. 97.

GENERAL NOTES ON USAGE

It is unfortunately impractical to give, in the description of a particular compound, the concentration to be used in pest control for, with each pest, the lowest effective concentration is recommended for reasons of both economy and safety. Such particulars must therefore be reserved for the disease or pest to be controlled. But even there the grower has to use a particular formulation so it might appear more convenient if the concentration of the formulation were given rather than the concentration of active ingredient. However with the older pesticides and those for which patent protection has expired, the grower is faced with a range of competitive formulations and it was to help him in his selection that the Agricultural Chemicals Approval Scheme was established. Also the concentration of a.i. in these formulations is seldom uniform and to phrase recommendations in terms of formulation would require lists of alternatives far longer than the somewhat tiresome lists of a.i. alternatives which he will meet in later pages.

Acephate (1)*. This organophosphorus compound was introduced in 1971 as an insecticide of moderate persistence and some residual systemic activity, effective against a wide range of fruit pests. It is not included in the Health and Safety (Agriculture) (Poisonous Substances) Regulations and has been cleared for use on hops and plums provided that an interval of at least 3 weeks elapses between application and harvest.

Aldicarb (56). This carbamate, introduced in 1965, is an anticholinesterase designed to resemble cholinesterase in molecular structure. It is included in the Health and Safety (Agriculture) (Poisonous Substances) Regulations as a Second Schedule, Part II substance (see p. 87) and, because of hazards to operators, is recommended for use only in granular form. (The Poisons Rules apply to this substance.) In addition to its insecticidal properties it is an effective nematicide. Uses cleared under the Pesticides Safety Precautions Scheme include: as a pre-

*The numerals refer to the Appendix, p. 392.

planting soil-incorporated treatment for early potatoes at rates up to 2·3 kg/ha, on main-crop potatoes up to 3·4 kg/ha; on sugar beet, soil-incorporated applied with the seed at drilling at rates up to 1·15 kg/ha; on bulb onions, as a soil-incorporated treatment, applied with the seed at drilling, at rates up to 1·7 kg/ha; on ornamentals grown outdoors or under glass at rates up to 11·4 kg/ha applied up to 3 times each crop.

Aldrin (41). This chlorinated hydrocarbon is of the cyclodiene group, so-called because cyclodiene derivatives are used in its manufacture. Its insecticidal properties, discovered in 1948, render the compound a potent stomach poison and contact insecticide, particularly effective against soil insects. It is of low solubility in water and somewhat volatile so that, of itself, it is not very persistent. However, aldrin is converted by biological oxidation to its epoxide, dieldrin, which is extremely persistent. It is compatible with almost all other pesticides, though certain minerals, used as diluents, induce a catalytic dehydrochlorination avoided by the incorporation of an inhibitor in w.p. and dust formulations.

Because of its conversion to dieldrin and the uncertainty of the long-term effects of the presence of small amounts in the animal body, the uses of aldrin have been limited by agreement under the Pesticides Safety Precautions Scheme. The currently agreed uses of the dust and liquid formulations are: control of wireworm in potatoes, of leatherjackets on DDT-susceptible varieties of spring barley, of cabbage root fly on brassicas, of narcissus bulb fly, of hop root weevil and of vine weevil.

2-Aminobutane (84). This fungicidal fumigant, introduced in 1962, has been approved for the treatment of seed potatoes, a process to be carried out only in premises subject to the Factories Act or in a process specifically cleared by Government Departments under the terms of the Pesticides Safety Precautions Scheme and under licence from the National Research Development Corporation. These recommendations are to be read in conjunction with the official Code of Practice entitled 'Fumigation of Potatoes with 2-aminobutane', obtainable free from the suppliers of the chemical. It is not in the Health and Safety (Agriculture) (Poisonous Substances) Regulations 1975 and its use should not present a hazard to operators provided they observe the precautions given on the label.

Azinphos-methyl (2) is an insecticidal and acaricidal organophosphorus compound first introduced in 1954 and since used for the control of a range of phytophagous insects and mites. Being relatively stable and of low vapour pressure, it is moderately persistent, and it is included in Part III of the Regulations, see p. 87, and the Poisons Rules apply to these products. The least time interval permitted between application as a spray and the harvesting of edible crops is 3 weeks in the case of apples, cherries, pears and plums, 2 weeks (3 weeks for azinphos-methyl plus demeton-S-methyl sulphone) in the case of blackberries, blackcurrants, gooseberries, loganberries, raspberries, strawberries, peas, mustard, brassica crops grown for seed and Brussels sprouts. Not more than 5 applications, each at 50 g a.i./100 litres, may be made in any one season to apples and pears; 4 applications each at 44 g a.i./100 litres to strawberries; 3, each at 38 g a.im/100 litres, to plums; 1, at 2ig a.i./100 litres, to cherries; 2, each at 44 g a.i./100 litres, to blackcurrants and gooseberries; 1, at 19 g a.i./100 litres, to blackberries and loganberries, applied at bud burst; up to 4 applications, each at 320 g a.i./1100 litres/ha to peas; 2, each at 44 g a.i./100 litres, to mustard and brassica crops grown for seed, applied before the flowers open *or* up to 3 applications, each at 28 g ami./100 litres, 2 applications to be made pre-blossom; up to 3 applications, each at 320 g a.i./680–1100 litres/ha, to Brussels sprouts. If applied LV, the amount of a.i. used per hectare should not exceed that which would have been applied HV. Stock and poultry should be kept from treated areas for 2 weeks.

Azinphos-methyl may also be used as a drench for brassicas, a single application at 50 g a.i./100 litres, applied at 77 cc/plant at transplanting.

The usual formulation is the 22 per cent w/v e.c. A w.p. formulation is available containing 25 per cent w/w azinphos-methyl plus 7·5 per cent w/w demeton-S-methyl sulphone for the control of the pre-blossom pest complex on top fruit and pests of soft fruit, mustard brassicas and peas; the minimum pre-harvest interval for this mixed product is 3 weeks.

Benodanil (85). This compound has recently been introduced for use as a systemic fungicide against the rust diseases of certain crops. It is systemic in action and appears to be without serious hazard to operators or to consumers of treated crops.

Benomyl (52). This systemic fungicide is one of the benzimidazole group and was introduced in 1967. It has been approved for use against a wide range of fungus diseases, particularly of fruit; it has acaricidal properties. Its toxicity hazards are slight and it is generally non-phytotoxic though the fruit colouring of certain apple varieties may be affected and some lenticel pitting has occurred on Egremont Russet. The usual formulation is the 50 per cent w.p. A seed dressing containing 30 per cent w/w benomyl and 30 per cent w/w thiram is available to control loose smut and other seed-borne diseases of winter wheat, *Ascochyta* and pre-emergence damping-off and seed decay of pea and field bean and neck rot of onion.

Binapacryl (73). This dinitrophenyl compound, introduced in 1960, has proved an effective acaricide and fungicide for use on apple against fruit tree red spider mite and powdery mildew. Up to 7 applications each at 50 g a.i./100 litre HV, may be applied for this purpose, or the equivalent amount LV. It should not be used on undercropped or intercropped fruit as blackcurrant, strawberry and gooseberry may be damaged. The interval between application and harvest should be at least 7 days and livestock should be kept from treated areas for at least 4 weeks.

Bordeaux mixture. The original copper fungicide for the protection of foliage and first devised in 1885. It is prepared by mixing a solution of copper sulphate with a lime suspension, a common recipe employs 1 kg copper sulphate crystals (bluestone) and 1·25 kg hydrated lime per 1000 litre (for further details, see Chapter 6, page 128). The copper sulphate is specified in MAFF Technical Bulletin 1, as containing not less than 98 per cent $CuSO_4.5H_2O$.

Fungicidal activity is associated with the slow formation, from the Bordeaux deposit, of water-soluble copper compounds, the ultimate toxicant being the cupric ion. The survival of the mixture for so long is largely due to the excellent adherence to foliage of the freshly prepared precipitate. The adherence of the precipitate deteriorates rapidly with time, so the spray should be used as soon as possible after preparation.

Bordeaux mixture is relatively safe to use, though care must be taken in its preparation as soluble copper salts are poisonous and the solution of copper sulphate should *not* be made in metal containers, otherwise the latter will receive a copper coating.

Bromophos (3). This organophosphorus compound was introduced in 1964 as a non-systemic contact and stomach insecticide which, at the recommended rates of application, persists on sprayed foliage for 7 to 10 days. Its mammalian toxicity is low enough for it not to be included in the Health and Safety (Agriculture) (Poisonous Substances) Regulations though it may be harmful to livestock and wildlife. It has been cleared for use on edible crops growing outdoors provided that the interval between use and harvest is at least 7 days; for use as a smoke on glasshouse crops and mushrooms with a minimum interval between treatment and harvest of 2 days; for use as a seed dressing on seeds of brassicas, carrots, parsnips, onions, beetroot and beans (broad, runner and dwarf).

Burgundy mixture is a variant of Bordeaux mixture in which the lime is replaced by washing soda. The proportion of washing soda used is slightly greater, a typical recipe being 1 kg bluestone, 1·25 kg washing soda per 1000 litres. The resultant spray is more likely to cause spray damage than is the similar Bordeaux.

Calomel (85) is the old name for mercurous chloride, long used both as an insecticide and as fungicide, though its strong phytotoxicity precludes its use on foliage except grass. Its action is thought to be due to its slow dissociation to metallic mercury, a reaction hastened by alkali, hence it is incompatible with lime and other alkalies. As an insecticide, its main use is against root maggots and, as a fungicide, it is used for the treatment of turf disease and against clubroot of brassicas and white rot of onion. For these purposes it is formulated as a 4 per cent dust, though the undiluted chemical is often recommended as a dip or overall spray.

Although a salt of mercury, it presents, unlike the related mercuric chloride (corrosive sublimate), no serious toxicity hazard though its powerful purgative properties renders care in handling imperative.

Captafol (87). This fungicide is closely allied, chemically, to captan but has found preference against certain diseases.

Captan (88), a protective fungicide, introduced in 1949 and since widely used for foliage protection. Though of pungent odour, the compound is practically non-volatile and being stable and virtually insoluble in water, it is of long persistence and compatible with most other pesticides. The reasons for its fungicidal activity are unknown but it is somewhat specific, being ineffective against powdery mildews. The common formulation for spray preparation is the 50 per cent w.p. and, for seed treatment, a 75 per cent dressing. The compound is of low toxicity to warm-blooded animals and presents only the hazard of skin irritation. It is, however, harmful to fish hence care is necessary to avoid contamination of ponds, waterways and ditches with the chemical or used containers—see 'Code of Practice for the disposal of unwanted pesticides and containers on farms and holdings'.

Carbaryl (57). This contact and stomach insecticide was introduced in 1956 and is used for the control of codling, tortrix and winter moth caterpillars, capsids and earwigs on apples, against pea moth and for the control of earthworms in turf.

Being of low vapour pressure and but slightly soluble in water, it is of moderate persistence but it is incompatible with alkaline pesticides such as Bordeaux mixture or lime sulphur.

Being a carbamate, its insecticidal action is probably due to the inhibition of the insect cholinesterases though the compound is not unduly toxic to warm-blooded animals. It is, however, toxic to fish and contaminations of streams and ditches must be avoided. An interval of at least 1 week must elapse between application and the harvesting of apples, pears, gooseberries, outdoor grapes, brassicas, lettuces, peas and tomatoes; 14 days in the case of raspberries and 6 weeks on blackcurrants and strawberries.

Carbendazim (53). This systemic fungicide, one of the benzimidazole group, was introduced in 1970 and is effective against a wide range of fungal diseases including powdery mildews, *Botrytis* spp. and apple scab. Its recommended uses should present no hazard to operators, animals and bees. It is available as a w.p. 50 per cent a.i. w/w or 60 per cent a.i. w/w; solutions of its salts are also available for injection into the trunks of elms to combat Dutch Elm disease.

Carbophenothion (4). This organophosphorus insecticide and acaricide was introduced in 1955. It is not included in the Regulations (see p. 87) but, under the Safety Precautions Scheme, is approved only for the treatment of cereal seed to be used in premises subject to the Factories Act or in a process specifically cleared by Government Departments under that Scheme.

Carboxin (58). This systemic fungicide, introduced in 1966, is available in Great Britain, only in admixture with an organomercury compound (and sometimes also gamma-HCH) or with thiram for use as seed dressings. It should not be used on cereal seed with a moisture content more than 16 per cent; treated seed should not be treated with any other seed dressing and should not be sown straight after treatment but the delay should not exceed a week or so. Livestock should be kept from the area sown with treated seed for at least 7 weeks. The Poisons Rules and usual precautions apply to the organomercury products.

Carbon tetrachloride. CCl_4, was first used in 1908 for the fumigation of nursery stock but, on the whole, its insecticidal potency is low. It has anaesthetic effects and is a potential liver and kidney poison. Any liquid spilt on the skin must be washed off at once and adequate protection must be provided to operators using the compound for fumigation. The vapour decomposes on hot surfaces to produce highly toxic gases, including phosgene, so smoking and the use of naked lights should be prohibited. At least a week's airing should be allowed before fumigated materials are used for human or animal consumption.

Cheshunt compound. A cuprammonium fungicide introduced in 1922 specifically for the control of damping off of seedlings. It is prepared by the intimate mixing of 2 parts by weight of bluestone (crystalline copper sulphate) and 11 parts of ammonium carbonate, the latter first being exposed to air to ensure the formation of the bicarbonate. Solutions of the mixture are used for the watering of soil and seed boxes.

Chlordane (42). This chlorinated hydrocarbon, introduced in 1945, is now of limited use as an insecticide and, in Great Britain, is recommended only for the control of earthworms in turf and lawns. Label instructions must be observed and livestock and poultry kept from treated areas for at least 2 weeks after treatment.

Chlorfenvinphos (5). This organophosphorus compound, was introduced in 1963, particularly for use against soil insects. It is included in Part III of the Regulations (see p. 87) and the Poisons Rules apply to formulations other than granules. It is cleared, under the Pesticides Safety Precautions Scheme, for use on brassicas, turnips, swedes, carrots, onions, maize, beans, celery, parsley, parsnips and cereals at rates up to 2·3 kg a.i./ha. On fen soils only, carrots may be treated at rates up to 5·7 kg a.i./ha per application. At least 3 weeks should elapse between application and harvest; on celery this interval should not be less than 12 weeks.

Its use as a seed dressing for cereals sown in the autumn is also acceptable but it should not be used on poor seed or on seed containing more than 16 per cent moisture.

Formulations are available for use in mushroom composts at rates not exceeding 110 g a.i./tonne compost and 50 g a.i./tonne casing soil.

Chloropicrin (89). An insecticidal fumigant first used as a pesticide in 1908. Because of its extreme toxicity, its use by inexperienced operators is seldom recommended but the intense irritation it causes to the eyes and mucous membranes enables it to be used as a warning agent in other fumigants. It is included in Part I of the Regulations (see p. 87) and the Poisons Rules apply. It should be used as a soil fumigant only in accordance with the Code of Practice issued by the MAFF for the safe and efficient fumigation of soil. Unprotected persons should not enter treated glasshouses and adjacent premises until advised otherwise by the fumigator.

Chlorothalonil (90). This fungicide, introduced in 1963, is effective against a wide range of crop diseases and, though irritant to skin and mucous membranes, is not included in the Health and Safety (Agriculture) (Poisonous Substances) Regulations. It has been cleared for use as a spray on the following crops provided that the interval between use and harvest is, on apples and pears 2 weeks, on strawberries, blackcurrants, redcurrants, gooseberries, grapes, raspberries, blackberries and logan berries 3 days. In glasshouses and minimum intervals when used as a spray are 12 hours for cucumbers, capsicum and tomatoes, 14 days for lettuce; when used as a smoke, 2 days for cucumbers and lettuces, 12 hours for tomatoes.

Chlorpyrifos (6). This non-systemic organophosphorus insecticide was introduced in 1965. It is not included in the Health and Safety (Agriculture) (Poisonous Substances) Regulations and has been cleared for use on apples and pears subject to an interval of at least 14 days before harvest and for use, in granule formulations, on leaf brassicas at planting out.

Copper fungicides. Many formulations of copper compounds have been introduced for application as sprays or dusts and to avoid the troublesome preparation of Bordeaux and Burgundy mixtures. These products are prepared from a variety of copper compounds, usually from copper oxychloride, made by the aeration of scrap copper in cupric chloride-sodium chloride solutions. The resultant product approaches in composition to $3Cu(OH)_2.CuCl_2$, though, if calcium chloride is present, the precipitate will approximate to $3Cu(OH)_2.CaCl_2$. Alternatively, curprous oxide may be used.

The copper dusts may have contents of 6–35 per cent metallic copper, though the 6 per cent dust is usual. W.p. formulations range from 20–50 per cent metallic copper.

The so-called colloidal formulations consist of suspensions of copper oxychloride in a dispersing agent and subjected to an intense grinding by which the particles are reduced to a size such that they remain permanently in suspension both in the paste and when diluted to spray strength.

Copper compounds present few hazards though they may be injurious to sensitive varieties of plants. Livestock, and particularly sheep, should be kept from treated areas for at least 3 weeks.

Corrosive sublimate (92). At one time this compound, mercuric chloride, was used for soil treatment against clubroot of brassicas and pests such as root maggots, but it has been superseded by the less poisonous calomel. Though not included in the Health and Safety (Agriculture) (Poisonous Substances) Regulations, it is scheduled under the Pharmacy and Poisons Act, 1933 and Poison Rules, 1960. The only accepted use now is by groundsmen as a turf fungicide in formulations containing not more than 15 per cent w/w mercuric chloride.

Cresylic acid. This name is applied to a coal tar derivative which consists mainly of the 3 isomeric cresols. The specified product contains

40 per cent w/w of phenols when analyzed by the steam distillation of the acidified product. On treated soil, planting should be delayed for at least 3 weeks, 4 to 5 weeks on heavy soils. It should be kept off the skin and away from the eyes, food and food containers. It is hazardous to fish. Keep greenhouses fully ventilated during application and for at least 24 hours afterwards.

Cufraneb is the common name approved by BSI for an ethylenebisdithiocarbamate complex containing not less than 8·15 per cent zinc, not less than 8·05 per cent manganese, not less than 5·5 per cent copper and not less than 1·0 per cent iron. It is effective against hop downy mildew and potato blight. Hops, apples, pears and blackcurrants should not be harvested within one week of its application; if used on mushrooms, 2 clear days should elapse before picking.

Cycloheximide (93). This antibiotic is produced in culture by *Streptomyces griseus* and is recovered as a by-product of streptomycin manufacture. It is included in the Regulations (see p. 87) as a Part II substance and has been accepted for use in forest nurseries for the control of Needle Blight of Western Red Cedar. Instructions for use must be carefully followed.

Cyhexatin (79). This organotin compound was introduced in 1968 and finds its main use for the control of phytophagous mites. It is not included in the Health and Safety (Agriculture) (Poisonous Substances) Regulations and has been cleared for use in up to 2 applications on apples, pears and plums at rates up to 550 g a.i./ha and on raspberries at rates up to 600 g a.i./ha provided that at least 4 weeks elapse between application and harvest; on strawberries after the crop has been harvested; on glasshouse cucumbers and tomatoes in up to 2 applications at rates up to 750 g a.i./ha with a minimum interval between treatment and harvest of 2 days.

Dazomet (65) is a soil fumigant introduced in 1952 and found successful against soil-borne diseases, eelworms, insects and weeds. The compound breaks down in soil to a dithiocarbamate which, in turn, yields methyl isothiocyanate. It is available as a prill containing 98 per cent a.i. and is of low toxicity to mammals though precautions should be taken to avoid inhalation or undue contact. Soil treatment indoors should be carried out at least 8 weeks before planting; outdoors, 4 to 5 weeks. Check that the soil to be planted is free from phytotoxic residues using the cress germination test (p. 253).

DDT (43). The first protective insecticide for outdoor use, the insecticidal properties of which were discovered in 1939 and which, by 1945, had attained a production of about 14 000 tonnes. This success encouraged a search for insecticidal properties in related compounds giving rise to a series usually called the chlorinated hydrocarbon group.

The name DDT is an abbreviation for *d*ichloro*d*iphenyl*t*richloroethane though the accepted chemical name for the compound is 1,1,1-trichloro-2,2-di-(4-chlorophenyl)ethane. The technical product is, however, a mixture of compounds and may contain up to 30 per cent of the *op'*-isomer. The specification approved by WHO calls for a content of at least 70 per cent of the *pp'*-isomer, which is the more effective insecticide.

The persistence of DDT is due to its low volatility, low solubility in water and to its good stability though, when in solution, it is readily decomposed by alkalies with loss of hydrochloric acid and insecticidal activity. A similar dehydrochlorination by an enzyme is thought to occur in some strains of insects from which the continued use of DDT has resulted in the selection of resistant strains.

DDT is both a stomach poison and a contact insecticide. It is harmless to plants, except to cucurbits and many varieties of barley. To man it is innocuous enough to be used for the control of body lice though it is not rapidly degraded in the body and may persist in the body fat. The ubiquitous contamination of the environment by DDT and by other persistent organochlorine pesticides which has resulted from their widespread use is deplorable and it should be used only where a persistent insecticide is really necessary. Livestock should be kept from treated areas for at least 14 days and the same interval should elapse between application and the harvesting of edible crops, except when used as an aerosol or smoke, when 2 days is long enough.

Dust formulations for agricultural use range in DDT content from 1–10 per cent, the usual being 5 per cent. W.p. formulations range from 20–50 per cent, the usual being 50 per cent. Solutions for e.c. or for aerosol use likewise have various DDT contents and a mixture with malathion is also available. DDT is sufficiently heat-stable to be used in smokes either of itself or in combination with gamma-BHC.

Demephion (7). This systemic organophosphorus insecticide was introduced in 1968, mainly as an aphicide and acaricide for use on fruit and vegetable crops. The name is applied to a mixture of *OO*-dimethyl *O*-2-methylthioethyl phosphorothioate and the corresponding thiolate, *OO*-dimethyl *S*-2-methylthioethyl phosphorothioate. It is placed in Part III of the Regulations (see p. 87) and the Poisons Rules apply. An interval of at least 3 weeks should elapse between application and harvest; livestock should be kept from treated areas for at least 2 weeks.

Demeton-S-methyl (8). This systemic organophosphorus insecticide is an improved form of demeton-methyl, introduced in 1950 as an aphicide and acaricide for fruit, vegetable and ornamental crops. Demeton-methyl consisted of a mixture of *O*-2-ethylthioethyl *OO*-dimethyl phosphorothioate and the related thiolate *S*-2-ethylthioethyl *OO*-dimethyl phosphorothioate but was, in 1963, replaced by preparations containing mainly the latter compound. Its insecticidal activity is due to its inhibitory action on the enzyme, cholinesterase, and a similar action in the animal body renders it poisonous to mammals. It is placed in Part III of the Regulations (see p. 87). Hence a strict adherence to the requirements of the Regulations and to instructions on the label is imperative. At least 3 weeks should elapse between application and harvest (2 weeks in the case of wheat and barley) and 2 weeks before allowing access of stock or poultry to treated areas. Mangolds should not be clamped within 10 days of treatment to avoid risks to workers; no part of the treated crop should be used as fodder until 3 weeks have elapsed. Brassica crops should not be treated after the end of October. Some ornamentals are sensitive to sprays.

Demeton-S-methyl sulphone (9). Oxidation of demeton-S-methyl leads first to the sulphoxide, and then to the sulphone, which was introduced in 1967 as a component of a formulation of azinphos–methyl. Safety precautions to be taken in the use of this formulation are accordingly those given for azinphos-methyl though the pre-harvest interval is extended to 3 weeks on edible crops.

Derris (92). The use, as an insecticide, of the ground root of certain species of Derris was patented in 1912, since when it has been established that the active components are a series of compounds called rotenoids, of which the main insecticide is rotenone. The rotenoids have also been found in other species of tropical plants used by the natives as fish poisons, including *Lonchocarpus* spp. from South and Central America. The reasons for the insecticidal activity of derris remain obscure but the insect respiratory system and heart are slowly paralyzed. The rotenoids are of low toxicity to most warm-blooded animals but, on edible crops, at least 1 day should elapse between use and harvest. Their extreme toxicity to fish, however, makes care necessary especially in the disposal of empty containers, see 'Code of Practice for the disposal of unwanted pesticides and containers on farms and holdings'. Although extracts are used in some formulations, it is not usual to extract the rotenoids from the root, which is used for dust formulations, generally containing 0·2–0·5 per cent rotenone. W.p. and e.c. formulations are available with rotenone contents ranging from 1–6 per cent.

Diazinon (10). This non-systemic organophosphorus insecticide and acaricide was first introduced in 1952 and has found particular use against aphids, thrips, flies, mites and some species of soil pests. It is of low water solubility and but slightly volatile. It is compatible with other pesticides but should not be combined with copper fungicides. Although an inhibitor of the enzyme cholinesterase, the compound does not present serious poison hazards to man and it is not included in Regulations though care in its handling is important. Two weeks should elapse between use and harvest, though with aerosols 2 clear days is long enough. Livestock and poultry should be kept from treated areas for at least 2 weeks. Cucumbers and tomatoes should not be treated before mid-May as there is risk of damage; maidenhair fern is sensitive. The usual formulations are the 16 per cent e.c. and 20 per cent atomizing solution; 5 per cent granules and a 40 per cent w.p. are available for soil use, the latter as a drench.

Dichlofluanid (94). This substituted sulphamide was introduced in 1966 as a wide-range fungicide, especially for the control of powdery mildews and of *Botrytis* spp. on soft fruit, vegetable crops and ornamentals. It is not included in the Regulations but an interval of at least 2 weeks should elapse between its application and the harvesting of strawberries, raspberries and dwarf beans, and of 3 weeks in the case of blackberries, blackcurrants, gooseberries, loganberries, outdoor-grown grapes, autumn-sown salad onions, outdoor lettuces, leaf brassicas under glass. When used on tomatoes under glass as a directed spray on flower trusses, 3 days is an adequate interval. It should not be used on strawberries under glass or plastic because of possible phytotoxicity.

p-Dichlorobenzene (49). This chlorinated hydrocarbon, introduced in 1915, finds its main use as a domestic fumigant against clothes moth, for which purpose it is marketed as crystals of high purity. It is also used against bulb-mite of narcissus.

Dichloropropane-dichloropropene mixture. This by-product of plastics manufacture was first used as a soil fumigant in Hawaii in 1943 and has since proved an effective nematicide in cooler climates. It consists of a mixture, roughly 54 per cent mixed 1,3-dichloropropenes and about 23 per cent of 1,2-dichloropropanes. Because of its phytotoxicity, a delay of at least 6 weeks is necessary before planting in treated soil. Its odour serves to warn operators against its harmful properties but care in handling is essential. On no account should the mixture come into contact with eyes, nose or mouth. Any liquid splashed on the skin must be washed off with paraffin and copious quantities of soap and water. Avoid breathing the concentrated fumes, do not open containers in confined spaces or apply in a closed house. Contaminated clothing, including polyethylene boots and gloves, must be washed immediately with soap and water and must not be used as long as the odour remains. The mixture must be kept away from foods, feedingstuffs, and food containers and contamination of ponds, streams and water courses must be avoided. As it is corrosive to many metals, equipment must be thoroughly washed out with paraffin immediately after use.

Dichloropropene (44) is available as a separate entity and is used for the same purposes as the mixture; similar precautions must be taken in its use.

Dichlorvos (11). Although this organophosphate was introduced in 1955 mainly for use against houseflies, it has recently been adapted for horticultural use in the form of e.c. and treated plastic strips from which the compound slowly volatilizes. It differs from other organophosphorus insecticides in its comparatively high vapour pressure and its ready decomposition. These properties reduce hazards to operators to the extent that, although the compound is placed in Part III of the Regulations and it is subject to the Poison Rules, it is safe to enter treated areas unprotected after 12 hours. Edible crops should not be harvested within 24 hours of use. It should not be applied to cucumbers or roses as damage may occur; certain chrysanthemum varieties are also sensitive. On mushrooms it should not be used at rates greater than 18 g a.i./100 m³;

on other edible crops the maximum rates are 7 g/100 m³ for indoor crops and 1·1 kg/ha for outdoor crops.

Dicloran (95). This fungicide, introduced in 1959, has been successfully used for the control of *Botrytis* spp., especially on lettuces. Its mode of action differs from that of most other protective fungicides in that it has little effect on spore germination but appears to intervene in cell division. Being practically insoluble in water, stable and of low volatility, it is persistent and compatible with other pesticides. Its mammalian toxicity is negligible though commonsense precautions should be taken in handling and the disposal of containers. The usual formulation is the 4 per cent dust and crops should not be harvested within 3 weeks of the last post-planting application. On lettuce the maximum rate and frequency of application, in terms of the 4 per cent dust, are: pre-planting 1·35 g a.i./m²; post-planting 0·34 g a.i./m², except for crops to be cut before Christmas when the rates are: pre-planting 0·67 g a.i./m²; post-planting 0·34 g a.i./m² applied once only at 2 to 3 weeks after planting.

Dicofol (50). This acaricide was introduced in 1955. If its chemical structure is compared to that of DDT, it will be seen that the only difference is that the hydrogen (H) of the bridge of DDT is replaced in dicofol by the hydroxy (OH) group, a change which converts the highly insecticidal but non-acaricidal DDT to an acaricidal but poorly insecticidal dicofol. The latter is effective against all stages, including eggs, of phytophagous mites.

Dicofol should not be used on seedlings or young plants under glass before mid-May or in bright sunshine as damage may occur. On apples, pears, plums, cherries and strawberries at least 7 days should elapse between application and harvest but, on edible indoor crops, this time may be reduced to 2 days. On hops and melons the interval should be at least 21 days and on grapes the interval between successive sprays and between the last application and harvest should be at least 7 days.

Dieldrin (45). This cyclodiene derivative was first employed against insects in 1948, since when it has been widely used mainly as a soil insecticide. Though its molecule contains oxygen it is grouped with the chlorinated hydrocarbons for it is the epoxide produced from aldrin; indeed it is manufactured by the oxidation of the latter. The name is reserved for a product containing at least 85 per cent of the epoxide. It is of high chemical stability and is compatible with all other pesticides. It is not included in the Health and Safety (Agriculture) (Poisonous Substances) Regulations, but great care should be taken to avoid skin contact for the compound is suspected of a hazardous toxicity. For the reasons given under aldrin (see p. 30), the uses of dieldrin have been limited by agreement under the Pesticides Safety Precautions Scheme. Liquid and w.p. formulations may be used for the control of cabbage root fly on brassicas to a rate of 1·5 kg a.i./ha. Seed dressings should not be used on poor seed or seed with a moisture content of more than 16 per cent because of phytotoxic risks. Other uses are on rubbed or graded sugar beet seed for precision drilling up to 2 g a.i./kg seed; on onion seed for the control of onion fly at a rate up to 38 g a.i./kg seed; on seed of french or kidney beans and runner beans up to 1·7 g a.i./kg seed; on spinach seed up to 2 g a.i./kg; as a dust on compost used for the potting of ornamentals against vine weevil at rates up to 19·5 g a.i./m³ of compost.

Dimefox (12) is one of the earliest of the systemic organophosphorus compounds discovered in Germany during the 1939—34 war. It was introduced in the UK in 1949 but its range of usefulness is limited by its high toxicity to warm-blooded animals, for which reason it is included in Part I of the Health and Safety (Agriculture) (Poisonous Substances) Regulations and the Poisons Rules apply. The common formulation is a 50 per cent e.c. for use on hops, at least 4 weeks before picking, at up to 2 applications each at 1·6 kg a.i./ha. Other accepted uses are on strawberries and Brussels sprouts, one application at 2·2 kg a.i./ha, on strawberries before flowering begins; on mangolds, fodder and sugar beet, up to 2 applications each at 3·3 kg a.i./ha.

Dimethirimol (97). This substituted pyrimidine was introduced, in 1968, as a systemic fungicide effective against powdery mildews, particularly of cucurbits. Its use should present no hazard to operators but 2 days should elapse between application and harvest. It is available as stock solutions of the hydrochloride for use as soil drenches.

Dimethoate (13). This systemic phosphorus compound was introduced in 1956, simultaneously in USA and in Italy, for the control of certain fruit flies but its use has now extended to many other sap-feeding insects. It is rather soluble in water and has a comparatively short life in plant tissues. Moreover it is of low mammalian toxicity permitting it to be tested for the systemic control of cattle grubs. As it is readily degraded by alkalies, it is incompatible with pesticides of an alkaline reaction, such as lime sulphur or Bordeaux mixture. An interval of at least 7 days should elapse between application and harvest when it is used as a spray, 4 weeks when used as granules. Livestock and poultry should be kept from treated areas for at least 1 week. It should not be applied to chrysanthemums, hops (except the variety Fuggles) or to ornamental *Prunus* spp.

Dinobuton (74). This dinitrophenyl compound was introduced in 1963 as a non-systemic acaricide and direct fungicide of value for field and glasshouse use for the control of red spider mites and powdery mildews. Care is necessary for it may be phytotoxic to tomatoes and to some cultivars of chrysanthemum and rose. It should not be mixed with alkaline pesticides nor with carbaryl. It is not included in the Health and Safety (Agriculture) (Poisonous Substances) Regulations but its use on edible crops should be restricted, on apples and pears, to up to 4 applications each at 2·3 kg a.i./ha *or* up to 10 applications each at 50 g a.i./100 litres HV at 10 to 14 day intervals; on strawberries up to 3 applications each at 50 g a.i./100 litres HV at 10 to 14 day intervals; on cucumbers up to 5 applications each at 50 g a.i./100 litres HV *or*, as an aerosol, up to 5 applications each at 6·7 g a.i./100 m³ glasshouse space. An interval of at least 3 weeks should elapse between use and the harvesting of apples, pears and strawberries, and 3 days between application and picking cucumbers. Livestock should be kept from treated areas for at least 3 weeks.

Dinocap (75). This compound arose from a search among dinitrophenyl derivatives for an acaricide less phytotoxic than DNOC. In 1949 it was found to be fungicidal, specifically against powdery mildews, for which purpose it finds its main use today. Originally thought to be a simple compound, it is now known to be a mixture of 2,4- and 2,6-dinitro-isomers, with the C_8 group joined to the aryl ring through the 2'-, 3'- or 4'- carbon atom of the chain. It is an ester and therefore unstable in the presence of alkalies so it should not be mixed with lime sulphur or other alkaline pesticides; nor should it be used with oil-containing sprays on account of its oil solubility.

Although closely related chemically to the poisonous DNOC, dinocap is less toxic to man though care should be taken to avoid inhalation and the skin and hair should be protected from its staining properties. Dinocap is toxic to fish and care is therefore required in the disposal of empty containers. A minimum interval of 1 week must be observed between last treatment and the harvest of outdoor crops and 2 days in the case of crops under glass. It should not be used on some cultivars of chrysanthemum.

Dioxathion (14). Introduced in 1954, this organophosphorus insecticide and acaricide is included in the Health and Safety (Agriculture) (Poisonous Substances) Regulations as a Second Schedule Part III substance (see p. 87) and the Poisons Rules apply. Cleared uses, apart from on non-edible crops, are on apples, pears and peas, provided there is an interval of at least 7 days between its last application and harvest. Livestock must be kept from treated areas for at least 2 weeks.

Disulfoton (15). This organophosphorus systemic compound, introduced in 1956, has found its greatest use for the protection of young plants against aphids. It is almost insoluble in water and stable except to strong alkalies. Being a toxic chemical, it is placed in Part II of the Regulations (see p. 87) and the Poisons Rules apply. The clearance maxima under the Pesticides Safety Precautions Scheme are as follows: on leaf brassicas 3·4 kg a.i./ha; on carrots and parsnips 2·3 kg a.i./ha; on potatoes, mangolds, fodder and sugar beet and strawberries 1·7 kg a.i./ha; on broad, horse, field and tick beans, peas, red beet, parsley and coriander 1·2 kg a.i./ha. If a single application is made within 7 days of planting, granules to a rate of 1·2 kg a.i./ha to french and runner beans and marrows, and to a rate of 2·3 kg a.i./ha to celery are cleared. There should be an interval of at least 6 weeks between last application and harvest of an edible crop. Granular formulations of various a.i. content and with various inert carriers are available.

Dithianon (98). This compound was introduced in 1962 as a protective fungicide of promise for the control of many diseases of pome and stone fruit and blackcurrants, though not effective against powdery mildews. In the UK it is used for the control of apple and pear scab, for which a colloidal formulation is available. This use should not present a hazard to operators.

DNOC (76), 4,6-dinitro-*o*-cresol also known as DNC, was the active component of a product introduced in 1892 for the control of the Nun moth. In a systematic examination of substituted aromatic compounds, carried out at the Rothamsted Experimental Station in 1925, it was found to be an excellent ovicide. Unfortunately its action on foliage is sufficiently drastic for it to be used as a herbicide, hence its use as an insecticide is limited to dormant trees. For this purpose it proved useful, for its toxicity to aphid and sucker eggs enables its use as an alternative to tar oils in the winter wash. DNOC-petroleum oil washes are specified, the MAFF specification requiring a content of not less than 60 per cent neutral oil and a content of DNOC not less than 1/45th of the weight of neutral oil in the product. The usual formulation, generally of the stock emulsion type, contains 65–75 per cent oil and 1·35–3·5 per cent DNOC.

DNOC-petroleum washes should not be applied to the plum variety Myrobalan nor to the red current variety Raby Castle.

The reasons for the ovicidal action of DNOC are unknown but the compound is also toxic to man and Statutory Regulations apply to the use of products containing more than 5 per cent DNOC. Although the winter wash is thereby excluded, protection of the skin and hair is advisable to avoid the staining properties of the cresol.

DNOC is itself specified in MAFF Tech. Bull. 1, but is mainly used as a weedkiller and its sodium salt, DNOC-sodium, is sometimes used for the disinfection of mushroom houses. This compound is subject to Part II of the Regulations and an added hazard is its inflammability for which reason DNOC-sodium is marketed with water as a paste. The MAFF specification requires that the product should contain between 97 and 103 per cent of the amount of DNOC stated on the label, but is applicable only to herbicides.

Dodemorph (99). This systemic fungicide was introduced in 1967 and is effective against the powdery mildews. It is not included in the Regulations (see p. 87) but care is necessary in its use for it is a skin irritant. It is phytotoxic to begonias, cinerarias and certain cultivars of roses. It is normally used to control powdery mildew on roses, both outdoor and under glass.

Dodine (100). This protective fungicide, introduced in 1956, has proved of special value against apple scab because of its ability to eradicate recent infections. Chemically, the compound has the features of a cationic wetter (see p. 25) and therefore should not be mixed with formulations of pesticides containing an anionic wetter except with the approval of the manufacturer. Its toxicity to warm-blooded animals is low but it may produce severe skin irritation, hence care is necessary in handling the concentrate which is normally a 65 per cent w.p.; a 20 per cent liquid formulation is also available.

Drazoxolon (101). This fungicide was discovered in 1960 and was introduced for use against powdery mildews and as a seed treatment against damping-off fungi. It is scheduled in Part III of the Regulations (see p. 87) and the Poisons Rules apply. Its use as a seed dressing is permitted only in establishments registered under the Factories Act 1961. Its use on foliage is limited to blackcurrants for the control of powdery mildew and leaf spot, at least 4 weeks before picking. Livestock should be kept from treated areas for at least 4 weeks. It is also used as a soil drench to control root-invading fungi. The normal formulation is a 30 per cent w/v col; a mixture with pirimiphos-ethyl is available for dressing seed.

Endosulfan (46). This cyclodiene insecticide, introduced in 1956, is sometimes included in the chlorinated hydrocarbon group though its molecule includes a sulphite grouping. Chemically the product is a mixture of 2 isomers which do not, however, appear to differ greatly in insecticidal activity. Being of a high mammalian toxicity it is included in Part II of the Health and Safety (Agriculture) (Poisonous Substances) Regulations and the Poisons Rules apply. Its use on edible crops is limited to blackcurrants, hops and strawberries. On fruiting blackcurrants to which up to 2 applications are permitted, one at first open blossom stage and one 3 weeks later, each at 50 g a.i./100 litres HV (or the equivalent LV), at least 6 weeks should elapse between the last application and picking; stock and poultry should be kept from sprayed areas for at least 3 weeks. On hops up to 6 applications each of 50 g a.i./100 litres are cleared. Its use on strawberries should be limited to the period between the completion of picking and the next season's flowering. Uses on non-edible crops are currently restricted by agreement to the control of bulb scale mite on narcissus, and of Tarsonemid mite on ornamentals grown under glass, of big bud mite on non-fruiting blackcurrant bushes for propagation and of pollen beetle on oil-seed rape and mustard not destined for use in feeding stuffs. It is available as e.c.s. containing 20 per cent a.i. w/v and 35 per cent a.i. w/v.

Endrin (47). This insecticide, introduced in 1951, is an isomer of dieldrin and, like the latter, is of long persistence. But as it finds use for foliage rather than soil application, the danger that its repeated use will encourage the selection of resistant strains is less significant than with dieldrin. Its mammalian toxicity is, unfortunately, higher than that of dieldrin and it is scheduled in Part II of the Regulations (see p. 87). On fruiting blackcurrants or blackberries the last application must not be made later than immediately before flowering; and on strawberries use should be limited to the period between the completion of picking and the next season's flowering. Stock and poultry should be kept from sprayed areas for at least 3 weeks. Formulations include a 19·2 per cent w/v e.c.

Apart from the limited uses on edible crops described above, uses on non-edible crops are currently restricted to the control of Tarsonemid mite on narcissus and ornamentals grown under glass.

Ethirimol (102). This systemic pyrimidine fungicide was introduced, in 1968, specifically for the control of powdery mildews of cereals, used either as a seed dressing or as a spray. It is not included in the Regulations and the following uses have been cleared: as a liquid seed dressing for spring-sown barley or applied to the soil when drilling barley; as sprays applied to growing winter wheat oats and barley. Treated seed should not be used for human or animal consumption nor should sacks which have contained dressed seed be used for grain for milling, malting or feeding stuff.

Etridiazole (103).† This compound has recently been introduced as a fungicide against certain soil-borne diseases of ornamentals and turf. Until its use has been cleared under the Health and Safety (Agriculture) (Poisonous Substances) Regulations, its recommendation must be regarded as provisional and experimental. Label instructions must be strictly followed.

Fenitrothion (16). This non-systemic organophosphorus compound, introduced in 1959, is of low mammalian toxicity and is not included in the Regulations. At least 2 weeks should elapse between use and harvest, 7 days for raspberries. Livestock should be kept from treated areas for at least 7 days.

Fentin acetate (79). This organotin compound was introduced in 1954 particularly for the control of potato blight. It is included in Part III of the Regulations (see p. 87) and the Poisons Rules apply. Provided the rate of each application does not exceed 280 g a.i./ha it should present no hazard to consumers. Livestock should be kept for at least 1 week from treated areas. In the UK it is available in combination with maneb as a w.p. containing 60 per cent fentin acetate and 20 per cent maneb. It should not be mixed with oil-based formulations.

Fentin hydroxide (81). This organotin fungicide was introduced in 1954, particularly for the control of potato blight. It is included in the Health and Safety (Agriculture) (Poisonous Substances) Regulations as a Second Schedule Part III substances and the Poisons Rules apply; provided that the rate of each application does not exceed 280 g a.i./ha, its use on potato foliage should present no hazard to consumers. Livestock should be kept out of treated areas for at least 1 week.

Folpet (104) is a fungicide closely related, chemically, to captan which it resembles in physical and biological properties but it may be found better than captan for certain purposes. It is currently available only in w.p. formulations in combination with dinocap for use on roses.

Fonofos (17). This organophosphorus insecticide, introduced in 1967, is particularly suited for the control of soil insects but is sufficiently hazardous to be scheduled in Part II of the Regulations (see p. 87).

---

†Proposed BSI common name for 5-ethoxy-3-trichloromethyl-1,2,4–thiadiazole.

Provided that at least 6 weeks elapse between application and harvest, its use on cereals, potatoes, leaf brassicas, swedes, turnips, carrots, sweetcorn and maize, at a rate not exceeding 4·5 kg a.i./ha, should not present a hazard to consumers. It is available in granular formulations.

Formaldehyde. The commercial solution of this compound is known as formalin and was, at one time, used for the treatment of cereal seed but it now finds a wider use as a soil disinfectant. The Poisons Rules apply to it. It is specified by the MAFF as a clear solution containing not less than 37·5 g and not more than 40·5 g formaldehyde (H.CHO) per 100 ml at 20°C. The solution may become cloudy through a polymerization of the formaldehyde, a process hastened by low temperatures; hence long storage in the cold should be avoided. Treated soil should not be sown or planted until all smell of the chemical has gone which takes about 3 weeks on light soils but up to 6 weeks on heavy clay soils.

Formothion (18). This systemic organophosphorus aphicide and acaricide was introduced in 1964 and is suitable for application to most crops except chrysanthemums and hops. It is not included in the Regulations but an interval of at least 1 week should elapse between application and harvesting or allowing access of stock or poultry to the treated area.

Gamma-HCH (48), is the most active insecticide present in HCH (*q.v.*) from which it is prepared by crystallization. Although practically insoluble in water and of low volatility, it is sufficiently lipoid-soluble to act in a pseudo-systemic manner on insects within the plant tissue adjacent to the deposit. It is compatible with other pesticides though, like DDT, it is dehydrochlorinated by alkali when in solution. An interval of 2 weeks should elapse between application and harvest or permitting the access of stock or poultry to the treated area. Though not of the strong tainting propensities of HCH, the hazard is there if applied to susceptible crops or if potatoes or carrots are planted where gamma-HCH has been used during the previous 18 months.

Dust formulations range from 0·2−0·65 per cent; e.c. products range from 12−20 per cent. Solutions for aerosol use are available and the compound is sufficiently heat-stable to be used pelletted with a pyrotechnic mixture or in smoke generators. When so used the interval before harvesting should be at least 2 days. Gamma-HCH is also incorporated, in amounts of 20−40 per cent, in certain organomercury seed dressings to extend activity to soil insects such as wireworm and the wheat bulb fly. The precautions to be taken are given under organomercury seed dressings.

HCH (formerly **BHC**). The insecticidal properties of this product were first put to use in the early 1940s. It is a mixture of several of the isomers of hexachloro*cyclo*hexane, a chlorinated hydrocarbon, of which the *gamma* isomer, see gamma-HCH, is the most toxic to insects.

The use of HCH is restricted by the readiness with which many treated crops, in particular, potato and blackcurrant, acquire an unpleasant taint, a risk which is taken if susceptible crops are grown on land treated more than once in the current of the 2 preceding years. This period indicates that HCH is not rapidly decomposed in soil and a build-up may occur if it is applied annually on the same area. A period of at least 2 weeks should elapse between its use and harvest or the access of stock or poultry to the treated area. Both e.c. and w.p. formulations with various contents of gamma-HCH are available for spray preparation. The heat stability of the product permits its use in smokes, which may be enriched with gamma-HCH.

Lead arsenate (105) (diplumbic hydrogen arsenate, $PbHAsO_4$) is the classic stomach poison first used in 1892. It is specified by the MAFF both as powder and as paste. The powder should contain not less than 62 per cent lead oxide (PbO) and not less than 32 per cent arsenic, calculated as arsenic oxide (As_2O_5). As water-soluble compounds of arsenic are generally phytotoxic, their content must not exceed 0·25 per cent As_2O_5. The corresponding figures for the paste are 28·4 per cent PbO, 14 per cent As_2O_5 and 0·25 per cent As_2O_5 in water-soluble form, with a content of water not more than 52 per cent. Though the Poisons Rules apply to lead arsenate, hazards to operators are not serious if commonsense precautions are taken, but those to consumers require an interval of at least 6 weeks between application and harvest. Stock and poultry should be kept from treated areas for at least 6 weeks though, if heavy rain intervenes, this period may be reduced to not less than 3 weeks. It should be noted that, under the Arsenic in Food Regulations, S.I. 1959, No. 831 and the Lead in Food Regulations, S.I. 1961 No. 1931, it is illegal to sell, deliver or import any food containing arsenic in amounts exceeding 1 ppm, expressed as elementary arsenic, or lead in amounts exceeding 3 ppm Pb.

Lime sulphur (106) has been used as a fungicide for over a century. It is made by boiling sulphur and lime water and, chemically, is a solution of calcium polysulphides ($CaS.S_x$) with a smaller content of calcium thiosulphate. It is specified by the MAFF, the specification requiring that the solution should be clear and free from sludge, should have an apparent density at 20°C of $1·30 \pm 0·1$, expressed as g/ml and should contain not less than 24 per cent w/v of polysulphide sulphur (the S_x of the above formula). On dilution with water the solution acquires an alkaline reaction which limits its compatibility with other pesticides. On exposure to air, the solution decomposes and the polysulphide sulphur is liberated as elementary sulphur. Lime sulphur may therefore be regarded as a formulation of sulphur and its biological properties are accordingly discussed under that title. Lime sulphur is, however, more effective than sulphur against scale insects, no doubt because of the alkalinity of the undecomposed solution.

Malathion (19). This non-systemic organophosphate was introduced in 1950 and is used for the control of sap-feeding insects and mites. It is slightly soluble in water and is of short to moderate persistence. Being susceptible to decomposition by alkalies it should not be used with alkaline spray materials. It is apt to injure certain ornamentals. Consumer hazards are low but an interval of at least 1 day should elapse between application and harvest though, to avoid taint, it is advisable to allow 4 days.

Mancozeb (66). This complex of zinc ion and maneb containing 20 per cent manganese and 2·5 per cent zinc was introduced in 1961 as a dithiocarbamate fungicide particularly for use against potato blight. The following intervals should elapse between application and harvest: 7 days on apples, blackcurrants, celery, cereals, gooseberries, hops, lettuce and pears; 2 days on mushrooms and tomatoes under glass. It is formulated as an 80 per cent w.p. and also in combination with zineb.

Maneb (67). This dithiocarbamate fungicide was introduced in 1950 and has been found useful, particularly for the control of downy mildews. There is evidence that the manganese present accelerates decomposition to the active fungicide, thought to be etem. Maneb is of low toxicity to animals, whether man or insect, but care should be taken to avoid inhalation or skin contact. Treated outdoor edible crops should not be harvested within 7 days of application, indoor edible crops 2 days.

Manganese and zinc dithiocarbamate complex. This dithiocarbamate fungicide is used for the control of leafspot and rust of blackcurrant, scab of apple and pear and potato blight. At least 1 week must elapse between treatment and harvest.

Manganese, Zinc, Iron Dithiocarbamate complex. This dithio-carbamate fungicide was introduced, in 1966, for the control of potato blight and of tulip fire. It can be irritating to the skin, eyes and nose. Crops should not be harvested within 1 week of treatment.

Menazon (20) was introduced in 1961 as a systemic organophosphate with selective insecticidal properties, being a potent aphicide yet without marked toxicity to other animals. It is sparingly soluble in water and with a negligible vapour pressure yet is taken up by the plant which is kept free from aphids for long periods. Its heat stability is poor but it is compatible with all but strongly alkaline pesticides. It is not included in the Regulations but livestock should be kept from treated areas for at least 3 weeks, and edible crops should not be harvested until at least 3 weeks after treatment. It is marketed as a 40 per cent liquid formulation and as a 70 per cent w.p., the latter being used for dips and soil drenches. Dry seed dressings are available for the protection of sugar beet from early aphid attacks.

Mephosfolan (21). This systemic organophosphorus insecticide, introduced in 1963, has been cleared for the control of aphid on hops by soil application. It is included in the Health and Safety (Agriculture) (Poisonous Substances) Regulations as a Second Schedule Part II substance and the Poisons Rules apply. At least 4 weeks should elapse between application and hop-picking and at least 10 days before allowing the access of livestock to treated areas.

Mercuric oxide, HgO, has been cleared for use in canker paints and as a protective seal on bark injuries and pruning cuts. When formulated as a ready-to-use paint containing not more than 5 per cent w/w a.i., it is not included in the Health and Safety (Agriculture) (Poisonous Substances) Regulations.

Metaldehyde (108). This compound, the slug-killing properties of which were discovered about 1936, is specified by the MAFF, the specification requiring a content of not less than 95 per cent by weight of metaldehyde calculated as acetaldehyde ($CH_3.CHO$). Its molluscicidal properties are usually attributed to an anaesthetic action which prevents the slug from regaining shelter with consequent desiccation; it is also toxic to warm-blooded animals. The compound has a domestic use as a solid fuel and must be kept out of reach of children. If used on edible crops, 10 days should elapse before harvesting. Poultry should be kept from treated areas for at least 7 days. For use against snails and slugs the crushed tablets are mixed with a protein-rich milling offal such as bran, but a number of approved slug baits are available with contents of 2·5–6 per cent a.i. and liquids with a higher a.i. content.

Metham-sodium (68). This soil fumigant was introduced in 1954 and has been found useful for the partial sterilization of soil and for the control of soil nematodes and fungi. Chemically it is a dithiocarbamate readily soluble in water and, though stable in concentrated solutions, is decomposed in the soil to form methyl isothiocyanate which is the active agent. Users should follow label instructions carefully for the compound is irritant to the eyes and mucous membranes. Metham-sodium is also so strongly phytotoxic that it has been used as a weedkiller. Treated soil should not be planted until decomposition is complete and all smell of the chemical gone. When used under glass the first following crop should only be carnation, tomato or lettuce. A cress germination test (see p. 253) is advisable and care must be taken to ensure that the fumes do not reach other plants.

Methidathion (22). This organophosphorus compound was introduced, in 1966, as a non-systemic insecticide and acaricide and is scheduled in Part III of the Regulations (see p. 87) and the Poisons Rules apply. For use in apples, up to 5 applications each at 50 g a.i./100 litres HV or the equivalent LV and, on hops, up to 10 applications at the same strength are allowed, at least 3 weeks before picking. Livestock should be kept from treated areas for at least 2 weeks.

Methiocarb (59). This carbamate, introduced in 1965 as a non-systemic insecticide and acaricide, has found a special use as a molluscicide. It is not included in the Regulations but it should not be used on land growing edible crops within 7 days of harvest. Poultry should be kept from treated areas for at least 7 days; containers and compound should be kept from children and birds; operators should wash hands and exposed skin before meals and after work. For the control of slugs and snails, the usual formulation is as pellets containing 4 per cent w/w, broadcast at 220 g a.i./ha.

Methomyl (60). This carbamate, introduced in 1966 as a systemic insecticide and acaricide, is included in the Health and Safety (Agriculture) (Poisonous Substances) Regulations as a Second Schedule, Part II substance and the Poisons Rules apply. It has been cleared for use as a foliar spray on hops, up to 10 applications per season provided there is an interval of at least 7 days between the last application and harvest. Unprotected personnel should be kept out of treated areas for at least 24 hours; livestock for at least 7 days.

Methyl bromide (109), was first used as an insecticide in 1938 and has since proved of great value in the fumigation of stored products, as a soil disinfectant and nematicide, and as a herbicide. It is a colourless gas, almost insoluble in water and with a chloroform-like odour which is usually masked by the chloropicrin added at 2 per cent as warning agent. Methyl bromide is an insidious poison, included in Part I of the Poisons List. It should be used for soil and compost fumigation only by contract servicing companies in accordance with the recommendations given in the pamphlet 'Methyl Bromide: Precautionary Measures, 1972', published by Her Majesty's Stationery Office.

Mevinphos (23). This systemic organophosphorus insecticide was introduced in 1953. It is of brief persistence for it is water-soluble and is decomposed by water. It has a high mammalian toxicity which places it in Part II of the Regulations (see p. 87) and the Poisons Rules apply. The maximum rate of application per season is 410 g a.i./ha on mangolds and fodder beet, 275 g a.i./ha on apples, cherries, pears, plums, currants, gooseberries, potatoes, spinach and brassicas, 206 g a.i./ha on hops, lettuce, peas and sugar beet, 137 g a.i./ha on beans and strawberries. At least 3 days should elapse between application and harvest, except 3 weeks in the case of potato and 1 week for currants and gooseberries. Unprotected personnel, livestock and poultry may enter treated areas after 1 day but wildlife should be flushed from the area before mevinphos is applied.

Nabam (68). When the fungicidal activity of this compound was discovered in 1943, it aroused interest as the first example of a water-soluble compound exerting protective action. The explanation was later found to be a decomposition on aeration to form the insoluble and fungicidal etem. But in the field, nabam proved unsatisfactory partly because of phytotoxicity and it was found necessary to add zinc sulphate to precipitate the nabam as the zinc salt, since called zineb. Later the manganese salt, maneb, was found better for some uses. Accordingly, nabam is mainly used, apart from soil application, for the tank-mix preparation of these two fungicides. Nabam is available both as a powder containing at least 93 per cent of the anhydrous salt and as a stock solution containing at least 22 per cent of the anhydrous salt.

It is of low toxicity but when used on edible crops, either alone or in conjunction with zinc or manganese salts, an interval of at least 7 days should elapse between use and harvest of outdoor crops, 2 days for crops under glass.

Naled (24). This organophosphate, introduced in 1956 in the United States, has useful fumigant properties as an insecticide and is of comparatively low mammalian toxicity. Provided that there is an interval of at least 24 hours between use and harvest, its use as a fumigant on mushrooms and tomatoes at a rate not exceeding 10 g a.i./100 m³ of glasshouse or mushroom house space should not present a hazard to consumers. Unprotected personnel should be kept out of treated glasshouses for at least 12 hours and the house should be ventilated for at least 30 minutes before it is entered.

Nicotine (109) is the insecticide of the tobacco steep which, 50 years ago, was about the only available general aphicide. Nowadays nicotine is extracted from tobacco waste and is specified by the MAFF as the mixed alkaloids of tobacco containing not less than 95 per cent by weight of nicotine. It is readily soluble in water and is sufficiently volatile to act as a fumigant, for which reasons it is non-persistent. Its toxicity to mammals warrants its inclusion in the Regulations as a Part III substance though formulations containing less than 7·5 per cent w/w are exempt. A particular hazard is due to its rapid absorption through the skin; splashes of the concentrate must be at once washed off with cold water. Edible crops should not be harvested for at least 2 days after treatment, except as smokes when the minimum interval is 1 day. Unprotected personnel should not enter treated glasshouses for at least 12 hours; livestock should be kept from treated areas also for 12 hours.

Formulations include dusts, usually containing 3 per cent nicotine, and a number of liquid formulations of various nicotine contents. The concentrate generally used is the 95 per cent material though it may be noted that the sale of this material is not practised in many other countries where the solution of nicotine sulphate containing 40 per cent nicotine is the rule.

Omethoate (25). This organophosphorus compound, introduced in 1965 as a systemic insecticide and acaricide, is included in the Health and Safety (Agriculture) (Poisonous Substances) Regulations as a Second Schedule Part III substance and the Poisons Rules apply to it. It has been cleared for use on non-edible crops; as a spray on winter-sown cereals; as a spray on apples and hops subject to an interval of at least 3 weeks between application and harvest. Livestock must be kept from treated areas for at least 7 days.

Organomercury seed dressings. The fungicidal components of these pesticides will not be dealt with alphabetically for, nowadays, the dressings are usually applied by the seed merchant, a practice to be encouraged. A number of compounds are used and among those of cereal seed treatment for the control of seed-borne fungi may be mentioned 2-methoxyethylmercury silicate and phenylmercury urea, used in dry treatment, and methylmercury dicyandiamide, used in liquid treatment. The dry dressings contain from 0·6–1·5 per cent mercury, expressed as metal, and the liquid dressings 0·6–2·0 per cent mercury. The latter are supplied only to authorized seed merchants and are subject to the Factories Act, 1961. Growers using dry dressings must observe the following precautions: wear rubber gloves and protect the wrists and forearms, for some organomercury compounds are strong vesicants; protect the nose and mouth from any dust; wash with soap and water, immediately if the skin is contaminated, after dressing the seed and before eating, drinking or smoking. Similar precautions should be taken when drilling treated seed. Treated seed must never be used as food or fodder and sacks used for treated grain should never be used for millable grain and should be well shaken and washed before use for feeding stuffs. The practice of reinforcing organomercury seed dressings with gamma-HCH is referred to under the heading of the latter compound.

For the treatment of sugar beet seed the compound generally used is ethylmercury phosphate; that for the steeping of seed potato tubers is 2-methoxyethylmercury chloride. Both treatments should only be carried out in premises subject to the Factories Act or by a process specifically cleared by Government Departments under the Pesticides Safety Precautions Scheme.

Oxamyl (61). This carbamate, introduced in 1969 as a systemic insecticide and nematicide, is included in the Health and Safety (Agriculture) (Poisonous Substances) Regulations as a Second Schedule Part II substance and the Poisons Rules apply to it. It has been cleared for use against potato cyst eelworm (potato), stem eelworm (onion), Docking disorder, pygmy beetle, millepedes, early leaf miner (sugar beet) and potato root eelworm (tomato, at planting).

Oxycarboxin (62). This compound, introduced in 1966 as a systemic fungicide for the control of rust diseases of cereals and vegetables, is not included in the Health and Safety (Agriculture) (Poisonous Substances) Regulations. It has been cleared for use as a spray on barley, wheat and on ornamentals outdoors or under glass.

Oxydemeton-methyl (26). As indicated by its common name, this organophosphate is a derivative of demeton-methyl which is oxidized in the animal body to the sulphone with oxydemeton-methyl as an intermediate product. Accordingly the latter, which is also known as demeton-S-methyl sulphoxide, was introduced in 1960 as a systemic insecticide and acaricide. Because of its mammalian toxicity, it is placed in Part III of the Regulations (see p. 87) and the Poisons Rules apply to it. A minimum interval of 3 weeks should elapse between application and harvest, except in the case of wheat and barley when 2 weeks is stipulated. When used on mangolds or fodder beet there should be an interval of at least 10 days before clamping and the treated crop should not be used as fodder for at least 3 weeks. It should not be used on brassica crops after the end of October. Livestock should be kept from treated areas for at least 2 weeks.

Parathion (27), the first organophosphorus compound to find insecticidal use, was introduced in 1944. It is of slight water solubility and is sufficiently volatile to exert a pseudo-systemic action on insects, such as leaf miners, feeding near the plant surface. It is therefore of moderate persistence. As it is rapidly decomposed by alkalies, it is not compatible with alkaline pesticides. Its insecticidal action is due to an inhibition of the cholinesterases causing a failure of nervous function though, to accomplish this, parathion must first be oxidized to the corresponding phosphate. As this oxidation proceeds in the animal body faster than detoxication, parathion is extremely toxic to man and is therefore placed in Part II of the Regulations (see p. 87) and the Poisons Rules apply. When used as a spray the maximum rate and frequency of application per season are: for peas, 1 application at 19 g a.i./100 litres; for fodder and sugar beet and mangolds, 2 applications each at 19 g a.i./100 litres; on tomatoes and cucumbers under glass, up to 3 applications each at 12·5 g a.i./100 litres. For soil application as a spray on cucumbers, 1 application at 6·25 g a.i./100 litres; on tomatoes, 1 application at 38 g a.i./100 litres *or* up to 3 applications each at 16 g a.i./100 litres are specified. If sprayed LV the amount applied per acre should not exceed that which would have been applied HV. For soil application as granules on cereals and brassicas, 1 application at 5·7 kg a.i./ha is accepted before or soon after planting. In glasshouse use as an aerosol, up to 5 applications each at 5 g a.i./100 m^3; as a smoke, up to 5 applications each at 14 g a.i./100 m^3 are accepted. There should be an interval of at least 4 weeks when used as granules or spray, 2 days when used as an aerosol and 24 hours when used as a smoke, between use and the harvesting of an edible crop. Unprotected personnel should not enter treated areas for at least 1 day; livestock should be kept out for at least 10 days.

Paris green (111). This pigment was used, a century ago, against Colorado beetle but, for foliage application, it was soon replaced by lead arsenate. But it still finds use as the toxic component of baits for the control of slugs and soil insects for which purpose it is mixed with dried blood or a protein-rich milling offal. It is specified by the MAFF as a fine green powder containing not less than 30 per cent copper, expressed as cupric oxide; not less than 55 per cent arsenic, calculated as As_2O_3; not more than 1·5 per cent arsenic, as As_2O_3, soluble in water; not less than 10 per cent acetate, expressed as acetic acid. The powder should pass completely through an 80 mesh BS sieve and not more than 5 per cent should be retained on a 300 mesh BS sieve. The compound is a powerful poison and domestic animals should be kept from the bait for at least 3 weeks. Moreover, it causes suppuration of open wounds which should be protected from contact.

Petroleum oils. Distillates from petroleum, often called mineral oils, have in one form or another been used for pest control since the days when paraffin or kerosene was first used as an illuminant. Nowadays their use for crop protection is limited to certain high-boiling fractions from the lubricating oil range. Chemically such oils are predominantly hydrocarbons but, to ensure safety to foliage, a high degree of refinement is necessary; in technical terms the percentage of unsulphonated residue must be high. Their main use is acaricidal either on the dormant tree or on foliage. For these purposes, 4 types are specified by the MAFF.

(1) Winter washes of the e.c. (miscible oil) type should contain not less than 75 per cent by weight of neutral oil with the following characteristics: a specific gravity of between 0·96 and 0·93 at 15·5°C/15·5°, a distillation range such that less than 5 per cent by volume will distil over at an oil temperature of 350°C and an unsulphonated residue not less than 65 per cent by volume. As phenols are sometimes used as mutual solvents in this method of formulation and tend to be phytotoxic, their content must not be greater than 6 per cent by weight.

(2) Winter washes of the s.e. type should contain not less than 60 per cent by weight of neutral oil of the same characteristics as under (1).

(3) Summer washes of the s.e. type for orchard use should contain not less than 60 per cent by weight of neutral oil of the following characteristics: a specific gravity of between 0·84 and 0·92 at 15·5°C/15·5°, a distillation range such that 10 per cent of the oil distils between 310 and 340°C, 50 per cent between 350 and 375°C and 80 per cent between 380 and 400°C, a viscosity between 100 and 200 sec. Redwood at 70°F and an unsulphonated residue not less than 92 per cent by volume.

(4) Summer washes for glasshouse use should be of the s.e. type and contain not less than 60 per cent by volume of neutral oil of the same characteristics as (3), except that the specific gravity should be between 0·84 and 0·89 at 15·5°C/15·5°, and an unsulphonated residue not less than 95 per cent by volume. In all 4 types, the specifications require that the concentrated and diluted emulsions shall pass prescribed stability tests.

The action of the oil on the mite is presumably physical for oils of these characteristics could hardly be toxic in the usual sense, indeed oils of the glasshouse type approach medicinal paraffin in blandness. To extend their usefulness the incorporation of other pesticides is commonly practised, either to the formulation (see DNOC) or to the diluted emulsion.

Phenylmercury compounds. Phenylmercury chloride found use as a foliage fungicide because of its ability to eradicate recent infections of apple and pear scab, for which purpose it was introduced in 1942. It is virtually insoluble in water and, unlike the related phenylmercury acetate which is used for seed treatment, is non-volatile. Certain apple and pear varieties are liable to damage especially in wet weather and the label instructions on varieties must be obeyed. Apples and pears should not be sprayed after the end of July and an interval of at least 6 weeks is advised between application and picking whereas stock and poultry should be kept out for at least 2 weeks. Formulations may include lead arsenate to extend activity to codling moth and leaf-eating caterpillars or sulphur to add fungicidal properties against apple mildew. Phenylmercury chloride is available as a 2·5 per cent ($\equiv$1·5 per cent Hg) w.p. used at 200 g product/100 litres HV; other organomercury scab fungicides are phenylmercury dimethyldithiocarbamate in 3 per cent ($\equiv$1·5 per cent Hg) w.p. used at 100 g product/100 litres HV; phenylmercury nitrate in 2·5 per cent ($\equiv$1·6 per cent Hg) w.p. used at 200 g product/100 litres HV; and a liquid formulation of phenylmercury salicylanilide at 6 per cent ($\equiv$4 per cent Hg) used at 160 litres product/ha, HV or LV. Phenylmercury salicylate is used in greenhouse aerosols, the concentrate containing 0·3 per cent Hg. This product comes within Part IV of the Regulations and requires an interval of at least 12 hours between use and the harvesting of an edible crop, of which only tomato is recommended. The number of applications is limited to 5 at intervals of at least 7 days, the dosages at each application should not exceed 140 mg organically-combined Hg/100 m^3.

2-Phenylphenol (113). This disinfectant finds use as a paint for the control of apple canker. It should be used only in the dormant period and not on maiden trees. If used for the prevention of rots in harvest fruit, the dip or spray should contain not more than 2 per cent w/v, otherwise there may be a risk of contravening the Preservatives in Food Regulations, 1962.

Phorate (28). This systemic organophosphate, introduced in 1954, has found its main use for seed or soil treatment to protect young plants from attack. For this purpose it is formulated as granules containing 10 per cent a.i., whereby the hazards arising from its mammalian toxicity are reduced. Phorate is practically insoluble in water and is unstable in solution; in the plant sap it is thought to be oxidized, ultimately to a stable sulphone, but the intermediate products are unstable. It is placed in Part II of the Regulations (see p. 87) and the greatest amounts which should be applied in any one season are: on brassicas, except root crops, 4·5 kg a.i./ha; on carrots, celery at transplanting, mangolds, fodder beet, red beet and sugar beet, parsnips and potatoes, 3·3 kg a.i./ha; on brassica root crops, dwarf, french and runner beans, peas, 2·2 kg a.i./ha; on broad, horse, field and tick beans, strawberries and sweet corn, 1·7 kg a.i./ha. There should be an interval of at least 6 weeks between application and harvest or before allowing livestock to enter treated areas.

Phosalone (29). This non-systemic organophosphorus compound was introduced in 1963. It is not included in the Regulations but its recommended use is restricted, on apple and pear, to not more than 5 applications per season at 675 g a.i./ha, not more than 4 postblossom; on plums, to 3 at 675 g a.i./ha of which not more than 2 should be postblossom; on brassica seed crops, 3 applications at 450 g a.i./ha at green bud, early yellow bud (if weevil attack is severe) and yellow bud; on grape vine outdoors, 5 applications on preblossom and 4 postblossom at 340 g a.i./ha. An interval of at least 3 weeks should elapse between use and harvest and livestock should be kept from treated areas for at least 4 weeks.

Phosphamidon (30). This systemic organophosphate, introduced in

1957, is readily soluble in water and, in solution, is rapidly decomposed; its half-life in plant sap is estimated to be 2 days. It is effective against sap-feeding insects and mites. It is sufficiently toxic to be included in Part III of the Regulations (see p. 87) and the Poisons Rules apply to it. Its use on non-edible crops is not restricted but, on edible crops, the number of applications and amounts applied are defined, under the Pesticides Safety Precautions Scheme: on apples and pears, up to 5 applications at 32 g a.i./100 litres, on hops, up to a total of 2·2 kg a.i./ha, the last application not being greater than 550 g a.i./ha; on sugar beet, fodder beet and mangolds, up to 3 applications at 350 g a.i./ha; on brassica crops, up to 4 applications at 300 g a.i./ha; on carrots and peas, 1 application at 300 g a.i./ha; on broad and field beans, up to 2 applications at 220 g a.i./ha; on potatoes, up to 4 applications at 325 g a.i./ha; on blackcurrants, 1 application at 20 g a.i./100 litres; and on strawberries, up to 2 applications at 20 g a.i./100 litres. A minimum interval of 3 weeks between application and harvest is recommended. Livestock should be kept from treated areas for at least 2 weeks.

Pirimicarb (63). This carbamate was introduced in 1969 as a systemic insecticide of particular promise for the control of aphid strains resistant to organophosphate aphicides. It is not in the Regulations (see p. 87) and the following uses have been cleared under the Pesticides Safety Precautions Scheme: on non-edible crops; as a spray on apples at the green cluster stage; as a spray on the following crops provided that the interval between application and harvest of cucumbers and peppers under glass is at least 2 days; tomatoes, strawberries, 2 days; Brussels sprouts and dwarf beans, 3 days; lettuces (outdoors), cabbage, cauliflowers and peaches, 7 days; peas, broad beans, potatoes and lettuces (under glass), 14 days; sugar beet, 8 weeks; as a smoke on lettuces under glass, 14 days. No interval is needed when pirimicarb is used as a smoke on cucumbers, tomatoes and peppers under glass, nor when it is used as a soil drench on glasshouse tomatoes and cucumbers. Livestock should be kept from treated areas for at least 7 days.

Pirimiphos-ethyl (31). This organophosphorus compounds, introduced in 1971, as a broad-range insecticide mainly for soil use, is included in the Health and Safety (Agriculture) (Poisonous Substances) Regulations as a Second Schedule Part III substance and the Poisons Rules apply to it. It has been cleared for use as a granular formulation incorporated in mushroom compost or casing and as a seed dressing for dwarf beans and runner beans. Its use as a seed dressing is permitted only in establishments registered under the Factories Act, 1961, for which purpose it is currently available in the UK only in admixture with drazoxolon.

Pirimiphos-methyl (32). This organophosphorus compound was introduced in 1970 as a fast-acting insecticide and acaricide, not truly systemic but capable of a translaminar action. Unlike its ethyl analogue (see above) it is not included in the Regulations. It has been cleared for use as a fog applied through a fogging machine in mushroom houses and to cucumbers, tomatoes, peppers and ornamentals growing under glass; and as a spray on carrots, sugar beet, and winter wheat provided at least 7 days elapse between treatment and harvest.

'Polyram' is the trade mark of a dithiocarbamate fungicide, a complex of zineb and polyethylenethiuram disulphide, once given the BSI common name of metiram, a name withdrawn when it was shown that the 'complex' is a mixture. It is also known as Zinc PETD and was introduced, in 1961, mainly for the control of potato blight. Inhalation and skin contact should be avoided.

Propineb (70). This dithiocarbamate fungicide, once named mezineb, was introduced, in 1966, mainly for the control of downy mildews and potato blight. Regular use will also suppress red spider mite. It is not included in the Regulations and has been cleared for use on apples, blackcurrants, celery, gooseberries, hops, outdoor grapes and potatoes.

Propoxur (64). This carbamate insecticide was introduced in 1959. It is not included in the Regulations (see p. 87) but when used on edible crops growing outdoors, an interval of at least 7 days should elapse before harvest; on glasshouse tomatoes and cucumbers, 7 days, but when used as a smoke, an interval of at least 2 days is adequate. Livestock should be kept from treated areas for at least 7 days.

Pyrazophos (33). This organophosphorus compound, introduced in 1971, is a systemic fungicide effective against powdery mildews. Unlike the other organophosphorus compounds described in this Chapter, it has a limited insecticidal and acaricidal action. The uptake by roots when applied to the soil or as a seed dressing is insufficient for an adequate systemic control. It is not included in the Regulations and its use as a spray has been cleared on non-edible crops, on apples, brassicas, cereals, hops and raspberries provided that there is an interval of at least 2 weeks between use and harvest; of 3 days on cucumbers grown under glass. Livestock should be kept from treated areas for at least 2 weeks.

Pyrethrum (82). The ground flower heads of certain species of pyrethrum, in particular *Chrysanthemum cinerariaefolium,* have been used as a domestic insecticide for over a century. The active components are a group of esters known as pyrethrins, present in amount from 0·7 to 3 per cent, which are potent contact insecticides with a rapid paralytic effect ('knockdown'). But these esters are susceptible to decomposition especially in sunlight and their persistence in the open is brief. Moreover they are rapidly decomposed in the insect body though a group of compounds, pyrethrum synergists, are known which intervene in the detoxication. Certain of the synthetic pyrethrins are less subject to detoxication and of higher insecticidal activity. One is resmethrin which has been cleared for agricultural and horticultural use.

Both pyrethrum and its synergists are virtually harmless to warm-blooded animals though the flower heads contain an oil which may cause dermatitis on sensitive persons. For dust use, the ground flower heads are diluted with a non-alkaline carrier; for spray or aerosol use, a synergist is usually added and often derris or DDT to ensure that the insects knocked down do not recover.

Quinomethionate (114). This compound, once named oxythioquinox, was introduced in 1965. It is a cyclic carbonate and has both fungicidal and acaricidal properties, the latter of use against mites resistant to organophosphorus compounds. Its main uses are as an acaricide on apples and against powdery mildews and leafspot on currants and gooseberries. It is not included in the Regulations (see p. 87) but skin contact and inhalation should be avoided. On cereals it should be used at least 2 months before harvest. An interval of 3 weeks should elapse between application and the harvesting of apples and pears, of 2 weeks for currants, gooseberries, and strawberries. On cucumbers and tomatoes under glass, 2 days is long enough but 1 week for outdoor marrows.

Quintozene (77). Although this fungicide, also known as PCNB, was first introduced in the late 1930s, it has but recently become widely used as a soil fungicide and for the control of Rhizoctonia. As it is practically insoluble in water, of low volatility and stable in soil, it is persistent; it is of low toxicity to man and other animals. Certain crops are sensitive to it and cucurbits may be damaged when grown in treated soil; tomato seed should not be sown in treated compost until 2–3 weeks after treatment; potatoes are tolerant. The usual formulation is the 20 per cent dust.

Sodium pentachlorophenate. Pentachlorophenol has been widely used for timber preservation since 1936 and its sodium salt, being water-soluble, is a handy form in which to employ its potent fungicidal properties. Unfortunately it is rather toxic to mammals, the acute oral $LD50$ for rats being 210 mg/kg; solutions stronger than 1 per cent may cause intense skin irritation and its dust or vapour provokes intense sneezing. Hence the greatest caution is needed in its use for the disinfection of wooden structures such as mushroom houses.

Streptomycin. This antibiotic, isolated in 1944, is of wide medicinal use but has been applied for the control of certain bacterial diseases of stone fruit. It is sometimes mixed with another antibiotic, oxytetracycline, and such products for crop protection uses containing one or both compounds must, by law, be rendered unpalatable and unfit for medical or veterinary use. Uses on edible crops are as a spray on cherries at a rate not in excess of 22 g streptomycin with 2·1 g oxytetracycline per 100 litres HV, or the equivalent LV, applied at a period not between petal fall and harvest. Streptomycin by itself may be used on hops, 2 applications each at 110 g a.i./100 litres but not within 8 weeks of picking. Protective gloves should be worn when handling or stringing hops within 10 days of treatment. The use of an emulsion paint containing 1 per cent streptomycin on stone fruit after harvesting should not lead to a consumer hazard but the operator should wear rubber gloves, rubber boots, face-shield and a mackintosh or overalls with sleeves rolled down and collar fastened when using the paint. If extensive use is to be made of these antibiotics, general practitioners and hospitals in the area should be asked to look out for signs of untoward effect.

Sulfotep (34). This organophosphorus compound was introduced in 1944 as a contact insecticide of low persistence but its main use today is as a fumigant in closed spaces. Its high mammalian toxicity places it in Part II of the Regulations (see p. 87). Although these regulations do not require operators to wear protective clothing when lighting a smoke generator, protective clothing must be worn if it is necessary to re-enter the house to relight a generator. Any person entering a treated glasshouse during the 12 hours following treatment is required to wear an overall, hood, rubber gloves and respirator. Its use as a smoke on any edible indoor crop should be at a rate not greater than 0.3 g a.i./100 m^3 and an interval of at least 24 hours should elapse between application and harvest.

Sulphur (115), has been in long use as a fungicide and acaricide but the reasons for its toxicity to these organisms is still little understood. It is insoluble in water and almost non-volatile, though, with suitable equipment, it is used for the fumigation of glasshouses. Chemically it is stable and is compatible with other pesticides. It is non-toxic to man and warm-blooded animals and to plants, though certain varieties are peculiarly sensitive to sulphur; the label instructions on 'sulphur shy' varieties must be carefully followed. Dusting sulphurs usually contain 95 per cent sulphur. W.p. formulations are specified by the MAFF as dispersible sulphurs and the specification requires a content of not less than 70 per cent sulphur with a particle size such that less than 40 per cent of the sulphur shall be present in particles of 6µm or less diameter and not less than 9 per cent in particles of 2µm or less diameter. Paste formulations are specified as colloidal sulphurs of which the sulphur content should not be less than 40 per cent and the particle size of this sulphur should be such that not less than 90 per cent should be of particles of 6µm or less diameter and not less than 55 per cent in particles of 2µm or less diameter.

Tar oils, derived by the distillation of coal tar, have for years been used as fungicides for wood preservation. But tar distillate washes for dormant use on fruit were first introduced in the UK in 1921, and at once became popular for the control of aphids, scales and suckers and for the removal of lichens. The tar distillates used are those from which anthracene has been recovered, hence the anthracene oils, which consist mainly of aromatic hydrocarbons with small amounts of derivaties soluble in alkali, tar acids or phenols, or soluble in acids, tar bases. The ovicidal compounds, specially against aphid and psyllid eggs, reside in the neutral fraction (insoluble in acids or alkalies). Strong phytotoxicity limits the use of tar oils to fully dormant plants but they should not be applied to the plum variety Myrobalan nor to the currant variety Raby Castle. The Poison Rules apply and contact may cause dermatitis.

Tar oil washes are specified by the MAFF; the specification for the e.c. (miscible oil) type calls for a content of not less than 80 per cent by weight total oil, not less than 52 per cent by weight of the wash should be soluble in dimethyl sulphate and not more than 10 per cent by weight of the wash shall be phenols. The neutral oil soluble in dimethyl sulphate (i.e., aromatic hydrocarbons) shall have a distillation range such that at least 40 per cent by weight of the wash should distil above 230°C, not less than 22 per cent above 290°C, and not less than 10 per cent above 335°C. The specification of the s.e. type requires a content of not less than 60 per cent by weight of neutral oil; not less than 54 per cent by weight of the wash should be neutral oil soluble in dimethyl sulphate and not more than 5 per cent by weight of the wash shall be phenols. The neutral oil soluble in dimethyl sulphate should have a distillation range such that the fraction distilling above 230°C should constitute at least 48 per cent by weight of the wash, that distilling above 290°C at least 27 per cent by weight of the wash and that distilling above 335°C, at least 11 per cent by weight of the wash. In both types, the undiluted and the diluted wash should pass certain stability tests.

To extend the ovicidal properties of the wash to such insects as capsid bugs and to mites, the addition of petroleum oils to the concentrate is possible but the amount of petroleum oil needed is roughly twice that of the neutral tar oil. But the combined tar-petroleum wash, with its 10 per cent oil content, is seldom used today and it seems unnecessary to cite the specifications which are to be found in the MAFF Techn. Bull. 1.

Tecnazene (76). This fungicide, also called TCNB, was introduced in 1947, and is, nowadays, mainly used for the control of the dry rot of potatoes caused by *Fusarium* spp. Its fungicidal properties are similar to those of the related quintozene but it is more volatile and hence of lower persistence in the open. Moreover its ability to inhibit the sprouting of potato tubers enhances its usefulness for the storage of ware potatoes. It is marketed either as a dust of 3–6 per cent a.i., as a 6 per cent e.c., or for use as a smoke. The latter are used for the control of *Botrytis* spp. in glasshouses and gamma-HCH may be added to extend activity; but the mixture should not be used on roses. At least 2 days should elapse before smoked edible crops are harvested.

Tetrachlorvinphos (35). This organophosphorus compound was introduced in 1966 as a selective insecticide and acaricide. It is non-systemic and is not included in the Regulations. Its use on apples, brassicas, peas, blackcurrants and gooseberries should present no hazard to consumers provided that the interval between use and harvest is at least 7 days.

Tetradifon (51). This acaricide was introduced in 1954 and is effective against eggs and young active stages of phytophagous mites and renders the adult females sterile. It is virtually insoluble in water, non-volatile and of a stability which renders it of good persistence and compatible with other pesticides. Yet it is capable of a pseudo-systemic effect and is toxic to mites on the side of a leaf opposite to that sprayed. It is of low toxicity to insects and mammals and harmless to plants except young cucumbers or plants that are wet. It is available as a concentrate of 8 per cent a.i.

Thiabendazole (54). The fungicidal properties of this benzimidazole derivative, in use since 1962 as an anthelminthic, were discovered in 1964. It acts systemically against a wide range of foliar diseases and is also effective for the post-harvest treatment of fruit and vegetables against storage diseases. It is not included in the Regulations and has been cleared for use on edible crops and as a post-harvest dip for top fruit, walnuts and grapes. Its use as a fog on ware potatoes is only acceptable if in combination with an iodorphor provided that there is an interval of at least 21 days between treatment and removal for sale or processing.

Thiometon (36). This systemic organophosphorus insecticide and acaricide was introduced in 1953 and is used mainly as an aphicide on vegetable, farm and fruit crops. It is included as a Part III substance in the Regulations (see p. 87) and the Poisons Rules apply to it. When used on edible crops, at least 3 weeks should elapse between application and harvest. Livestock should be kept from treated areas for at least 2 weeks. It is available as a liquid concentrate and as a dust, the latter containing not more than 10 per cent a.i.

Thionazin (37). This systemic organophosphorus compound, introduced in 1959, has found its main uses as a nematicide for the control of stem and bulb eelworm of narcissus and tulip and of leaf and bud eelworm of chrysanthemum; as an insecticide against cabbage root fly and as an acaricide against bulb scale mite. It is included in the Regulations (see p. 87) as a Part II substance and the Poisons Rules apply to it. Its uses, apart from bulb treatment, are restricted to use as a soil drench on leaf brassicas to which it should be applied within a week of transplanting; to tomatoes which should not be harvested within 11 weeks of treatment; on mushroom compost at rates up to 20 g/tonne. Livestock should be kept from treated areas and from dipping premises for at least 8 weeks. Bulbs for dipping should not be presoaked nor should pre-cooled bulbs be dipped if roots have developed. Drenches should not be applied to tulips and brassicas should not receive drenches before April.

Thiophanate-methyl (55). This compound of the benzimidazole group was introduced in 1970 as a systemic fungicide effective against a wide range of foliar diseases. It is not included in the Regulations and has been cleared for use as a spray on non-edible crops, on apples, pears, blackberries, blackcurrants, strawberries, raspberries, outdoor grapes, field, navy, broad and dwarf beans, mushrooms, lettuces, tomatoes and cucumbers and cereals; as a post-harvest dip on apples and pears; as a paint for canker wounds and pruning cuts on fruit trees, ornamental trees and shrubs; as a seed treatment for onions. Its use on strawberry runner beds is not recommended nor should it be applied to dessert apples from petal fall to fruitlet stages. Soil drenches may be used for glasshouse crops.

Thiram (71). This dithiocarbamate was introduced as a fungicide in 1934 and has since been used both on foliage and seed. It is practically insoluble in water and non-volatile and is compatible with most other pesticides. Its action on fungi is not fully understood but it is different from the other dithiocarbamates listed here for, being the derivative of a secondary amine, it is therefore not susceptible to the reactions thought to be associated with the fungitoxicity of these primary amine derivatives. It is harmless to plants at the concentrations used and is of low toxicity to warm-blooded animals, except perhaps poultry for it can be used as a bird repellent. But it should not be applied to fruit intended for canning or deep freezing because of the possibility of taint and a discoloration of the tins. Care in handling is necessary because of irritation to the skin and mucous membranes. The usual formulations for spray use are in 80 per cent w.p. and a 50 per cent colloidal suspension; for seed treatment a 50 per cent powder is available but the usual method of treating seed is the thiram soak treatment described on p. 314. Edible crops should not be harvested within 7 days of treatment and treated seed should not be used for human or animal consumption.

Triazophos (38). This organophosphorus compound was introduced in 1971 as a broad-range insecticide and acaricide with some nematicidal properties. It is included in the Health and Safety (Agriculture) (Poisonous Substances) Regulations as a Second Schedule Part III substance and the Poisons Rules apply to it. It has been cleared for use as a spray on ornamentals, apples, pears, cucumbers, peas, carrots, brassicas and maize; as a winter wash on top fruit trees; also for use in granular formulations at planting time on the above crops. At least 4 weeks should elapse between application and the harvest of these crops, except for peas when the interval should be at least 3 weeks. Livestock should be kept from treated areas for at least 7 days.

Trichlorphon (39). This phosphonate was introduced in 1952 mainly for the control of DDT-resistant house flies but it has found some use in crop protection. It is moderately soluble in water but is decomposed by alkalies and hence incompatible with alkaline pesticides. Its mammalian toxicity is low enough for it to be excluded from the Health and Safety (Agriculture) (Poisonous Substances) Regulations but the usual precautions must be taken especially by those who have to handle other organophosphates. It exerts a pseudo-systemic action which renders it effective against leaf-miners and cabbage root fly larvae attacking Brussels sprouts in the month before harvest for only 2 days need elapse between application and harvest.

Tridemorph (116). This systemic fungicide was introduced in 1969 for the control of mildew on spring barley. It is not included in the Regulations and is cleared for use on growing cereals and grass seed crops and on ornamentals. Livestock should be kept from treated areas for at least 14 days.

Triforine (117). Introduced in 1969 as a systemic fungicide, this compound is effective against powdery mildews and other leaf diseases. It is not included in the Regulations and has been cleared for use on non-edible crops, on cereals up to growth stage 10; on blackcurrants and gooseberry bushes after the fruit has been picked.

Urea, is used to protect tree stumps. It is not included in the Health and Safety (Agriculture) (Poisonous Substances) Regulations.

Vamidothion (40). This systemic insecticide and acaricide was introduced in 1962 and is included in Part III of the Health and Safety (Agriculture) (Poisonous Substances) Regulations and the Poisons Rules apply. Under the Pesticides Safety Precautions Scheme, the maximum rates and frequency of application are given as: on apples and pears, 1 application at 20 g a.i./100 litres between green cluster and pink bud and up to 2 applications later at 40 g a.i./100 litres HV or the equivalent LV; on cherries, plums and raspberries, 1 application at 40 g a.i./100 litres HV or the equivalent LV; on hops, up to 3 applications each at 710 g a.i./ha; on brassicas, fodder beet, sugar beet, mangolds and potatoes, up to three applications each at 570 g a.i./ha; on grapes up to 2 applications each at 925 g a.i./ha. Brassicas should not be treated after the end of September. There should be an interval of at least 6 weeks between application and the picking of apples, pears, cherries, plums or raspberries; and of 4 weeks in the cases of brassicas, fodder and sugar beet, mangolds, hops and potatoes. Livestock should be kept from treated areas for at least 4 weeks. The usual formulation is the 40 per cent s.c.

Zinc, manganese, copper, iron dithiocarbamate complex. This dithiocarbamate fungicide, was introduced in 1967 for the control of potato blight, downy mildewe and other foliage diseases. It is marketed as an 80 per cent w.p. and is compatible with most insecticides and with copper oxychloride, but should not be mixed with lead arsenate, lime or lime sulphur.

Zineb (72). When, in 1943, it was found that the addition of zinc sulphate improved the fungicidal efficiency of nabam, zineb was prepared in the tank in the field (see p. 50). But nowadays the ready-prepared compound is available and finds use against the downy mildews. It is practically insoluble in water and non-volatile and is compatible with other pesticides. Its fungicidal action is thought to be that of etem and it is harmless to all, except zinc-sensitive, plants. Its toxicity hazards are low but inhalation and skin contact should be avoided. On edible crops at least 7 days should elapse between application and harvest except that for crops under glass and mushrooms, 2 clear days is adequate.

Chapter 3
The application of pesticides

In protecting crops against pest and disease it is necessary to distribute a small amount of pesticide over the surface of the host plant, a task much more difficult than the treatment of seed or soil. On the crop, the pesticide may be applied as a dust or, more usually, diluted with water and sprayed on. A variety of methods are available for making the application and these must be selected to suit the particular requirements of the crop or pest.

3.1 Spraying

3.1A Methods and terminology

Volume. In almost all cases in Great Britain, water is used as the diluent of the spray. The chemical is supplied as a concentrate, the several types of which have been described in Chapter 2. Various amounts of spray may be applied to each area of crop and the volume used may affect the distribution of the chemical. A large volume increases the chances of full cover and may assist the wetting of difficult targets. When the volume is reduced and the same level of pesticide is required, it is necessary to increase the concentration of the spray and to use a finer drop 'spectrum' (a term for the distribution of the drops of a spray into size groups). The cover may vary from a complete film of spray on the target to a distribution of discrete droplets.

The volume applied will give different degrees of cover depending on the area of foliage of the crop and the following terms are commonly applied:

	Litres/ha on bushes and trees	Litres/ha on ground crops
Very low volume	under 200	under 50
Low volume	200–500	50–200
Medium volume	500–1000	200–700
High volume	over 1000	over 700

Volumes as little as a few litres per hectare have been used, overseas, in certain insect control measures for many years. The insecticide in this case is in a diluent which is much less volatile than water and this is formed into a very fine spray of drops just large enough to settle on the target. The technique, which has become known as ultra low volume (ULV) application, is now being tried on a variety of crops in Britain with specially designed equipment and formulations.

It has long been clear that control of certain specific insects and fungi can be achieved by this method but it is by no means generally applicable. The nature of European agriculture, with its close juxtaposition of various crops, places severe restrictions on the use of very fine sprays of low volatility because of drift hazards, and in some countries their use is illegal. Because the fine droplets are almost invisible, application by this technique is difficult to monitor, and trials should not be undertaken except in collaboration with the companies providing the formulations or with official bodies able to determine the distribution of material being achieved.

Concentration. The amount of chemical to be applied to ground crops will usually be recommended as amount per hectare; and recommendations for fruit tree spraying may be quoted as amount per 1000 litres for high volume application or, for all types of application but usually for low volume, as amount per hectare. The volume of spray may be adjusted to suit the density of the crop and, in all cases, label instructions must be followed.

Dose and cover. Dose and cover in a spraying context are usually taken to mean respectively the amount of chemical on a unit area of the target and the proportion of that area covered. Fortunately for the grower it is unnecessary to pursue these concepts and it may be accepted that dose is roughly governed by the amount of active chemical applied per hectare of ground and that cover is often dependent on the amount of water used. Where high or medium volume is used, the addition to the spray of surface-active 'wetters' or 'spreaders' may be recommended in order to increase the cover obtained. Such materials can markedly affect the spread and retention of spray on the plant and they should only be used as recommended. The distribution of the spray, i.e. the uniformity of dose, can obviously be influenced by the type of the spraying equipment and its operation.

Requirements. It is difficult to generalize since some pests and diseases can be readily controlled by chemicals even when the latter are in-differently applied, whereas, with other pests and diseases, it is essential to obtain as complete a cover as possible. Generally, the application of insecticides and fungicides and very selective herbicides must be carried out with more precision than herbicides like MCPA and 2,4-D and foliar nutrients. Systemic compounds, which are translocated by the plant, do not require complete cover to be effective but this property does not mean that they may be carelessly applied.

3.1B   Equipment

Types of machine

The ground crop sprayer. The ground crop sprayer most widely used by the farmer and smallholder is tractor mounted and comprises a tank of 200–700 litres capacity, a pump driven directly from the tractor power take-off and a horizontal boom with a span up to about 10 m. The boom is adjustable for height above the crop and is fitted with nozzles so placed to ensure a complete swath of uniform deposition. This type of machine is widely used at low and medium volumes for weed control.

Tank contents are agitated by fluid returned to the tank from the pump or by a mechanical stirrer. The working pressure of the pump can range between 3 and 20 bar depending on the type used and it can be regulated within the available range by adjustment of a relief valve. In order to reduce spray drift, this pressure is often reduced to the lowest consistent with a satisfactory nozzle performance, but the precaution is of more importance in the application of herbicides than in the application of insecticides and fungicides for which higher pressures are recommended. The pump capacity is selected to suit the volume of spray to be applied per hectare, the swath width and the speed of the tractor.

The spray contractor or the farmer with a large acreage uses a similar but larger machine. The boom may be up to 20 m in length and these machines often employ a trailed tank or have saddle tanks mounted on a large tractor. The tank capacity and the size of the pump must be matched to the requirements of the machine. The larger tanks are sometimes provided with mechanical agitation of the contents. For extensive use with toxic materials it is advisable to provide air-conditioning in the tractor cab. In all applications it is necessary for the operator to take necessary precautions as detailed in Chapter 4, pp. 83–99.

Sprayers for bush and cane crops. Ground crop sprayers, when fitted with vertical spray bars, can be used for spraying small trees, bushes and canes. In densely leaved crops like blackcurrants outputs up to 2000 litres/ha and pump pressures up to 40 bar may be needed. It is also advantageous to angle the individual nozzles up the height of the crop in a manner that will produce cross flows in the spray to move the leaves and achieve penetration into the crop and minimize underdosed areas of leaf and cane.

Tree sprayers. Although the trend in apple orchards is towards intensive plantings of small trees which can be sprayed with spray bars of hydraulic nozzles, the great majority of plantations will for some years to come be treated by machines using an airstream to carry the spray to the tree. While some jobs (control of certain insects or the application of foliar nutrients) may require large volumes of spray the majority of treatments can be made satisfactorily at not more than 600 litres/ha if air is used to distribute a fine spray through the foliage.

The 1960s saw a general trend towards airblast sprayers with physically smaller fans than were used earlier. This meant smaller volumes of air at higher pressures. In many instances the fan pressure was increased to 0·7 to 1 m w.g. so that air shear nozzles could be used in adjustable outlets to give a simple and relatively cheap machine suitable for a wide range of crops. Most machines used air at around 200 to 250 mm w.g. either from twin centrifugal fans directed one to each side, or from axial fans similar to, but smaller than, those used on older sprayers. The smaller fans gave a relatively compact fan and pump mechanism which could be mounted on the tractor linkage leaving only the tank to be trailed. Tanks were usually of around 700 to 1000 litres and the same mechanism was sold both for tractor mounting or built onto the tank in a conventional trailed machine. For smaller scale operations a few completely mounted units were sold having an integral tank of about 300 litres capacity.

Recent research has shown that penetration of spray into trees and the uniformity of spray deposition is improved if it is carried in a large volume of slowly moving air. Such airstreams are also much less affected by the travelling speed of the sprayer so that even in small trees which are easy to penetrate they may be used to give the advantage of faster spraying. For this reason the trend in tree sprayers is towards the use of trailed machines with fans and outlets as large as overall machine dimensions allow.

Machines having low pressure fans require conventional hydraulic nozzles working at a pressure of 20 bar or more to give the fine sprays needed, or compressed air nozzles may be used. Agitation of the tank contents on all but the smallest of these machines is by mechanical stirrer.

Orchard sprayers, sometimes with adjustment of the nozzle and outlet arrangement, are used for the treatment of hops in Kent. Smaller machines, suited to the European vineyard market, are often used in the West Midlands where tree crops do not have to be treated on the same holding.

Aerial spraying equipment. Aircraft spraying in the United Kingdom is restricted by the small acreages of crops and the hazards presented to the pilot. Its main use is in potatoes and other horticultural crops which require treatment when they are at a stage when tractor wheels would cause damage. Special marking and navigational aids are required but the basic spraying equipment is no different in principle from that used on ground equipment.

Manual knapsack sprayers. Five to 15 litres of spray can be carried in a knapsack sprayer and the contents of such a machine may be pumped out either by the operation of an attached hand pump, the second hand being used to direct a nozzle, or by a charge of pressurized air above the liquid in the container. The first type is now made mainly in plastic materials and is mass-produced at relatively small cost. A small pressurized chamber is included with the pump unit in all but the smallest and cheapest designs so that continuous pumping is not necessary unless a very large, or several small nozzles, are being used.

The older types which do not require pumping while in use are still available. They have metal containers for the spray which also hold a charge of compressed air. The air capacity is usually adequate to discharge the liquid contents without repumping and, if pressure-regulating valves are fitted, the liquid may be sprayed out at constant pressure. In one type of this appliance, the spray is put into the tank and air is then pumped in above by a fitted pump or a compressed air supply may be used. In a second type the air is retained and the sprayer is recharged by pumping in the liquid to be sprayed. Pressures of 1·5 to 5 bar are normally used.

These machines, with a single nozzle or with three or four small nozzles on a short boom, are capable of excellent spraying where the scale of work involves the application of only a few litres of spray and are particularly useful on small areas where the lay of the land precludes the use of vehicles.

Although the pressurized machines are manufactured with an adequate safety factor, it is well to subject old, fatigued or corroded sprayers to a periodic hydraulic pressure test. The recommended pressure should never be exceeded.

Motorized knapsack sprayers. In recent years lightweight two-stroke engines driving high speed fans have been fitted to knapsack sprayers. The fan creates a stream of air into which an air-shear nozzle is fitted and the spray container is pressurized to supply the nozzle at a liquid pressure of only a fraction of a bar. They are useful where a few tall trees, out of range of manual sprayers, require treatment.

3.2   Specification of main components

Tank. In all spray operations it is important to have equipment which is adequate for the job. Unless the tank capacity is suitable, more time may be spent on mixing and filling than on spraying. A simple calculation will show how far a tankful will go at the normal rate of application (divide tank capacity by volume/ha) and how long it will last (divide capacity by output of machine). The operation will become tedious unless the tank provides for 15 or 20 minutes' spraying. If repeated fillings are needed, it is important that the water supply should be nearby and adequate, the water flow should be enough to fill the tank in about 5 minutes.

Galvanized steel is the traditional material for the construction of the tank but, after a number of years, corrosion may be troublesome. As the result of developments in plastics technology, and also in welding stainless steels, these materials are becoming more widely used. Plastic tanks are now often made in two layers, an outer structural skin with a solvent resistant liner. Machines using plastic tanks have chassis members designed to cradle the tank so that stresses at fixing points are reduced. Tanks should have a strainer in the filler hole which should be of generous dimensions. They must also be readily drained and cleaned.

Agitation. It is necessary to maintain, in the spray tank, a uniform dispersion of suspensions and emulsions throughout the time that the machine is in operation. Mechanical agitation is generally more efficient than recirculation though, with both, it is necessary to ensure that the floor of the tank is thoroughly scoured without areas of 'dead' liquid. Hence the tank should be of regular shape with rounded corners. A danger from agitation dependent on recirculation is inadequacy when the nozzle output is near the maximum of the pump capacity. Some formulations cause frothing particularly if the stirrer or return flow is above the level of the tank contents. Also cavitation at the suction outlet must be avoided as this will cause frothing and reduce the output of spray.

Pumps. The capacity of the pump must be adequate to maintain a constant working pressure and should therefore be selected for the duty it has to perform. A variety of types is available.

Plunger pumps were used on most of the older tree sprayers and worked at pressures up to 45 bar. These draw in liquid via a valve on the suction stroke; this valve closes on the pressure stroke when the outlet valve opens and the spray is forced onto the delivery circuit. This type of pump must be stoutly made and has a number of component parts in contact with the spray liquid which must be made of corrosion resistant material; it is therefore expensive. Two or three cylinders are commonly used and an air vessel is needed in the outflow to even out the pulses of pressure.

Diaphragm pumps operate on the same principle as plunger pumps; pressure is applied to the rear of the diaphragm by a reciprocating plate which moves like the plunger. Since the spray fluid is sealed from the moving parts, abrasion and corrosion of close fitting parts are eliminated. The performance of these pumps is limited by the strength and movement of the diaphragm, usually of synthetic rubber, and output per 'cylinder' is accordingly limited. Development of these pumps in recent years, and especially of the diaphragm materials, has resulted in their replacing plunger pumps for pressures up to 35 bar.

Gear and vane pumps: most ground crop and many orchard machines are equipped with continuous delivery pumps, which may be either positive displacement or centrifugal. By positive displacement is meant any pump which, like the plunger pump, takes in a definite volume of liquid from the inlet and transfers it, without possibility of escape, to the outlet. The centrifugal pump, by contrast, takes in liquid at the axis and throws it by centrifugal force to the periphery where it is delivered. If the outlet of such a pump is restricted or blocked the outflow decreases or stops and there is no risk of damage to the pump by the build-up of pressure. In like circumstances with the displacement pump, damage may result unless a relief valve comes into operation or a safety plug blows.

There are three types of continuous positive displacement pumps, namely, gear, vane and roller vane. In the gear pump, a pair of gears run together in mesh in a casing; liquid enters between the teeth as they come out of mesh and is carried round between casing and teeth to be discharged at a later point before the teeth enmesh once more. Such pumps are relatively less expensive but are of limited output and are subject to heavy wear particularly if used for dispersions of wettable powders.

Vane pumps operate similarly but with a single rotor. The space between the rotor and casing is divided into sections by vanes in the rotor which reach to the casing. Liquid enters in at one portway in the casing to fill each sector and is transferred round to another point where it is ejected. The simple vane pump has most of the drawbacks of the gear pump but the roller vane, which is a development of it, has certain advantages. Vanes are replaced by rollers which greatly reduce the wear from abrasive materials. Most small sprayers now have this type of pump which is suitable for direct drive and mounting on tractor power take-off shafts. Outputs are of the order of 50 litres/minute at 3 or 4 bar. Performance of these pumps falls off as clearances become excessive. Some may be damaged if run dry for any prolonged period for liquid is relied on to reduce friction.

Centrifugal pumps are usually free from wear except where the shaft of the impeller enters the casing. For a given output they are relatively small and cheap. They need to be run at a fast speed and high pressures can be obtained only by the use of multi-stages. They are often used to deliver the spray at low pressure to the air-shear nozzles of orchard sprayers, and in separate, motor driven units for tank filling.

Nozzles—fan and cone. The hydraulic nozzles used on sprayers are of two types. For most ground spraying, fan nozzles are used. These depend on a shaped orifice through which the spray is emitted under pressure to give a fan-shaped pattern of droplets.

Cone or swirl nozzles give a cone-shaped pattern and consist of two basic elements, the disc and the swirl plate. The disc is a circular ceramic or hard metal plate with a central hole. The spray, before passing through this orifice, is given a swirling motion in a chamber which it enters by angled slots in the swirl plate. In most designs throughput and pattern can be altered by change of disc and swirl plate; cone nozzles are thus more versatile but more expensive than fan nozzles and are generally used on orchard hydraulic sprayers.

British Standard 2968:Part I:1971 suggests standards for fan spray and cone spray nozzles for attachment to booms on ground crop sprayers. The specification relates to performance and to the means of attachment of nozzles, the main purpose being to provide for interchangeability of nozzles of like size and type made by different manufacturers. Variation of discharge rate of nozzles is permitted within a range of $\pm 2 \cdot 5$ per cent of the nominal amount. It is not yet possible to lay down standards for the performance of nozzles in terms of the sizes of drops to be produced or the pattern of deposition required where a three-dimensional crop is the target.

The angle of the fan or cone and the spray spectrum are important when adjusting the nozzles on a boom. Fan nozzles and small cone nozzles give a distribution across the swath greatest under the centre and tailing off toward each edge, and this must be borne in mind when adjusting overlap. Swirl nozzles, as used on booms, give a hollow cone of spray and the bulk of the output is around the periphery.

The output from hydraulic nozzles varies as the square root of the applied pressure, i.e. doubling or halving pressure will result in roughly 40 per cent increase or 30 per cent decrease in flow. It is therefore necessary to change the nozzle (or the disc where provision is made) to obtain big changes of output, though pressure regulation will give a final adjustment.

A required output can be obtained from a large aperture working at low pressure or a smaller one at a higher pressure; the choice depends on the size of droplets required, higher pressures form smaller drops. Drop size affects cover and the liability to drift (see Operation, p. 73).

The simplest nozzle used in airblast sprayers is an orifice across which air is blown at high speed to shear off the emitted liquid as small droplets. Drop size is dependent on air speed but the spray becomes coarser with increased throughput.

Rotary nozzles give a more compact range of droplet size than the nozzles so far described. With them, drop size is dependent on the speed of rotation of a disc or drum from which the liquid is thrown. Small drops are obtainable but only at high rotational speeds of upwards of 10 000 rev/minute. When more liquid is supplied than can leave the periphery of the atomizer in discrete drops, it is thrown off in sheets and performance rapidly deteriorates. This type of nozzle is often used on aircraft, and for ULV treatments from ground equipment, when only small volumes are applied as a close spectrum of small droplets. Developments are under way which may result in the wider use on field sprayers of this type of nozzle running at lower speeds.

Fan. On airblast orchard sprayers, the fan is a critical component and for a given power it is generally better to produce air at 100–150 mm w.g. pressure (about 40 to 50 m/s in the outlet) than at the higher pressures of more than 500 mm w.g. needed for shear nozzles. Large volumes of slowly moving air mean a physically large axial flow fan rotating at moderate speeds. Outlet design is critical if such fans are to be used efficiently, but the crop penetration is so much improved by the use of large volumes of slowly moving air that the trend should be in this direction. Faster air streams and air shear nozzles may continue to be used for small-scale operations, where spraying can be limited to favourable wind conditions, because of cheapness.

Compromise designs of machine, with centrifugal fans working at about 300 mm w.g. and discharging directly to one or both sides, have been widely used. These employ hydraulic pressure nozzles operating at rather less than 7 bar and rely on the airstream to break up the larger drops so produced. Recognition of the better penetrating properties of the lower speed air, and the development of good and relatively cheap liquid pumps capable of higher pressure, will very much favour the use of larger axial-flow fans occupying, with their outlets, nearly the full width of the machine.

With all designs of fan the output and power requirements vary very steeply with the rotational speed. It is important that the fan be run at its rated speed and that it matches the power output of the tractor. On the matching of machine to tractor power and power take-off drive speed, the sprayer manufacturer or his agents should be consulted.

3.3   Operation

Maintenance. The manufacturers' instructions for lubrication, cleaning and adjustment should be learnt and followed. Since much of the deterioration due to corrosion occurs when the machine is *not* in use, it must be washed, drained and dried before putting into store and stored where it will keep dry. Hoses should be stored in the dark and not left kinked. If the outside of the machine is wiped over with oil before use, the deposit of spray chemical can more easily be cleaned off; this oil film should be restored after cleaning before putting the machine into store. Wearing parts such as pump glands, valves, agitator shaft seals and bearings should be checked at the end of each season and replacements arranged well in time for the next.

Planning. It is first necessary to know what pests and diseases are likely to be met and the pesticides to be used for each, the acreage to be sprayed for each, the volume to be applied and the timing of the applications. The records of past results are so helpful in furnishing this information that details of all pesticide operations should be recorded for future reference. The Health and Safety (Agriculture) (Poisonous Substances) Regulations 1975 made under the Health and Safety at Work etc. Act (1974) require that certain such records be made in respect of certain pesticides (see p. 88) and these should be extended to all pesticides and include: amount of pesticide used, volume applied, the date, time and crop stage. The infestation or infection conditions and the degree of control achieved should be, if possible, noted. Interruptions by weather or breakdown should also be noted.

Simple arithmetic will often provide useful guidance in planning spray operations. As an example, consider the times taken to spray 5 ha at 2000 litres/ha and at 400 litres/ha with a machine of tank capacity of 600 litres. Let the time taken to refill be 20 minutes (10 minutes for mixing and filling and 5 minutes each way to and from the filling point), and assume that the working rate is 25 minutes per hectare at low volume and 40 minutes per hectare at high volume (i.e. the pump capacity is 50 litres/hour). The schedule in each case will be:

2000 litres/ha			400 litres/ha		
Time minutes		Area (ha)	Time minutes		Area (ha)
0– 20	filling		0 – 20	filling	
20– 32	spraying	0·3	20 – 57·5	spraying	1·5
32– 52	refilling		57 – 77·5	refilling	
52– 64	spraying	0·6	77·5–115	spraying	3·0
64– 68	refilling		115 –135	refilling	
and so on for 16 loads					
500–512	spraying	4·8	135 –172·5	spraying	4·5
512–532	refilling		172·5–192·5	refilling	
532–540	spraying 0·2 ha	5·0	192·5–205	spraying 0·5 ha	5·0

The time taken under these conditions is therefore 9 hours at high volume whereas, at low volume, it is only 3½ hours. Similar calculations may be made to estimate the value of increased tank capacity for various circumstances.

If a large programme of work is involved, it is advisable to consider the complete routine, including herbicide work if this will involve the same operator, to reach a decision on machine capacity. At the peak of the programme it is usual to budget for the treatment of the required area in about 5 working days if a 10-day routine is aimed at, which leaves an adequate margin for breaks at weekends and for unsuitable weather.

Warnings. The choice of spray materials depends to some degree on surroundings and the risk of spray drift. Drift is generally less of a problem with insecticides and fungicides than with herbicides so far as damage to neighbouring crops but the hazards described in Chapter 4 must be taken into account when choosing materials and methods of application. Drift from tree spraying is unavoidable, and in the first ninety metres downwind it will be appreciable whatever application method is used. On ground crops it can sometimes be considerably reduced by using pressures of 0.7 or 1 bar to give a coarse spray, but not all treatments are effective when applied as a very coarse spray.

The operation of the machine. The subsequent notes relate to field and orchard operations. For small scale work, all that is usually needed is the application of a measured volume to a given area; both volume and area are easily measured and the machine manipulated to give the required dose. One litre per hectare is the equivalent of one millilitre on 10 square metres; this makes field recommendations easily applied to the treatment of small areas.

For field work, a tractor engine speedometer is essential for accurate spray application and should always be fitted. Nozzle and machine manufacturers give instructions for the adjustment of the machine for a desired output, but this should be carefully checked the first time the machine is used. To get the correct volume per hectare, it is necessary to obtain the right combination of output and speed.

Measuring nozzle output. Tanks should be fitted with dipsticks near the centre and measurements should be made with the machine as level as possible, particularly if the dipstick is not central. To check nozzle output: (i) fill the tank with water with machine stationary at the set tractor engine speed, (ii) spray for a fixed time long enough to reduce the tank contents by an amount which can be accurately measured, say one quarter, (iii) read amount used, (iv) reduce the tank contents by spraying or draining to between one quarter and half full, (v) repeat (ii) and (iii). The average of the two results may be taken if they are in reasonable agreement, but if there is a serious reduction in output with the reduced head in the tank, the cause must be ascertained.

Checking rate of work. With ground spraying the area covered is the swath width multiplied by the speed of travel. In tree spraying the row to row distance is taken as the swath when spraying double-sided but it is half of this for the single-sided operation. An engine speed indicator will usually be used to set tractor speed but this setting should be checked, preferably under actual spraying conditions. Measure the speed at set engine revolutions and in the gear or gears which will be used by timing over a measured distance. The calculation of the rate of work in hectares/hour from the speed in miles/hour or kilometres/hour, and the swath in metres can be done by the use of scales 1, 2 and 3 of the nomogram on p. 75.

Setting volume per hectare. If the output of the machine has been fixed by the selection of nozzles and pump pressure, volume is adjusted by changing the speed of travel. Suppose, for example, that an application of 300 litres/ha is needed with a machine using 15 nozzles on a 5 m boom, with a nominal output of 85 litres/ha operated at 2·5 bar. The output of the machine, measured as described above, is 78 and 84 litres in two 5-minute periods. The actual output is therefore 16.2 litres/minute. Now refer to the nomogram Fig. 3.1, p. 75. Find 300 litres/ha on scale 5 and 16.2 litres/minute on scale 4. Lay a ruler across these two points and read where this ruler crosses scale 2, i.e. very close to 3·1 ha/hour. Now find the swath width on scale 3. Join this to the point found (3·1) on scale 2 and read where the ruler crosses scale 1 which is at 6·2 km/hour or 3·8 miles/hour; this is the required speed.

For a second example, suppose that the tractor engine speed is fixed by fan requirements and, having selected the appropriate gear, it is necessary to adjust nozzle output by changing liquid pressure to set to a particular volume per hectare. The trees to be sprayed single-sided are in rows 9·2 m apart and 600 litres/ha are to be applied from an airblast machine with

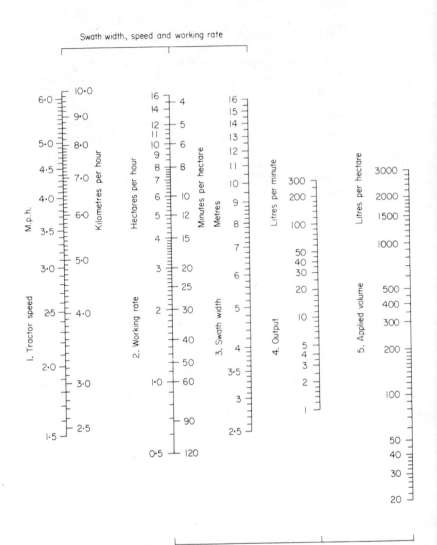

Swath width, speed and working rate

Fig. 3.1. Nomogram.

engine speed set at 1800 rev/minute at which revolutions the ground speed is 4·4 km/hour or 2·7 miles/hour. The effective swath is therefore 4·6 m, the point to find on scale 3, join this point and 2·7 on scale 1 with a ruler which will be found to cut scale 2 at 2·1 ha/hour. Now join this point on scale 2 to 600 on scale 5 and read off the required output on scale 4, namely, 21 litres/minute. The machine should now be set for this output and the actual volume delivered checked by the procedure given on p. 74.

Checking application rate. Whenever a new formulation is used, it is advisable to check the rate of application. Calculate the distance driven to spray a hectare by dividing 10 000, the number of square metres in a hectare, by the swath width in metres. Check the volume of spray used over this distance and repeat as the tank empties. Rough checks on the first tankful will detect gross misapplication and smaller deviations will become apparent after a larger area has been sprayed. The amounts used should be recorded.

Filling and mixing. The label of the pesticide container must be read carefully, the precautions observed and any recommended mixing procedure should be followed. The statutory requirements for handling scheduled chemicals are given in Chapter 4 and in literature available at any MAFF office.

The necessity of an adequate and clean water supply is obvious. Precreaming is recommended, especially when preparing the spray from wettable powders and the spray tank should be partly filled with water before the precreamed concentrate is added. Readily dispersed materials may be placed in the spray basket of the tank and washed in with the water. When the mixing of different pesticides is required, make sure that they are compatible especially when mixing the products of different manufacturers for, even though the active ingredients may be compatible, the formulating agents may be incompatible (see p. 25).

Cleaning. After the spraying operation has been completed, the machine should be drained, the filters taken out and cleaned, the tank flushed out and water pumped through to clean the lines and nozzles. Finally drain the tank and wash down the machine. If possible, do not use, for the application of insecticides or fungicides, a machine which has been used for herbicides, especially of the growth-regulating type. If unavoidable, the machine should be decontaminated by the procedure described in the *Weed control handbook*.

3.4   Nozzle arrangements and procedures for particular crops

Field crops. The normal arrangement of the nozzles on boom sprayers is for them to point directly downwards though other arrangements, such as directing them rearwards at an angle to the vertical or fitting them to drop legs so that they are carried between the rows of plants, are sometimes used. The nozzle spacing on the boom is fixed by the nozzle design and the amount of overlap between the nozzle spray patterns is adjusted by altering the height of the boom above the crop. Nozzles with wide-angled patterns can be operated at smaller boom heights than those with narrow-angled patterns.

Two types of patterns are recognized, namely triangular and rectangular. In the first type the heaviest deposit is under the centre of the nozzle and the amount decreases gradually towards each edge. This is the most suitable for use on booms where patterns are to be overlapped and is readily produced by fan spray nozzles. The rectangular distribution, i.e. equal amounts deposited across the entire width, can be achieved, approximately, by a special design of fan spray nozzle or by cone nozzles. Such a pattern is suitable for band spraying, where a single nozzle is used to cover a strip of ground. Small changes of nozzle height can result in strips receiving double or no dose if this distribution is used for boom nozzles. Triangular pattern nozzles should be set to overlap so that the edge of one fan of spray reaches the crop or ground directly beneath the centre of the next nozzle and the distribution across the swath will then not be very sensitive to small changes in boom height. An equally serious source of unevenness is that of variation, both in throughput and pattern, between individual nozzles which nominally are the same. Where a good degree of uniformity is required, nozzles should be used conforming to the B.S. specification mentioned on p. 70.

When selecting boom height, it should be remembered that, under most field conditions, the boom will not always remain parallel with the crop because unevenness of the ground causes tilting, and bumps cause boom tips to flap. Hence the boom should be set high enough to ensure that its lowest end will not be lower than the height necessary to obtain overlap. Excessive height should be avoided because it increases the risk of spray drift. When it is important to reduce drift and fan nozzles are being used, the nozzles can be pointed rearwards at an angle to the vertical enabling overlap to be obtained from a lower setting of the boom height. Booms having nozzles spaced at 3 to a metre, rather than the usual 2, also allow the nozzles to be operated closer to the crop and make the spraying operation much less susceptible to wind. The cost of the extra nozzles will often be well justified.

Drop legs are used when it is necessary to obtain a better cover of the under parts of plants. One drop leg equipped with two nozzles is required for each row to be sprayed. When using drop legs, care must be taken not to set the boom in so low a position that the nozzles dig into the ground for blockage or damage will result. The nozzles should be set at an angle to the crop rows so that the spray cone may develop before the spray reaches the main part of the target.

Airblast sprayers are rarely used for ground crops because of the difficulty of estimating the effective swath width and the increase both in the drift hazard, and variability of deposit.

The main points to be borne in mind when spraying field crops are the adequate overlapping of adjacent swaths, and the avoidance of missed areas between one tank load and the next. When row crops are sprayed, the correct place to start the second and subsequent swaths is determined merely by counting the rows but, for cereal crops, the swaths should be measured and marked, or some swath matching aid be used, to enable the tractor driver to follow the correct track across the field. If possible, stops to refill should be avoided between the headlands but, if not, the stopping place should be clearly marked, for example, by immediately turning out of line into the previously sprayed area. After refilling, the return should be made along the last set of wheel marks and spraying should be resumed immediately before the point where they curve. Anti-drip devices are available and may be operated should a blocked nozzle or other cause make it necessary to stop during a bout, but their use is not quite as important in insecticide and fungicide application as in the spraying of herbicides.

Plantation crops present a greater diversity of target than field crops and the nozzle arrangement should be suited to the shape and composition of the target. The adjustment is usually a matter of experience but advice on difficult problems may be sought of the ADAS. The addition of a fluorescent dye to the spray to give a picture of the deposition under ultraviolet light is one of the techniques used by specialists in the Advisory Service to help with setting up machines for spraying various crops.

For smaller bushes, cane fruit and for apples and pears grown on intensive systems, simple arrangements of hydraulic nozzles spaced over the vertical height of the crop are satisfactory but, where the leaf canopy is dense, as in blackcurrants, gaps in spray cover are unavoidable because of the screening of leaf by leaf. High nozzle pressures and large volumes of spray, combined with angling of the nozzles to give spray at each height from more than one direction, are necessary to get good results in this crop.

For larger trees, the critical area is in the centre and top of a spreading canopy. The bulk of the spray should be directed to the top of the trees but, if the underside of leaves and the wood must be covered, extra spray should be directed, from as close as possible, in to central and lower parts of the tree. When airblast machines are used, both the airstream and nozzle position should be adjusted: the airstream should normally be centred on the diagonal up through the tree but the spray should be concentrated in the higher and lower part of the arc according to canopy requirements. When adjustment from the tractor seat is possible and there is a cross wind, the nozzles spraying downwind should be elevated and those spraying upwind should be depressed. With low volume applications, heavy doses on the skirt of the tree near the machine outlets are difficult to avoid.

The sprayer output must be fully on when the first tree or bush is reached and the spraying should be to a set pattern so that rows are not missed. A skilled operator, making repeated application to a block, will be able to save much time at high volume rates by learning how far a full tank will go and so avoiding running-out of spray at points which involve a long journey to the filling point. It is often quicker to refill a tank not quite empty, especially in long rows affording no opportunity of turning.

One-third of the final deposit may be from spray settling on downwind rows. An additional traverse of the upwind headland at high speed to give, say, half the normal application rate, is therefore a wise precaution against failure to protect the upwind trees adequately.

Horticultural crops, when grown outdoors on a scale large enough to warrant tractor powered sprayers, are covered by the foregoing remarks. In nurseries and market gardens, the small ground crop sprayer can be adapted for some crops and spray pumps are available as accessories to some garden tractors. Automatic spraying from booms is sometimes possible. The alternative is the hand-held lance, more laborious and variable though capable of a good job if skilfully used.

Glasshouse crops may be sprayed, at high volume, with the hand-directed lance. Adequate lengths of hose are needed and the spray machine, if operated by a petrol motor, should be outside the door. Up to 10 000 litres of spray are needed per hectare of tomatoes or cucumbers to wet the foliage completely.

With certain crops, low volume applications may be made with a compressed-air paint spray-gun, a motorized mist-blower or other machine capable of dealing with concentrated aqueous dispersions of pesticides. From 40–100 litres/ha are required at low volume and more frequent applications are usually necessary than with HV spraying, the air compressor output and pressure must be adequate for the type of spray-gun used—normally $1/3$ m^3/minute at 3 to 4 bar.

Aerosols are generated from atomizable formulations using paint spray-gun type sprayers, preferably suction-fed types fitted with a calibrated glass container, or with integral-motored aerosol generators from which the atomized droplets are carried in a stream of air. The droplet stream is directed into the air of the glasshouse while the operator walks backwards the length of the glasshouse. The quantity of pesticide applied is regulated by adjusting the output rate of the machine or spray-gun to the speed of walking through the glasshouse. Inhalation of aerosols or low volume spray mists must be avoided.

3.5   Dusts and dry applications

Dry formulations may, for convenience, be divided into (a), dusts for use mainly in plantation crops and in situations where the tall growth retains the dust cloud; (b), granular pesticides. But before discussing the types of machine available for dusting operations, it will be well to outline the advantages and disadvantages of dusts as opposed to sprays, for often the alternative is open to the grower.

DUSTING VERSUS SPRAYING

The main advantages of dusting rest in the fact that the problems of water supply vanish. As a consequence, dusting appliances are lighter in weight and more easily handled in difficult terrain than sprayers; the dust must be purchased ready for use whereby mixing tanks become unnecessary and handling is reduced though toxicity hazards to operators may be especially serious because of the greater inhalation risks, see Chapter 4, p. 88. It may be said that, by and large, dusting is less costly in appliances, time and labour than spraying.

On the other hand, spraying usually gives better results than dusting for the dust deposit is less adherent to foliage than the spray, and by comparison with spraying only a small fraction of the expensive chemical applied is utilized. For this reason and because of the greater drift hazard of dusts, it is often preferable to dust only at dawn or dusk when there is moisture on the target and the air is still.

GRANULES

Insecticides formulated in the form of granules have been introduced in recent years for certain specific purposes. Their main use has been in the application of systemic materials on broad leaved crops where the granules can be lodged in the foliage to give a gradual release of insecticide over a period. Although broadcast application can be made from the air or by ground machine, special applicators have more generally been used to apply granules to each row of plants in crops like sugar beet and brassicae. These can be fitted on a tool bar to combine the insecticide application to the crop with cultivation of the soil between the rows. The increased cost of formulating in this way can thus be offset by economies from the greater persistence and more accurate placing of the insecticide.

TYPES OF MACHINES

All machines used for applying dry formulations consist essentially of a hopper which usually contains an agitator, an adjustable orifice or other metering mechanism and delivery tubes. In the case of dusters a fan or bellows provides the conveying air and the dust may join the airstream before or after the fan.

Although dusts are commonly applied from aircraft in some overseas countries, this type of operation is rare in the British Isles. Ground machines range in size from hand-held to tractor-powered and each type of sprayer has its counterpart with dust or granule outlets in place of spray nozzles. Also, coarse dusts and granules can be distributed broadcast by a spinning disc type of fertilizer spreader. Older machines were often driven from a ground wheel but this form of power is not often used now except sometimes to operate a metering mechanism. In this way application may be directly related to the ground covered and independent of travelling speed.

HINTS TO OPERATORS

1. Keep your powder dry! Even if the sacks are polythene-lined, store them clear of concrete floors.

2. Empty the hopper at the end of the day's work; otherwise store under cover or sheet down carefully.

3. Keep all foreign material such as string, pieces of paper, oil, grease, out of the dust; a coarse screen can sometimes be useful inside the hopper.

4. As with sprays, application rates must be accurate and constant throughout the treatment. To achieve this, the operator requires, in addition to a suitable machine, an accurate calibration chart to give him the initial setting, a means of checking the rate actually applied in the field, and an indication of his forward speed and of the speed of the mechanism so that, once set, both can be held constant. Many older types of duster, although tractor-mounted, were fitted with their own engine as a means of holding the fan speed constant. Now that most tractors are fitted with an engine speed tachometer or speedometer, the fitting of an engine is no longer necessary on these grounds. The checking of application rate of a dust or of granules is more difficult than that of a sprayer because once the material has passed into the airstream it cannot be easily recovered for weighing. Some machines are so designed that it can be caught in a container before it joins the airstream and the weight delivered in a given time can then be measured. Otherwise checks can only be made in the field by noting the level of material in the hopper, treating a known area, and weighing the amount required to refill the hopper to the previous level.

5. In addition to maintaining the engine or power take-off speed constant during work, it is important that no slip should occur in the drive to the fan. The drives to dust fans often run at high speeds and belt tension should be carefully watched.

6. Lubricate carefully, but not excessively, all bearings.

7. Tap all ducting at regular intervals to dislodge caked dust.

8. At one time, dusters suffered from a tendency for the delivery rate to fluctuate violently according to the amount of material in the hopper, a fluctuation amounting, in some cases, to as much as 2 or 3 times the required rate. Although this fault is less frequent in recent models, it is still wise to refill the hopper before the agitator in the base is fully exposed, because it is at this level that the greatest variation in rate occurs.

9. Adjust the height of the nozzles to suit crop and wind conditions.

IFH—4

10. When changing from one dust to another, clean out the machine and re-check the application rate, for the same setting will not necessarily give the same rate with the new material.

11. Take all the required precautions to protect yourself and others from toxicity hazards and, even when a dust has no poisonous ingredient, wear a mask especially when filling the hopper.

DOSE RATES

Granules are often used for treatment of areas, along the rows of rowcrops for example, less than the total field area, and application of rates must be related to the area of ground actually treated and not the field area.

With metrication of application recommendations there is no longer any difficulty in relating recommendations on a field basis, i.e. in kg/ha, to the treatment of smaller areas that may be measured in metres. One kilogram to the hectare is equivalent to 1 gram on 10 square metres.

Chapter 4
The safe and efficient use of pesticides

As might be expected, the large number of chemicals now available to the farmer and grower for dealing with their pest and disease problems includes some compounds which could be harmful if not properly used. The precautionary measures needed by users of toxic chemicals in agriculture were studied in detail by a Working Party appointed by the Minister of Agriculture and Fisheries in 1950, and as a result of recommendations, on promoting the safety of workers applying such materials contained in its first report, the Agriculture (Poisonous Substances) Act was passed in 1952. The first Regulations under this Act were introduced in the following year.

The second report of the Working Party, concerned with possible risks to consumers of treated crops, resulted in the formation of the Advisory Committee on Poisonous Substances used in Agriculture (now known as the Advisory Committee on Pesticides and other Toxic Chemicals) and its Scientific Subcommittee. The third report, in 1955, dealt with the possible effects of such chemicals on the natural fauna and flora of the countryside. Two years later, the Notification of Pesticides Scheme (now known as the Pesticides Safety Precautions Scheme) formally came into being as the result of negotiations between Government Departments and the Industrial Associations concerned. An account of how these safeguards to operators, consumers, and other parties are provided, forms the first part of this chapter. This is followed by a description of another Scheme—The Agricultural Chemicals Approval Scheme—which is concerned with the biological efficiency of proprietary products based on particular formulations of the active chemical ingredients.

4.1A The Pesticides Safety Precautions Scheme

The second report of the Working Party (Toxic Chemicals in Agriculture: Residues in Food, HMSO, 1953) contained a recommendation that manufacturers, importers and distributors should notify Government Departments of new toxic chemicals and formulations of pesticides before putting these on the market. As a result, the voluntary Notification of Pesticides Scheme was agreed between Government Departments and the Industrial Associations concerned (the Association of British Chemical Manufacturers: the Association of British Insecticide Manufacturers [now the British Agrochemicals Association] and the British Pest Control Association).

Negotiations were conducted by the Agriculture and Health Departments for England, Scotland and Wales, acting on advice from the Advisory Committee (which includes representatives of the Department of Industry, the Laboratory of the Government Chemist, the Agricultural, Medical Research and the Nature Conservancy Councils) and its Scientific Subcommittee, the latter being composed of official scientists selected for their specialized knowledge of various aspects of pesticides.

As experience with the safe use of pesticide products increased, it became clear that the Scheme needed to be brought up to date and its requirements strengthened. Accordingly, the Industrial Associations and those Government Departments concerned negotiated a revised Scheme which was finally accepted in May 1964 under its new title of 'The Pesticides Safety Precautions Scheme'.

Under this Scheme, a manufacturer, importer or distributor is required to notify a product, if it contains a new chemical or comprises a new formulation or new use of a chemical already on the market, for use as a pesticide in agriculture (including horticulture and home garden use) or food storage (including rodenticides).

Notification of products for agricultural use is made to the Director, Plant Pathology Laboratory, Ministry of Agriculture, Fisheries and Food, Hatching Green, Harpenden, Herts., or to the Director, Infestation Control Laboratory, Ministry of Agriculture, Fisheries and Food, Hook Rise South, Tolworth, Surbiton, Surrey, if it involves a rodenticide or a product for use in food storage practice. Each notification must be supported by extensive data and evidence to justify the claims concerning the safe use of the product. The information the notifier is asked to supply includes full details of the composition of the product, its proposed method of use, mode of action, toxicity, persistence and any other relevant data; and the hazards likely to result from its use, to those handling the product, consumers of treated produce, domestic animals and wildlife.

A notification may be processed in one of two ways. For those which the Director considers do not present special problems a 'quick procedure' is adopted and the notification is dealt with by the Scheme secretariat after consulting advisers including an official toxicologist. If they recommend that clearance should be given, the notification is sent to the appropriate Division of the Ministry of Agriculture, Fisheries and Food (which acts as the co-ordinating Department for the Scheme) for acceptance by Departments and the notifier is then informed of the decision.

The other way is for the notification to be processed through the 'committee procedure'. This generally applies to all new chemicals, new uses of the more toxic ones and those that fail to receive clearance by the quick procedure. Under the committee procedure, the information provided by the notifier and any other evidence available is considered by the Scientific Subcommittee and the notifier has the opportunity to attend the meeting to submit oral evidence. The Subcommittee makes recommendations to the Advisory Committee which, in turn, advises Departments. If Departments recommend precautions in the use of the chemical they will not be publicized until the notifier has accepted them and has confirmed that his proposal is no longer confidential. The recommendations are issued as supplements to a loose-leaf booklet entitled *Chemical compounds used in agriculture and food storage in Great Britain—user and consumer safety—advice of Government Departments.*

These Recommendations Sheets have a wide circulation and the substance of the recommendations must be included on the appropriate labels of pesticide products offered on the market. The advice given for a chemical cleared for agricultural use is concerned with three aspects: operator safety, consumer safety and wildlife safety.

On operator safety, Departments may consider that the chemical is either too toxic to be used at all or that its use must be regulated by law under the Health and Safety at Work etc. Act (see following section). On the other hand, it may be considered to offer insufficient hazard to justify regulating, but advice is given as to certain precautions that should be taken when handling it; or, again, that no special precautions are needed other than common sense general advice to treat all chemicals with care.

On consumer safety, the recommendations aim at ensuring that no unacceptable residues remain when the treated crop is harvested. The use of the chemical may be restricted to certain named edible crops and there might be limitations on the amount of active ingredient applied, and the time of such application(s). A minimum interval between the date of the last application and the time the crop is harvested is also usually stipulated. Unlike operator protection, which is provided under the Health and Safety at Work etc. Act, there is, with two exceptions, no direct legislation for consumer protection in relation to pesticides, although indirect protection is afforded by the Food and Drugs Act and Regulations made thereunder. The exceptions are arsenic and lead for which maximum levels in various food stuffs are prescribed under The Arsenic in Food Regulations 1959 (Statutory Instrument 1959 No. 831) and The Lead in Food Regulations 1961 (Statutory Instrument 1961 No. 1931), respectively.

The recommendations for the protection of livestock, wildlife, and the general public usually take the form of advice such as, for example, avoiding harm to bees by not applying the chemical when plants are flowering, and by ensuring that flowering weeds are kept down in orchards; avoiding harm to fish by not spraying near ponds or waterways, or contaminating them with concentrate, washings or empty containers; preventing animals, pets and children from having access to the chemical by storing full or part-full containers tightly closed in a safe place and thoroughly washing out empty containers. The recommendations may also take the form of restrictions on certain uses such as those for aldrin, dieldrin and DDT, or even complete withdrawal as for alkali arsenites, selenium and cadmium compounds.

Copies of the current Recommendations Sheet for any particular chemical can be obtained free of charge from the Ministry of Agriculture, Fisheries and Food (Pesticides Branch), Great Westminster House, Horseferry Road, London, SW1. In the Pesticides Safety Precautions Scheme, emphasis is placed on the advantages to be gained by early informal consultations between prospective notifiers developing a new chemical, or extending the uses of an established one, and the scientists who are concerned with the Scheme in an official capacity. Preliminary enquiries for further details relating to products for use on growing crops or as rodenticides, or intended for use in food storage, should be sent to the addresses given on p. 84.

The success of these arrangements depends on the co-operation and goodwill extended by the Industrial Associations, and there is every indication that this has been achieved since the formal inception of the original Scheme in 1957. Nevertheless, an additional safeguard exists should a product that is considered to offer a user, consumer or other hazard, appear on the market. Departments can ask the marketing agent to provide evidence in support of the claims that the product does not give rise to such risks when used according to the directions. From this stage the matter is then treated as a normal notification. Although most of the work occupying the time of the officials of the Scheme and the Committees is taken up in dealing with new chemicals, those already on the market are examined from time to time. As a result, new or revised Recommendations Sheets are issued to keep the recommendations for the safe use of pesticide products in step with the latest practices in agriculture.

Recent developments include the extension of the Scheme to cover pesticides used in forestry and other non-agricultural uses, in particular, wood preservation, and the participation of Northern Ireland as a Department granting clearances.

4.1B   The Health and Safety at Work, etc., Act, 1974

The Health and Safety at Work etc. Act places general obligations on:

  a. **employers** to ensure so far as reasonably practicable, the health, safety and welfare at work of their employees;

  b. the **self-employed** and **employees** to take reasonable care of their own health and safety at work;

  c. **employers,** the **self-employed** and **employees** not to put at risk, by their own activities, the health and safety of others;

  d. **manufacturers** and **suppliers** of articles and substances for use at work to ensure, so far as is reasonably practicable, that they are safe and without risk to health when properly used.

The Health and Safety (Agriculture) (Poisonous Substances) Regulations 1975 (Statutory Instrument 1975 No. 282) made under the Act supersede those made between 1966 and 1969 under the Agriculture (Poisonous Substances) Act 1952. The Regulations which cover users in England, Scotland and Wales and similar regulations made under the Acts applying to Northern Ireland, the Channel Islands and the Isle of Man are designed to protect personnel from harm in the use of the more dangerous chemicals employed as pesticides in agricultural operations. Users of such compounds specified in the Regulations are required by law to take certain precautions, including the wearing of the prescribed protective clothing specified for particular operations.

Inspectors are appointed to enforce the Act and any Regulations resulting from it. These inspectors have rights of entry to premises and can also enforce the production of certain documents, take statements and sample for independent analysis. The certificate issued by an approved analyst in such circumstances is a valid document in any legal proceedings which result, becoming admissible in evidence without the need for the analyst to appear in person as a witness. These inspectors will also give advice and assistance in connection with the precautions to be observed under the Regulations, and may be able to give vital help if cases of poisoning or suspected poisoning are reported to them without delay.

The chemicals to which the Health and Safety (Agriculture) (Poisonous Substances) Regulations 1975 apply are those listed in the Second Schedule, Part I, II, III and IV, as follows:

PART I chloropicrin[1]; demeton; dimefox; and mazidox;

PART II aldicarb; amiton and its salts; carbofuran; cycloheximide; dialifos; dinoseb and its salts[1]; dinoterb and its salts; disulfoton; DNOC and its salts[1]; endosulfan; endothal and its salts; endrin; fluoroacetamide[2]; fonofos; medinoterb and its salts; mephosfolan; methomyl; mevinphos; oxamyl; parathion; phorate; potassium arsenite[3]; schradan; sodium arsenite[3]; sulfotep; TEPP; and thionazin;

PART III azinphos-ethyl; azinphos-methyl; chlorfenvinphos; demephion; demeton-methyl; demeton-S-methyl; demeton-S-methyl sulphone; dichlorvos[1]; dioxathion; drazoxolon; ethion; fenamiphos; fenazaflor; fentin acetate; fentin hydroxide; formetanate; mecarbam; methidathion; nicotine and its salts; omethoate; oxydemeton-methyl; phenkapton; phosphamidon; pirimiphos-ethyl; thiometon; triazophos; vamidothion;

PART IV Any organo-mercury compound.

Note (1) The Regulations do not apply to (a) preparations used exclusively as insecticides and which contain not more than 5 per cent by weight of dinoseb or DNOC or the equivalent of their respective salts, and no other specified substance; (b) a preparation or mixture where the only specified substance contained in it is not more than 5 per cent by weight of chloropicrin; (*c*) substances which contain not more than 7·5 per cent by weight of nicotine or its salts and no other specified substance; (*d*) an aerosol where the only specified substance contained in it is not more than 1 per cent by weight of dichlorvos; (*e*) materials impregnated with dichlorvos for slow release.

(2) Under the Poisons Rules, 1968, fluoroacetamide may not now be purchased by farmers and growers.

(3) The use of potassium arsenite and sodium arsenite as potato haulm destroyers and weed-killers is banned by agreement with all the national organizations concerned and their sale for this purpose is prohibited under the Poisons Rules.

The Regulations are designed to take into account the fact that one method of using a chemical may be inherently more dangerous to the operator than another; thus, other factors being equal, soil or granular application is safer than ordinary spraying, which in turn is less hazardous than the use of aerosols under glass. It should be noted that in the Regulations, 'spraying' does not include 'soil application' when pesticides are applied to the soil in unbroken liquid form, nor does it include 'granular placement' when particles in granular form are deposited on or in the soil or on plants. The properties of particles in granular form are defined, see p. 27.

The Regulations specify 26 scheduled operations and list the type of protective clothing which must be worn, according to the chemical being used, depending on its classification as a Part I, II, III or IV substance (p. 87). Table 4.1 summarizes the present position for those chemicals in current use as insecticides and fungicides, with details of the clothing to be worn.

The Regulations impose obligations on the employer, who must provide the prescribed protective clothing, and make certain that the workers wears it, and on the employee, both of whom may be charged with infringements. Similar obligations are now imposed on the self-employed. Various other matters included in the Regulations are the maximum number of hours workers may carry out scheduled operations; the age at which they may be employed; precautions when working in greenhouses; the provision and maintenance of protective clothing; the provision of washing facilities for employees; the notification of sickness; the training and supervision of operators carrying out scheduled operations; the provision of drinking water and vessels; ensuring that tanks and containers for storing the substances are securely closed when not in use; and the keeping of a register containing details of all scheduled operations carried out.

An inspector may grant a certificate of exemption from some or all of the provisions of the Regulations if he is satisfied either that the worker can be protected adequately by other precautions or that the provisions are unnecessary under the proposed conditions of use. The certificate is granted only on specified conditions, binding upon employer and/or employee.

Table 4.1. **Protective clothing requirements**

Part I Substances	Jobs for which protective clothing must be worn	Clothing, etc. to be worn
Chloropicrin Dimefox	Except where the specified substance is in granular form: (a) opening a container containing a specified substance; (b) diluting or mixing a specified substance or transferring it from one container to another; (c) adjusting apparatus after filling the chemical tank or hopper with a specified substance; (d) washing out of containers which have held a specified substance.	Rubber gloves, rubber boots, a respirator, and either: (a) an overall and a rubber apron; or (b) a mackintosh.
	Opening a container containing a specified substance in granular form; transferring the contents from one container to another; adjusting apparatus after filling the chemical tank or hopper with a specified substance in granular form.	Rubber gauntlet gloves, rubber boots, a respirator and either (a) an overall and a rubber apron; or (b) a mackintosh; with the sleeves worn over the cuffs of the rubber gauntlet gloves.
	Washing or cleansing spraying apparatus, soil-application apparatus, or granule placement apparatus which has been used with a specified substance, other than chloropicrin, including apparatus used in connection with spraying from aircraft, such aircraft themselves and their chemical tanks or hoppers.	Rubber boots, rubber gloves, a face-shield, and either: (a) an overall and a rubber apron; or (b) a mackintosh.
	Washing and cleansing soil-injection apparatus which has been used with chloropicrin.	Rubber gloves, rubber boots, a respirator, and either: (a) an overall and a rubber apron; or (b) a mackintosh.
	Spraying any specified substance, other than chloropicrin, on any ground crop except where carried out: (a) from aircraft; or (b) in a greenhouse.	An overall, a hood, rubber gloves, rubber boots and a respirator.
	Spraying bushes, climbing plants, other than hops, or trees.	A rubber coat, rubber gloves, rubber boots, a sou'wester and a respirator.
	Spraying hops.	A rubber coat, rubber gloves, a sou'wester and a respirator.

Table 4.1.–*Cont.*

Part I Substances	Jobs for which protective clothing must be worn	Clothing, etc. to be worn
	Spraying in a greenhouse with any specified substance, other than chloropicrin, except where an aerosol dispenser, smoke-generator or smoke shreds are used.	Rubber gloves, rubber boots, a hood, a respirator and either an overall or a mackintosh.
	Granule placement by hand; or by means of hand-operated granule placement apparatus; or, except where carried out by aircraft, by means of granule placement apparatus operated otherwise than by hand; or, where such apparatus is being used for the purpose mounted on or drawn either directly or indirectly by a tractor, operating any other apparatus mounted on or so drawn by the tractor.	An overall, a hood, rubber gauntlet gloves, rubber boots and a respirator, with the sleeves of the overall worn over the cuffs of the rubber gauntlet gloves.
	Handling hops which have been sprayed within the previous 4 days.	Rubber gloves.
	Acting as a ground-marker in connection with the spraying of ground-crops from aircraft.	An overall, a hood, rubber gloves, rubber boots and a respirator.
	Acting as a ground-marker in connection with granule placement from aircraft.	An overall, a hood, rubber gauntlet gloves, rubber boots and a respirator, with the sleeves of the overall worn over the cuffs of the rubber gauntlet gloves.

	Soil-application, other than in a greenhouse, when carried out by:
	(a) the driver of:
	  (i) tractor-mounted soil-application apparatus; or
	  (ii) tractor-drawn soil-application apparatus (if the driver is un-accompanied);
		An overall, rubber gloves and rubber boots.

	(b) any operator on foot (including a person principally engaged as a tractor-driver whilst not engaged in tractor-driving).	An overall, a rubber apron, rubber gloves and rubber boots.
	Soil-application in a greenhouse.	An overall, a rubber apron, rubber gloves, rubber boots and a respirator.

Table 4.1.—*Cont.*

Part I Substances	Jobs for which protective clothing must be worn	Clothing, etc. to be worn

	Soil-injection of chloropicrin, other than in a greenhouse, when carried out by:
	(*a*) the driver of:
	   (i) tractor-mounted soil-injection apparatus; or
	   (ii) tractor-drawn soil-injection apparatus (if the driver is un-accompanied);
		An overall and rubber boots.

	(*b*) any operator on foot (including a person principally engaged as a tractor-driver whilst not engaged in tractor-driving).	An overall, rubber gloves and rubber boots.
	Soil-injection of chloropicrin in a greenhouse.	An overall, rubber gloves, rubber boots and a respirator.
	Removing the sheeting after soil-injection with chloropicrin out of doors.	An overall, rubber boots and rubber gloves.
	Removing the sheeting after soil-injection with chloropicrin in a greenhouse.	An overall, rubber gloves, rubber boots and a respirator.

Part II Substances	Jobs for which protective clothing must be worn	Clothing, etc., to be worn
Aldicarb Carbofuran Cycloheximide DNOC Disulfoton Endosulfan Endrin Fonofos Mephosfolan Methomyl Mevinphos Oxamyl Parathion Phorate Sulfotep Thionazin	Except where the specified substance is in granular form: (*a*) opening a container containing a specified substance; (*b*) diluting or mixing a specified substance or transferring it from one container to another; (*c*) adjusting apparatus after filling the chemical tank or hopper with a specified substance; (*d*) washing out of containers which have held a specified substance.	Rubber gloves, rubber boots, a face-shield, and either: (*a*) an overall and a rubber apron; or (*b*) a mackintosh.
	Opening a container containing a specified substance in granular form; or transferring the contents from one container to another; or adjusting apparatus after filling the chemical tank or hopper with a specified substance in granular form.	Rubber gauntlet gloves, a face shield and either an overall or a mackintosh, with the sleeves worn over the cuffs of the rubber gauntlet gloves.

Table 4.1.–*Cont.*

Part II Substances	Jobs for which protective clothing must be worn	Clothing, etc. to be worn
	Washing or cleansing spraying apparatus, soil-application apparatus or granule placement apparatus which has been used with a specified substance, including apparatus used in connection with spraying from aircraft, such aircraft themselves and their chemical tanks or hoppers.	Rubber boots, rubber gloves, a face-shield, and either: (*a*) an overall and a rubber apron; or (*b*) a mackintosh.
	Spraying any ground-crop except where carried out: (*a*) from aircraft; or (*b*) in a greenhouse.	An overall, a hood, rubber gloves, rubber boots and either a face-shield or a dust-mask.
	Spraying bushes, climbing plants (other than hops) or trees.	A rubber coat, rubber gloves, rubber boots, a sou'wester and a face-shield.
	Spraying hops	A rubber coat, rubber gloves, a sou'wester and a face-shield.
	Spraying in a greenhouse (except where an aerosol dispenser, smoke-generator or smoke shreds are used).	Rubber gloves, rubber boots, a hood, a face-shield and either an overall or a mackintosh.
	Spraying in a greenhouse where an aerosol dispenser, smoke-generator or smoke shreds are used.	An overall, a hood, rubber gloves and a respirator.
	Spraying in a livestock house where an aerosol dispenser, smoke-generator or smoke shreds are used.	An overall, a hood, rubber gloves and a respirator.
	Granule placement by hand or by means of hand-operated granule placement apparatus.	Rubber gauntlet gloves and either an overall or a mackintosh with the sleeves worn over the cuffs of the rubber gauntlet gloves.
	Granule placement (except where carried out from aircraft) by means of granule placement apparatus operated otherwise than by hand, or, where such apparatus is being used for the purpose mounted on or drawn either directly or indirectly by a tractor, operating any other apparatus mounted on or so drawn by the tractor.	Either an overall or a mackintosh.

Table 4.1.–*Cont.*

Part II Substances	Jobs for which protective clothing must be worn	Clothing, etc. to be worn
	Handling hops which have been sprayed: (*a*) within the previous 24 hours, with mevinphos; (*b*) within the previous 4 days with any Part II scheduled substance other than mevinphos.	Rubber gloves.
	Acting as a ground-marker in connection with the spraying of ground-crops from aircraft	An overall, a hood, rubber gloves, rubber boots and a face-shield.
	Acting as a ground-marker in connection with granule placement from aircraft.	A hood, a face-shield and either an overall or a mackintosh.

	Soil-application, other than in a greenhouse, when carried out by:
	(*a*) the driver of:
	  (i) tractor-mounted soil-application apparatus; or
	  (ii) tractor-drawn soil-application apparatus (if the driver is un-accompanied);
		An overall, rubber gloves and rubber boots.

	(*b*) any operator on foot (including a person principally engaged as a tractor-driver whilst not engaged in tractor-driving).	An overall, a rubber apron, rubber gloves and rubber boots.
	Soil-application in a greenhouse	An overall, a rubber apron, rubber gloves and rubber boots.
	Bulb dipping, bulb steeping, handling bulbs when wet from dipping or steeping, disposing of the thionazin solution and washing the apparatus used for dipping or steeping in the thionazin treatment	Rubber gauntlet gloves, rubber boots, an overall and a rubber apron.

Table 4.1.–*Cont.*

Part III Substances	Jobs for which protective clothing has to be worn	Clothing, etc., to be worn
Azinphos-methyl Chlorfenvinphos Demephion Demeton-S-methyl Demeton-S-methyl sulphone Dichlorvos Dioxathion Drazoxolon Fentin acetate Fentin hydroxide Methidathion Nicotine Omethoate Oxydemeton-methyl Phosphamidon Pirimiphos-ethyl Thiometon Triazophos Vamidothion	Except where the specified substance is in granular form: (a) opening a container containing a specified substance; (b) diluting or mixing a specified substance or transferring it from one container to another; (c) adjusting apparatus after filling the chemical tank or hopper with a specified substance; (d) washing out of containers which have held a specified substance.	An overall, rubber gloves and a face-shield.
	Opening a container containing a specified substance in granular form; transferring the contents from one container to another; adjusting apparatus after filling the chemical tank or hopper with a specified substance in granular form.	Rubber gloves, a face-shield and an overall.
	Opening a container containing smoke shred or a smoke-generator which contains not more than 40 per cent by weight of nicotine, or transferring the contents from one container to another.	Rubber gloves.
	Acting as a ground-marker in connection with the spraying of ground-crops from aircraft.	An overall, a hood, a face-shield and rubber boots.
	Applying nicotine to roosts, perches and other surfaces in a livestock house.	Rubber gloves, a face-shield and an overall.
Part IV Substances	Jobs for which protective clothing must be worn	Clothing, etc., to be worn.
Any organomercury compound	Applying aerosols in a greenhouse	An overall, a hood, rubber gloves and a respirator.

The Ministry of Agriculture, Fisheries and Food and the Department of Agriculture and Fisheries for Scotland have issued a leaflet APS/1 *The safe use of poisonous chemicals on the farm,* which includes a valuable summary in non-legal terms of the main provisions of the Regulations, as well as much general advice on the safe use of pesticides in relation to persons, livestock and wildlife; the cleansing and maintenance of respirators and dust-masks; notes on the symptoms of poisoning by various chemicals, and the necessity for constant medical supervision of operators. This leaflet also contains a list of addresses of safety inspectors appointed under the Act.

Chemicals included in the Health and Safety (Agriculture) (Poisonous Substances) Regulations are usually also subject to the provisions of the Pharmacy and Poisons Act, 1933 and Poisons Rules, which restrict their sale and impose certain labelling requirements and conditions under which listed poisons may be bought, packed, transported and stored on shop premises.

4.2   The Agricultural Chemicals Approval Scheme

(Insecticides, fungicides and herbicides)

This is a voluntary Scheme under which proprietary formulations of chemicals used in crop protection can be officially approved and its purpose is to enable users to select, and advisers to recommend, efficient and appropriate crop protection chemicals and to discourage the use of unsatisfactory products.

The chemicals covered by the ACAS are those used for the control of plant pests and diseases, for the destruction of weeds, for growth regulation and certain other crop protection purposes and for the control of insect and mite pests of farm stored grain. Those used as rodenticides or for the protection of stored products or for veterinary or domestic uses, are not included.

Participation is open to manufacturers and their agents and also to the authorized agents of overseas manufacturers. The Scheme is operated on behalf of the Agricultural Departments of the United Kingdom* by the Agricultural Chemicals Approval Organization, at the Plant Pathology Laboratory, Hatching Green, Harpenden, Herts., and a member of the Organization specializing in herbicides is attached to the ARC Weed Research Organization, Begbroke Hill, Kidlington, Oxford. Full support is given by the British Agrochemicals Association, the British Pest Control Association, the British Association of Grain, Seed, Feed and Agricultural Merchants, the National Unions and Associations of farmers and growers in the United Kingdom* and the National Association of Agricultural Contractors. Under this Scheme, it is intended that, where possible, products containing new chemicals will be given official approval at the time of marketing.

Approval is granted to products by the Organization for specific uses under United Kingdom* conditions when the recommendations made on

* For the purposes of this Scheme the United Kingdom includes England, Scotland, Wales, Northern Ireland, the Channel Islands and the Isle of Man.

Fig. 4.1. The official mark of the Agricultural Chemicals Approval Scheme.

the labels are supported by satisfactory evidence from the manufacturer's field trials, supplemented in appropriate cases by the results of work carried out by the Advisory Services, or by independent Research Stations.

A Certificate of Approval, which is subject to annual renewal, is granted to each product approved. The label of each approved product carries the identification mark shown in Fig. 4.1.

No product may receive approval under this Scheme until its safety in use has been considered under the Pesticides Safety Precautions Scheme (see p. 83) and recommendations for its safe use have been issued by the appropriate authorities.

A List of Approved Products is published in February each year. Copies of the current List may be obtained free of charge from the Ministry of Agriculture, Fisheries and Food (Publications), Block C, Tolcarne Drive, Pinner, Middlesex HA5 2DT, or from any of the Ministry's Regional and Divisional Offices. Copies can also be obtained from the main offices of the Agricultural Departments in Scotland, Northern Ireland, the Channel Islands and the Isle of Man. This booklet lists the proprietary names of approved products under the headings of the active ingredients they contain and is complementary to this Handbook which deals with pest and disease control solely in terms of active ingredients. To select a suitable approved proprietary product for a particular purpose, the List of Approved Products should be consulted.

Products approved too late for inclusion in the current List are announced in the trade press, before the next List appears.

Correspondence concerning the Scheme should be addressed to the Secretary, Agricultural Chemicals Approval Organization, Plant Pathology Laboratory, Hatching Green, Harpenden, Herts. AL5 2BD.

4.3   Pesticides—code of conduct

The official provisions for the safeguarding of users and public from the use of pesticides become ineffective if the pesticides are wrongly used. The Joint Association of British Manufacturers of Agricultural Chemicals/Wild Life Education and Communications Committee, representing manufacturers, distributors and users of pesticides, the MAFF, Nature Conservancy Council and voluntary conservation organizations, have issued a *Code of conduct* to promote this proper use. The responsibilities of the agricultural and horticultural users are summarized thus:

A   *Identification of the problem*

  i to become familiar with the appearance or symptoms of common pests, diseases and weeds;
  ii to make a balanced assessment of the economic benefits that may be expected from using the pesticide.

B   *Choice of product and method of use*

To choose, with the best available advice,

  i a correct pesticide to combat the pest;
  ii the effective dosage, time and conditions of application.

C   *Compliance with instructions and recommendations*

  i to become familiar with Government requirements and to observe them strictly;
  ii to READ THE LABEL and all relevant literature, observing especially the dosage rates recommended and the safety precautions; to adhere strictly to the interval that must elapse between applying the pesticide and harvesting the crop.

D   *Safety to operators*

  i to make certain that the operator knows the correct method of applying a pesticide and all the safety precautions that must be observed, including the use of protective clothing if necessary;
  ii to ensure that operators wash well immediately after working with pesticides and before eating, drinking or smoking;

iii to warn operators of the danger of clearing blocked nozzles by mouth;
iv to notify a doctor immediately if an operator shows any symptoms of illness during or after working with pesticides and, subsequently, to notify the safety inspector of one of the agricultural departments if a pesticide was shown to be the cause of the symptoms;
v to ensure that equipment and protective clothing is frequently cleaned and stored safely;
vi to check the condition of protective clothing and equipment and replace when necessary.

E   *Safety for others*

To apply pesticides carefully and to AVOID

 i spray drift on surrounding areas;
ii spraying blossom and flowers which would harm bees and other beneficial insects;
iii pesticides draining into water courses, which might harm fish or other wildlife, bees, or farm stock.

Some important safety points:

Train operators properly.
Keep machinery in good order.
Always read the product label.
Store both full and partly used containers in a locked store.
Make certain the partly used containers are tightly closed and clearly labelled.
Never transfer pesticide to other containers.
Wash out containers after use and return washings to spray tank.
Never wash or cleanse equipment in water courses.
Return, burn, or flatten and bury used containers.
Keep an accurate record of quantities and places where pesticides are used.

4.4   The disposal of unwanted pesticides and containers

Two other Acts of Parliament are of importance to the users of pesticides: The Deposit of Poisonous Waste Act, 1972 and The Control of Pollution Act, 1974. The purpose of this legislation is the protection of the environment from hazards arising through an improper disposal of poisonous waste. Regulations release farmers and growers from the notification procedure if they dispose of pesticides *in any safe manner* on agricultural land. To define 'any safe manner' the MAFF, in conjunction with the British Crop Protection Council, has issued a 'Code of Practice for the Disposal of Unwanted Pesticides and Containers on Farms and Holdings'. Observance of the advice given in this Code should avoid the creation of an environmental hazard and, thereby, the commission of an offence.

The advice given in the Code may be summarized as follows, though for details the original should be consulted.

Store unopened or partly filled pesticide containers in a well-ventilated place to which only authorized personnel should have access. Store empty containers to await disposal in a separate pound, preferably under cover, from which unauthorized people, especially children and stock and pets are excluded. Empty paper sacks should be placed in a special weatherproof container in the pound.

Empty containers should be well rinsed, the rinsings being added to the spray tank as spraying continues. When washing out the spray machine allow the draining to run into a soil soakaway well away from the catchment areas of boreholes, wells, springs or ponds.

Some local authorities will collect empty containers provided they are empty, rinsed and punctured. If not, engage a waste disposal contractor or bury or burn the containers as outlined below. Glass containers should be placed in a sack and crushed; metal containers should be holed and flattened. Both should be buried at least 450 mm deep in a marked, isolated place away from ponds, watercourses and boreholes. Paper sacks and plastic containers, but not empty aerosol containers, should also be buried. Alternatively, empty paper or plastic packs may be burnt in a roaring fire not within 15 m of a public highway nor where smoke may drift over people, stock, houses or covered crops. Do not puncture or burn empty aerosol containers or spent vaporizer strips; put them in ordinary refuse bins, amongst other rubbish.

In the case of pesticides no longer wanted, if the container is in good condition, unopened and with label intact, ask the supplier to collect it. When opened or partly empty containers are to be disposed of, the contents may, in certain cases, be applied to cereal stubble. Dilute in accordance with label instructions or at the rate recommended for HV application and spray at about 340 litres/ha. Such treated areas should be ploughed and barley sown in the following spring.

Pesticides suitable for stubble disposal are listed in Appendix IV of the Code and include most of those described in Chapter 2. Exceptions are the chlorinated hydrocarbons, except HCH sprays, and a number of pesticides recently introduced. If the pesticide is not listed in Appendix IV ask the local official agricultural adviser for guidance.

Chapter 5
Pest and disease control in cereals

Healthy seed, drilled at the right time and to the right depth in a firm and adequately manured seedbed, is most likely to give a satisfactory yield in spite of pests and diseases. The higher the potential yield, however, the more is to be gained from pest and disease control, and the greater are the chances that the use of insecticides and fungicides will be profitable. Good husbandry and chemical control should not be thought of as alternatives, but as working together to give the best results.

It should not be assumed that cultural control invariably costs nothing. The early ploughing of grass, which prevents or greatly reduces damage by the 'ley pests', may entail the forfeiture of several weeks' grazing or of a hay or silage cut. Crop rotation avoids trouble from many serious pests and diseases but, where economic conditions favour intensive corn growing, it may require the continuance of less profitable enterprises. The practice of minimal cultivations and direct seeding has advantages in the saving of labour and time, but is not without its drawbacks in that there is often an increased risk of damage by pests and perhaps to a lesser extent by diseases both soil and airborne.

Because of the large annual acreage involved, the risks of 'side effects' from chemicals applied to corn crops are especially important. For this reason, as well as for the low cost, the development of control by seed treatment is of the greatest importance. Overtreating can injure the seed and undertreating will give poor control. It is very important that the manufacturers' instructions are followed implicitly, and that the seed is of good quality, dry (under 16 per cent moisture content), and undamaged. Many of the compounds are extremely poisonous, are harmful to fish and need careful handling. Treated seed should *never* be used for feeding stock or humans. Sacks which have contained treated seed must *not* be used for feeding stuffs or millable grain.

The greater value of grain crops currently, the increasing importance of foliar diseases, and the difficulty experienced by plant breeders in producing cereal varieties which retain their genetical resistance to some pathogens for more than a few seasons has meant that spray applications of fungicides to the growing crop (particularly some varieties) can under some circumstances be beneficial or even essential. In addition to contact fungicides a whole new range of systemic fungicides is being developed some of which are specific to a particular disease group (e.g. rust fungi). By contrast the wide spectrum benzimidazole- and related-systemic fungicides (benomyl, carbendazim and thiophanate-methyl) sometimes give a yield increase when disease is thought to be insignificant.

Some pathogens confine their attack to one cereal, others may attack two or more. In the following list multiple pathogens are listed under each crop heading with appropriate cross references. Pests and diseases are listed alphabetically by the generally accepted common name and under the appropriate host. Symptoms which are not caused by pathogens, but whose appearance is similar, are described to aid more positive diagnosis. Fuller information about cereal diseases and pests and their control is given in MAFF Bulletins 129 'Cereal Diseases' and 186 'Cereal Pests'.

There is little or no information available on some aspects of pest and disease damage to cereals. So far as is possible, however, the following recommendations take into account both the economic and biological factors.

5.1   Wheat

Pests

**Aphids. Pests of the greenfly group may cause serious injury to corn crops, particularly after a mild winter has been followed by a fine spring. For description see 'Key for the field identification of apterous and alate cereal aphids with photographic illustrations' MAFF, 1975. They are important as carriers of virus diseases. The 4 most common species are:

Bird-cherry aphid (*Rhopalosiphum padi*), a greenish brown species which is usually restricted to the lower leaves. It often overwinters in the summer form on grasses and cereals.

Rose-grain aphid (*Metopolophium dirhodum*), a light green species with a darker green stripe down the back. It overwinters on rose and may be found on cereal leaves in summer.

Fescue aphid (*Metopolophium festucae*), a species very like the rose-grain aphid but lacking the darker stripe down the back. Overwinters on grasses and cereals and in occasional years causes severe and widespread damage.

Grain aphid (*Macrosiphum (Sitobion) avenae*), a large green or reddish species that is often noticed on the ears of corn. Usually it causes little or no damage, but attacks on wheat heads were severe and widespread in 1968, when many crops were sprayed from the air; there was also a less severe build-up in 1970.

Plants affected by aphids are dwarfed and of poor colour, the tips of the leaves being often reddish, purple, or brown and withered. Discoloration may be due to virus infection. In severe cases large areas of the crop may be completely withered, so that they can be fired with a match. Aphids and their cast skins are very numerous. Severe attacks are most likely in hot, dry weather.

Control is obtained with demeton-S-methyl 245 g a.i./ha *or* dimethoate 335 g a.i./ha *or* formothion 420 g a.i./ha *or* pirimicarb 140 g a.i./ha.

Treatment against direct injury to the crop is likely to be profitable when bird-cherry and/or fescue aphids are numerous enough to be noticed easily. When parasites and predators are also numerous, spraying may not be required.

See also Barley yellow dwarf virus, p. 108.

*Cereal cyst eelworm—see under Oats, p. 111.

*Frit fly (overwintering generation)—see under Oats, p. 111.

*Gout fly—see under Barley, p. 108.

**Leatherjackets (A.L. 179). Certain crane-flies (*Tipula* and *Nephrotoma* spp.) lay eggs in grassland during summer and autumn; these eggs hatch in the autumn. When grass is ploughed for cereals the larvae, known as leatherjackets, may feed on the seedling corn. These insects, unlike wireworms, complete the life cycle in one year, so that injury is normally restricted to the first year after ploughing temporary or permanent grass.

Cereal plants are damaged at or below ground level; injured tissues appear torn rather than cut. Identification should be confirmed by finding leatherjackets in the soil.

Cultural control depends on early ploughing of grassland (July or early August) before the main egg-laying period. In certain areas early ploughing may bring a risk of wheat bulb fly attack.

Chemical control is by bait or spray. (*a*) Poison bait. To 31 kg bran add fenitrothion (0·5 kg a.i.) *or* gamma-HCH (280 g a.i.) but *not* if potatoes due next year. Add water to make the mixture crumbly and broadcast at this rate per hectare; *or* use proprietary baits; DDT should be avoided if possible.

(*b*) Sprays. Chlorpyrifos 0·75 kg a.i./ha *or* gamma-HCH 1·1 kg a.i./ha where potatoes are *not* to be grown within 18 months. Aldrin 1·6 kg a.i./ha may be used on DDT-sensitive barley varieties, but DDT 1·1 kg a.i./ha should be avoided where alternatives are available.

***Slugs (A.L. 115). Several species damage cereals, the commonest being the field slug, *Agriolimax reticulatus*. The most serious injury occurs below soil level; germinating grains are hollowed out, leaving only the outer skin and the shoot may be damaged or severed. Above ground, slugs eat narrow longitudinal strips from the leaves and in a bad attack leaves may be completely shredded. Damage is most common on the heavier soils and after grass, clover, or a crop (e.g. peas) which leaves in the soil abundant partially decomposed organic matter.

Prepared baits are now commonly used for control e.g. 6 per cent metaldehyde minipellets broadcast at 960 g a.i./ha, *or* 4 per cent methiocarb pellets broadcast at 220 g a.i./ha. Some control of underground attack can be expected from the use of these baits pre-drilling *or* at sowing after thorough mixing with the seed. Post-drilling baiting is only useful in helping preserve an existing plant stand.

***Wheat bulb fly (*Leptohylemyia coarctata*) (A.L. 177). Eggs are laid in bare soil from mid-July until the beginning of September. They hatch from late January to early March and the young larvae enter the shoots of wheat, barley and sometimes rye. Oats are not attacked. In early spring the centre leaves of damaged plants become yellow and die. Small untillered plants may be completely killed. Shoots often have a very small hole at the base (pull off the outer leaves, *cf.* wireworms), and a proportion of damaged shoots contain a white maggot.

Damage is normally confined to certain areas in the eastern part of the country and is usually restricted to cereals after fallow, bastard fallow, or after a crop (e.g. potatoes) in which there is bare soil between plants during the egg-laying period. It may also follow patches of poor growth in any crop.

Crops sown before the middle of October in a good seedbed should be able to withstand any normal attack. Top dressing in spring helps a thinned crop. Spring sowing, on land where trouble is expected, should be delayed until mid-March.

Chemical control (*a*) Seed dressings. Aldrin and dieldrin are no longer permissible for this purpose. For earlier sowings, use carbophenothion at 120 g a.i./100 kg seed *or* chlorfenvinphos at 100 g a.i./100 kg seed. These dressings should not be used after 31 December. For sowings in December and later, use a gamma-HCH seed dressing formulated for Wheat bulb fly control.

(*b*) Sprays. On unprotected crops, damage may be curtailed by spraying in early March with dimethoate 0·7 kg a.i./200–400 litres/ha *or* formothion 2·8 kg a.i./200–400 litres/ha *or* omethoate 0·6–1·2 kg a.i./500–1200 litres/ha.

Forecasts of the likelihood of attack are obtainable from the ADAS and are usually available by the beginning of October.

*Wheat shoot beetle (*Helophorus nubilus*). Small larvae, often difficult to find, which cause damage somewhat similar to that of wireworms or wheat bulb fly larvae (small lateral hole at base of shoot). Attacks occur only after grass, usually ley, and commence in very early spring. Some control is obtained by early ploughing of grass for autumn-sown cereals.

***Wireworms (A.L. 199). These tough, yellow, long-lived larvae of certain click beetles (mainly *Agriotes* spp.) are normal inhabitants of old grassland. Damage may be expected during the first 4 years after ploughing up grass that has been down for 4 years or longer. On some farms, usually on moderately heavy soils, wireworms may be troublesome on old arable land, particularly if it has been allowed to get weedy. Attacked plants become yellow and die. Shoots are damaged below soil level, there being often a distinct hole in the side of the plant at the base or the shoots are chewed and frayed just above the old seed. Sometimes only the centre leaf becomes yellow. Damage is often less severe on headlands. Attacks are usually seen in autumn and spring but, unlike wheat bulb fly and wheat shoot beetle, not as a rule in very early spring.

Attack is reduced by sowing in a firm seedbed which may be rolled if an attack develops.

Gamma-HCH is used for chemical control but soil treatments are justified only against high populations. Spray e.c. at 1·1 kg a.i./ha *or* apply dust at 1 kg a.i./ha before sowing, followed by thorough incorporation into the soil, *or* use dust at 0·5 kg a.i./ha combine-drilled with the seed. Do not grow potatoes or canning carrots within 18 months of the above treatment.

For lower populations, gamma-HCH seed dressings at 36–45 g a.i./100 kg seed usually provide adequate protection against attack. Wireworm seed dressings do not protect crops against wheat bulb fly. *Do not use* wheat bulb fly dressings against wireworm alone.

Diseases

*Barley yellow dwarf virus. Symptoms are similar to those described under barley (see p. 108) but the leaves of affected plants of some varieties may develop a purplish tint. For control, see Barley, p. 108.

*Black mould (*Cladosporium herbarum*). This fungus, and other sooty moulds often cause blackening of the ears, stems and leaf sheaths, especially in wet seasons. Poor plants with poorly filled heads are most likely to show this effect, but the fungi are secondary. They cause respiratory troubles in man and are sometimes mistaken for the smut diseases. Sprays based on dithiocarbamate, benzimidazole type or related fungicides (benomyl, carbendazim or thiophanate-methyl) applied after heading for other reasons should also suppress this fungus.

*Black point (*Alternaria* spp.). In this condition (unlike black mould) the plumpest grains are usually affected. The embryo end of the grain shows a brown or black discoloration, and there may be some shrivelling of the grain. Discoloured grains lower the market value of the sample, but germination is usually unimpaired. Post heading spray applications of dithiocarbamate fungicides applied to control more serious pathogens should also have some effect on this fungus.

Brown foot root and **Ear blight (*Fusarium* spp.). Young plants may be killed outright (seedling blight) resulting in a poor braird. Mature plants show rotten roots, brown rotten stem bases, often with pink or white fungus pustules on the lesion. Leaves may show a light coloured spot with a dark red margin, often become yellowed and sometimes wither. Ears are poorly filled or empty (whiteheads). The grain may bear red or pink fungal masses on the surface, the ear blight stage, and some penetration of the grain may occur. The disease is common in some seasons and organomercury seed treatments give partial control, though the pathogens are also found in the soil. Sprays of benzimidazole type compounds applied just after growth stage 10 to control glume blotch (*q.v.*) may reduce the earblight stage, perhaps increase yield and reduce the amount of seedborne inoculum.

*Bunt or **Stinking smut** (*Tilletia caries*). Affected plants produce bluish ears on shorter stalks, and with shorter and plumper grains than healthy plants. The grains contain a mass of black greasy spores, smelling of rotting fish. The spores adhere to the seed and on germination penetrate into the young seedling. This disease is now rare and is readily controlled by organomercury, benomyl-thiram or carboxin-thiram seed treatments.

*Ergot (*Claviceps purpurea*)—see under Rye, p. 114.

*Eyespot (*Pseudocercosporella herpotrichoides*) (A.L. 321). Infection is seen in young plants as light necrotic brown bordered oval lesions on the leaf sheaths. If the attack is severe the young plants are killed outright, but otherwise the fungus works its way in to attack the lower part of the stalk, resulting in an eye-shaped brown bordered spot with a black spot in the middle, from which symptom the disease is named. Later in the season the stalk, weakened at the eyspot lesion, may topple over, giving the condition known as 'straggling'.

Yields are reduced and a lot of tail corn is produced. Whiteheads may occur and become secondarily invaded by black mould (*q.v.*). Whiteheads may also be due to an attack of take-all (*q.v.*) or other causes. It occurs widely in the British Isles especially on heavy fertile soils.

Good crop rotations, eradication of couch (*Agropyron repens*), moderate but not excessive manuring and adequate drainage effect some control. Stiff-strawed varieties are less affected and a large number of varieties are available with resistance to this disease. Champlein is rather susceptible and is best not used for other than a first white crop. There is evidence that benzimidazole type materials applied at about growth stage 6 are beneficial and could be combined unless there are compatibility problems with weedkiller spray applications. Winter wheat can be sprayed with benomyl 0·25 kg a.i./ha *or* carbendazim 0·25 kg a.i./ha *or* thiophanate-methyl 0·7 kg a.i./ha.

**Glume blotch (*Leptosphaeria (Septoria) nodorum*). Dark brown spots on the ears, leaves with pale yellow spots with brown margins, the spots often bearing small black fruit bodies (pycnidia). Browning of the stem nodes may occur in wet seasons. The disease is common and seedling losses may be serious, and it is not adequately controlled by seed treatments. Spray applications of benzimidazole type compounds applied just after growth stage 10, have a beneficial effect and lessen grain shrivelling especially on susceptible varieties and crops grown in *Septoria* prone areas such as western coastal areas where in wet seasons the disease can be serious.

*Grey leaf (Manganese deficiency). Not common, but sometimes seasonally and/or locally troublesome—see under Oats, p. 113. Of the NIAB recommended varieties, Champlein and Mega of the winter and Maris Dove and Maris Butler of the spring wheats are the most tolerant to manganese deficiency.

*Leaf spot (*Septoria tritici*). Symptoms are indefinite light coloured spots on the leaves bearing small black fruiting bodies (pycnidia). It is common some seasons, and of consequence. Control measures are usually ineffective.

Loose smut (*Ustilago nuda*). Black spore masses on the diseased heads are very conspicuous in a crop as the ears emerge. The spores are readily dispersed by the wind and eventually just a bare stalk remains.

Seed should be taken from smut free crops. Hot water treatment of the seed, at 32°C (90°F) for 4 hours and then 10 minutes at 52–54°C (126–129°F) or 2·5 hours at 46°C (115°F) is a cure. Sow 'certified' or 'field approved' seed. Carboxin with organomercury as a seed treatment is an effective but somewhat expensive control. Its use is perhaps only justifiable economically on crops grown for seed. Maris Huntsman, Maris Nimrod, Flinor, and Maris Dove, the last being a spring wheat, are the most resistant varieties in the current NIAB recommended list of wheats.

Mildew (*Erysiphe graminis*) (A.L. 579). Seen as off-white or brown patches of mycelium on the lower leaves. As the season and the disease progresses, the lower leaves become yellow and shrivel, and the upper leaves become affected. The mycelial patches become dark brown with age and small black spore cases (perithecia) become apparent on the patches. In a severe attack mildew is found on the ears and glumes and may discolour and shrivel the grain.

Although mildew attacks wheat, barley, oats, rye and pasture grasses, the races of mildew are specific to each crop and the disease does not spread from one type of corn to another. Excessive use of nitrogen is to be avoided, and it should be applied early and spring crops sown early. Some of the newer varieties of wheats show some resistance to mildew at present, but by past experience this quality is usually quickly lost. The advantage of controlling mildew in the wheat crop specifically by fungicides is not clear cut, there being little evidence of a sufficient return on capital outlay. Sufficient control is often obtained from the side effect of some fungicides used for other wheat diseases. When severe attacks occur winter wheat can be sprayed with tridemorph at 0·53 kg a.i./ha.

Scab (*Gibberella zeae*) is similar to brown foot rot and kills young plants; older plants show a foot rot and pinkish fungal masses on the ears. Small blue-black fruit bodies (perithecia) often occur on ears and stubbles. Grain is shrunken and affected grain should not be fed to pigs, dogs or humans. The disease is not common but isolated outbreaks are occasionally reported. Organomercury seed treatment and crop rotations usually prevent the disease becoming of economic importance. Foliar sprays of dithiocarbamates and benzimidazole type fungicides applied for other reasons may have a beneficial effect against scab.

Sharp eyespot (*Corticium (Rhizoctonia) solani*) (A.L. 321). This fungus often attacks cereals at the seedling stage causing the roots to turn brown, whilst distinct lesions may occur on the older thicker roots. The leaves are erect, narrow and rolled and the foliage may be purple in colour (purple patch). Badly affected plants succumb, others may survive but are stunted and yield poorly.

The fungus also attacks the lower part of the stalks producing a lesion somewhat like that of eyespot (q.v.). Sharp eyespot can be distinguished from eyespot in that the lesions may be much elongated, have a much sharper border and outline and do not show the black spot in the middle. The disease is often troublesome after a ley or a non-cereal arable crop especially in cold soils, so sow winter crops early and spring crops late. Sprays of benzimidazole type compounds applied for a possible beneficial effect on eyespot are unlikely to do much against this disease.

***Take-all** (*Gaeumannomyces graminis*) (A.L. 304). The fungus attacks the roots of the plant resulting in either the complete death of young plants or in yellow stunted plants. The disease is, however, usually noticed after heading when patches of affected plants are seen, with bleached empty heads—the whitehead condition. These whiteheads often become covered with a dark fungal growth, the black mould condition (*q.v.*) but this colonization is secondary. The plants with whiteheads often show rotten roots and the base of the stalk may be covered superficially with black mycelium of the take-all fungus. Barley is also attacked as is couch grass (*Agropyron repens*). Oats are unaffected and ryegrass is resistant. However, there is a strain of the take-all fungus (*G. graminis* var. *avenae*) common in oat-growing areas, which can attack wheat, barley, oats and couch grass. The disease is favoured by alkaline and light soils.

No chemical control method is known but these precautions should be taken: avoid over-liming; control perennial weed grasses which can harbour the disease; do not include wheat and barley too often in the rotation; follow autumn sown crops of wheat and barley with spring sown ones, and not vice versa; cultivations which make for quick breakdown of stubbles should be practised. There is a phenomenon known as 'take-all decline', when long runs of wheat or barley may result eventually in a decline rather than an increase of take-all infection. This phenomenon is not universal, however, and further information is best sought from a cereals pathologist.

Yellow rust (*Puccinia striiformis*) (A.L. 527). This disease can be very harmful. New strains of the fungus which are capable of attacking most varieties of winter and spring wheats soon become widespread. The fungus also occurs on barley, rye and some grasses, but as distinct strains, so that these do not pass from one kind of cereal or grass to another. Similarly not all wheat varieties are susceptible to the same wheat strains.

Symptoms of yellow rust are seen initially as parallel lines of lemon yellow pustules on the leaves, but the stems and ears may also become affected. Later in the season the pustules turn black. Mild winters and cool moist weather in the spring and early summer favour the disease, but a hot dry spell will often arrest the disease spectacularly. Control is by growing resistant varieties, but this is not always reliable. It is foolish to grow one variety of winter wheat or winter and spring wheat susceptible to the same strains of the fungus. Grow varieties of varying race spectra. Despite this, when climatic conditions suitable for an epidemic persist, a severe attack may result and spraying becomes economic. Dithiocarbamate sprays are protectant and may need more applications than the more rust-specific systemic fungicides which have a knockdown and protectant action. Spray with a mixture of tridemorph at 0·53 kg a.i./ha and 'Polyram' (80 per cent) at 2·2 kg product/ha in not less than 250 litres/ha *or* benodanil at 1·1 kg a.i. plus wetter in not less than 225 litres/ha.

5.2   Barley

Pests

Aphids—see under Wheat, p. 101.

*Cereal cyst eelworm**—see under Oats, p. 111.

*Frit fly** (overwintering generation)—see under Oats, p. 111.

Gout fly (*Chlorops pumilionis*) (A.L. 174). Larvae live in shoots of barley, wheat and rye (*not* oats). There are 2 generations each year, causing damage in the autumn and during early summer. Attacked shoots are usually swollen ('gouty') and at first of a rich green colour—later they rot away. Shoots attacked at a later stage of growth are not swollen but the ear may be damaged on one side and there is a groove down one side of the stem below the ear.

To reduce attack, avoid sowing autumn corn too early and spring corn too late.

***Leatherjackets**—see under Wheat, p. 102. Note that many varieties of barley may be injured by DDT (e.g. Deba Abed, Impala, Inis, Julia, Mosane, Vada, Zephyr, and the winter barley Senat) whereas others are not affected (e.g. Maris Badger, Maris Otter, Proctor and Sultan).

Aldrin may be used to control leatherjackets only on those varieties of spring barley sensitive to DDT; spray aldrin at 1·6 kg a.i./ha.

Slugs—see under Wheat, p. 102.

*Wheat bulb fly**—see under Wheat, p. 102.

***Wireworms**—see under Wheat, p. 103.

Diseases

Barley yellow dwarf virus symptoms vary in severity, depending upon the strain of the virus, the age of the plant on infection and the variety of the cereal. Sometimes the virus can be recorded from plants which appear to be exhibiting no recognizable symptoms. The virus is transmitted by some cereal aphids. Young plants on infection show a golden-yellowing of the leaf tips and gradually this colouring extends down the leaf. Plants are stunted and may be killed. Surviving plants, or those infected later in life, show golden-yellow colouring of the leaves, heads may not emerge, ears may be blasted, yield is reduced. It is common some seasons when conditions are favourable for aphids, i.e. a mild autumn and winter, especially if followed by a warm and dry spring. Some varieties show slight tolerance to the disease. When a crop, particularly an autumn sown one, is infested soon after brairding with the aphid, *Rhopalosiphum padi*, spraying with an insecticide, to minimize further spread of the virus, is probably worth while.

*Black point (*Alternaria* spp.)—see under Wheat, p. 104.

Brown foot rot and **Ear blight (*Fusarium* spp.)—Common some seasons—see under Wheat, p. 104.

**Brown rust (*Puccinia hordei*). This disease, seen as small brown spore pustules on the leaf and leaf sheaths, has gained in prominence in recent years. Attacks usually occur late in the season when the effect on yield is probably slight. Occasionally more especially in S.W., S.E. and E. England the disease can be severe when the weather is hot and dry. There are a number of varieties with some resistance to the disease. Sprays of benodanil at 1·1 kg a.i. plus wetter in not less than 225 litres of water/hectare applied as soon as an early attack is noted might be worth while if hot and dry conditions persist up to the ripening period.

*Covered smut (*Ustilago hordei*). At ear emergence, affected plants produce black ears containing masses of black spores enclosed by the coat of the grain. The coat generally remains intact, but at threshing it ruptures releasing the spores. The disease is no longer common except in the remote areas and is controlled by organomercury and carboxin-thiram seed treatments.

*Ergot (*Claviceps purpurea*)—Infrequent—see under Rye, p. 114.

**Eyespot (*Pseudocercosporella herpotrichoides*)—Common—see under Wheat, p. 105.

*Grey leaf (Manganese deficiency)—Locally common some seasons, leaves pale green, brown streaks and spots interveinally—see under Oats, p. 113.

**Halo spot (*Selenophoma donacis*) (A.L. 580). Brown spots with a purple border, a grey centre and bearing black fruiting bodies (pycnidia) occur on the leaves. Severe attacks result in the death of leaves and scorched patches in the crop. The disease is possibly seedborne, but is not controlled by seed treatment. It is more common in Wales and south-west England than elsewhere. Special measures are unnecessary.

**Leaf blotch (*Rhynchosporium secalis*) (A.L. 580). Dark grey lesions with dark brown margins occur on the leaves. The disease is frequent on both winter and spring barley, often damaging, especially in Wales and south-west England. It also occurs on rye. Destruction of volunteer plants and wide separation of autumn sown and spring sown crops helps control. In situations known to be prone to leaf blotch, susceptible varieties like Maris Otter, Maris Mink and Universe should be avoided. In these situations sprays of captafol at 900 g a.i./ha *or* captafol at 720 g a.i. plus ethirimol at 140 g a.i./ha (also controls mildew) *or* thiophanate-methyl at 700 g a.i./ha (also controls eyespot) *or* tridemorph at 530 g a.i.

plus carbendazim at 125 g a.i. in not less than 250 litres/ha (also controls mildew and yellow rust) may be beneficial in epidemic years.

*Leaf spot (*Septoria passerinii*). Affected leaves have brownish yellow striped areas bearing the black fruit bodies (pycnidia). The disease seems to be more widespread than hitherto thought. Sprays of benzimidazole type fungicides applied against other diseases may be beneficial against leaf spot.

*Leaf stripe (*Pyrenophora (Drechslera) graminea*). Emerging leaves show pale striping and seedlings are often killed. Plants which survive often have their leaves striped, first pale green, changing later to yellow and finally to brown and may split. Poorly filled and sometimes discoloured heads are produced, or the plants may not head at all. The disease is controlled by grading out the thin severely affected grains, organomercury or carboxin-thiram seed treatments and crop rotation.

**Loose smut (*Ustilago nuda*). Symptoms similar to loose smut of wheat (*q.v.*). This disease is not controlled by organomercury seed treatment and only seed from healthy crops should be sown. Infected seed can be hot water treated. Winter varieties—4 hours in cold water followed by 10 minutes in water at 51–52°C (124–126°F). Spring varieties—4 hours in water at 32°C (90°F) followed by 10 minutes in water at 51–52°C (124–126°F). Alternatively all varieties may be treated for 2·5 hours at 42·5°C (109°F). Sow 'certified' or approved seed. A test to establish the level of loose smut infection in seed is carried out at the Seed Testing Stations at Cambridge and Edinburgh on request, and payment of a fee. Carboxin/organomercury treatment is an effective but somewhat expensive control. Thiram-benomyl seed treatments are also effective.

***Mildew (*Erysiphe graminis*) (A.L. 579) see under Wheat, p. 106. It is often severe especially on late sown spring crops, or when near mildewed winter barley. Mildew is undoubtedly the major disease of spring barley. Some varieties are more resistant than others but new resistant varieties quickly succumb to new strains of the fungus. Control can be obtained by the use of sprays, but their cost is not generally recouped on winter barley except in special situations. Ethirimol at 350 g a.i./ha (see below re tolerant strains of mildew) *or* tridemorph at 530 g a.i./ha *or* triforine at 280 g a.i./ha sprayed on spring barley according to manufacturer's instructions are effective. Ethirimol may also be applied to spring barleys as a seed treatment but should not be used in this form on soils with a high organic matter content as disease control may be diminished. It should not be applied to seed which has already been treated with an organomercury seed treatment. For the control of seed-borne diseases a suitable dry organomercury seed treatment should be applied at the same time. Ethirimol is applied at 520 g a.i./100 kg of seed. As with other seed treatments do not use in seed with a moisture content of more than 16 per cent as damage may occur in some circumstances. Control may be reduced where strains of mildew tolerant to ethirimol occur.

*Net blotch (*Pyrenophora teres*) is seen on leaves as irregular brown blotches, rectangular or elongated. These blotches show a network of dark brown lines, on top of a paler background. The disease is of little consequence, and is controlled by organomercury *or* carboxin-thiram seed treatments.

**Scab (*Gibberella zeae*)—see under Wheat, p. 106.

**Sharp eyespot (*Corticium (Rhizoctonia) solani*)—see under Wheat, p. 106.

**Take-all (*Gaeumannomyces graminis*)—see under Wheat, p. 107.

*Yellow rust (*Puccinia striiformis*). This disease has become more prevalent on barley in recent years although generally the attacks develop late and thus the effects on yield are minimized. The strains of this fungus on barley are distinct from those on wheat, rye and certain grasses and spread does not occur from one kind of cereal or grass to another. The symptoms are similar and spray materials can be used as for yellow rust of wheat (p. 108) but are not often necessary except for susceptible varieties during epidemics. Some of the materials also control brown rust, but the two diseases are unlikely to be serious together as the weather conditions for their development are dissimilar.

5.3   Oats

Pests

**Aphids—see under Wheat, p. 101.

***Cereal cyst eelworm (*Heterodera avenae*) (A.L. 421). Microscopic nematodes enter the roots, interrupt the flow of plant food and water and cause excessive root branching; infected plants are pale and stunted. The crop most seriously affected is oats, but wheat and barley may also be damaged; rye is highly resistant. Mature female eelworms form pinhead-sized cysts, at first white but becoming dark brown. They are full of eggs and may remain in the soil for several years. Heavily infested fields should be grassed down for at least 3 years. Barley varieties resistant to races 1 and 2 of this eelworm are now available and will outyield susceptible varieties on moderately- and heavily-infested land. They have the added advantage of reducing the eelworm population in the soil.

***Frit fly (*Oscinella frit*) (A.L. 110). There are 3 generations each year; they arise from eggs laid in spring, summer, and autumn. Larvae of the first generation damage the shoots of spring sown oats, causing the centre leaf to become yellow and die. Small plants may be killed but larger ones usually survive and may produce an excessive number of short tillers. Each damaged shoot contains a very small white maggot. Except in Scotland the second generation larvae damage oat grains, causing the kernels to be shrivelled and black, but having little effect on the husks. Larvae of the third (overwintering) generation live in grasses but may transfer from ploughed grass to a following cereal crop. Symptoms resemble those of the first (spring oats) generation.

The first and second generation attacks are largely avoided by sowing oats early (before 1 April); damage from the overwintering generation is avoided by ploughing grass early and leaving an interval of at least 4–6 weeks between ploughing and sowing.

***Leatherjackets**—see under Wheat, p. 102.

Slugs—see under Wheat, p. 102.

Stem eelworm, Tulip root (*Ditylenchus dipsaci*) (A.L. 178). Microscopic nematodes enter shoots of oats and rye, causing swelling, distortion, and often death by rotting. Host plants of the oat race include mangolds, sugar beet, beans, onions and rhubarb, and also several weeds. There is no cyst stage (*cf* cereal cyst eelworm). Some oat varieties are highly resistant.

Control is possible by crop rotation, weed control and the use of resistant varieties (e.g. Manod, Peniarth, Maris Quest).

***Wireworms**—see under Wheat, p. 103.

Diseases

Barley yellow dwarf virus. Leaves of affected plants become purplish red. Other symptoms are similar to those seen with infected wheat or barley (*q.v.*).

Brown foot rot and **Ear blight** (*Fusarium* spp.) is common some seasons—see under Wheat, p. 104.

*Covered smut** (*Ustilago hordei*). At ear emergence, affected plants show black ears containing masses of spores, which sometimes remain within the coat of the grain. Now and again the spore masses are exposed, making differentiation between covered smut and loose smut of oats (see below) difficult except by microscopic examination. Sometimes black stripes occur on the topmost leaves. The disease has rarely been seen in recent years: organomercury *or* benomyl-thiram *or* carboxim-thiram seed treatments effect control.

Dark leaf spot (*Leptosphaeria (Septoria) avenaria*). Rounded or lozenge shaped orange bordered dark brown spots, with fruiting bodies (pycnidia) appearing translucent or faintly pink when held up to the light, occur on the leaves and leaf sheaths. Under wet conditions spotting of the panicle occurs and the stalk may be attacked, resulting in rotting and breaking of the straw. Under these conditions the glume can also be attacked and the underlying grain is infected and sometimes discoloured. The disease can then be carried over on the seed. It occurs widely in the oat growing areas of north England, Scotland, Ireland and Wales, and is sometimes damaging, particularly in wet seasons. No practical means of control are known, but sprays of benzimidazole-related fungicides might be beneficial.

*Ergot** (*Claviceps purpurea*). Very rare—see under Rye, p. 114.

*Eyespot (*Pseudocercosporella herpotrichoides*) is frequent, but not usually serious in Great Britain. It is common and serious, however, in the Republic of Ireland, see under Wheat, p. 106.

***Grey leaf (Manganese deficiency). Light green spots occur in the leaves, often about a month after emergence. These spots and sometimes streaks develop, becoming greyish or brown in colour and may have purple margins. When these spots meet across a leaf, the blade collapses at this point and the distal end hangs down. Infected plants are stunted and badly affected ones succumb. The early stages are often mistaken for halo blight (*q.v.*). To control the disease, avoid overliming and spray plants with 9–11 kg manganese sulphate/ha, plus a wetter, in 225–1120 litres/ha.

*Halo blight (*Pseudomonas coronafaciens*) is seen on leaves as a spot with a dead brown centre surrounded by a pale green area, circumscribed by a halo. The spot may fan out in different directions but the distinctive halo generally remains at the edges. This bacterial disease is widely distributed in northern and western areas. Sprinkling the grain with formalin (40 per cent formaldehyde), 310 ml/100 litres of water, gives partial control.

***Leaf (stripe) spot (*Pyrenophora (Drechslera) avenae*). Affected seedlings show narrow brown stripes on the leaves and the first leaf may be distorted and twisted. Plants can die at this stage, or even before they emerge from the soil (pre-emergence blight). Less severely attacked plants survive, showing brown spots on the leaves. Subsequently the younger leaves and the spikelets become infected and other plants in the crop can be secondarily infected. Organomercury seed treatments are not always effective as mercury tolerant strains of the fungus exist. In these cases use benomyl-thiram *or* carboxin-thiram seed treatments.

*Loose smut (*Ustilago avenae*). When the ears emerge, black spore masses are seen replacing the grain in the ears. The spore masses may sometimes be partly or completely enclosed by the coat of the grain, resembling covered smut of oats (see p. 112). Sometimes black stripes occur on the topmost leaves. The disease is rare for, unlike the loose smuts of wheat and barley, it is controlled by organomercury *or* benomyl-thiram *or* carboxin-thiram seed treatment.

***Mildew (*Erysiphe graminis*) (A.L. 579) is a common and often severe disease. For symptoms see Wheat—mildew (p. 106). Varieties of oats resistant to mildew quickly succumb to new races of the fungus. With susceptible varieties and in areas and sites where mildew can be severe, spraying with ethirimol or tridemorph at the same growth stage and concentration as recommended for mildew of barley (p. 110) is likely to be worth while.

**Scab (*Gibberella zeae*). Not common—see under Wheat, p. 106.

*Sharp eyespot (*Corticium (Rhizoctonia) solani*). Occasional, rarely severe—see under Wheat, p. 106.

**Take-all (*Gaeumannomyces graminis* var. *avenae*). This fungus is distinct from *G. graminis* (see under Take-all of Wheat) in that it attacks wheat, barley and oats, and is common in oat-growing areas but occasional elsewhere. To reduce the disease, control perennial grass weeds and avoid cereals too close in the rotation—see also under Wheat, p. 107.

5.4   Rye

The major pests are aphids, leatherjackets, slugs, wheat bulb fly and wireworms (see under Wheat); gout fly (see under Barley); stem eelworm (see under Oats). All varieties of rye may be injured by DDT. Of diseases, black point, brown foot rot and ear blight, which is common some seasons, bunt, which is rare, eyespot, and sharp eyespot are described under Wheat, pp. 104–107. For leaf blotch, which is frequent but not important, see under Barley, p. 109. Rye is fairly resistant to take-all (see under Wheat).

Ergot (*Claviceps purpurea*) (A.L. 548). The ergot is seen as a horn-shaped body protruding from a diseased spikelet, where it completely replaces the grain. It can be as much as 18 mm long. The same fungus is often to be found in the spikelets of grasses. The strains which attack grasses are not always distinct from those attacking cereals. Ergots contain toxic alkaloids derived from ergotine, many of which, though used in medicine, are capable of causing acute illness in animals and humans. The dangers of corn samples containing ergots must be emphasized. Some control is effected by rotation of crops and deep ploughing. Spraying at or just prior to anthesis with benzimidazole-related fungicides may be beneficial.

5.5   Stored cereal pests

***Saw-toothed grain beetle** (*Oryzaephilus surinamensis*) (A.L. 492) is the most important pest of farm stored grain. It destroys the germ and causes grain to heat and become caked and mouldy (A.L. 404). Only a few insects are required to start an infestation; they are not brought in from the field but from premises previously infested. Cooling of grain below 17°C (63°F) prevents breeding. Provided that the granary or bins are reasonably gas tight, infested grain can be fumigated. Ethylene dichloride and carbon tetrachloride are now available as a 1:1 mixture which should be used at the rate of 220 ml/100 kg grain. Farmers should not attempt to fumigate large bulks of grain without professional assistance. An alternative treatment, which must be carried out by a contractor, uses aluminium phosphide preparations to give 1 g phosphine/tonne grain.

Infested buildings should be thoroughly cleaned and sprayed with pirimiphos-methyl (emulsion) at 10 g a.i./litre *or* fenitrothion (w.p.) at 10 g a.i./litre *or* malathion (w.p.) at 15 g a.i./litre. If there is a risk of clean grain being infected it may be mixed with pirimiphos-methyl at 4 g a.i./tonne *or* malathion at 10 g a.i./tonne.

*Grain weevils (*Sitophilus* spp.) (A.L. 219), *S. granarius* being the commonest species. Larvae eat endosperm, hollow out grains and cause heating (A.L. 404). They cannot breed below 13°C (55°F). For fumigation of infested grain see under Saw-toothed grain beetle. For *Sitophilus* alone or when both *Sitophilus* and *Oryzaephilus* are present, spraying of infested buildings is best done using pirimiphos-methyl (emulsion) at 10 g a.i./litre *or* a mixture of malathion (10 g/litre) and gamma-HCH (5 g/litre).

*Flat grain beetles (*Cryptolestes* spp.). For damage and treatment, see under Saw-toothed grain beetle.

**Mites (Acarina). Several species are commonly found, the most important being *Acarus siro*. All species feed on the germ and some of them cause tainting. Mites are liable to develop in grain stored for any length of time at over 14 per cent moisture content. Drying is the best treatment for infested grain, but turning of bulk grain is helpful. Infested grain may be treated with pirimiphos-methyl at 4 g a.i./tonne *or* may be fumigated as under Saw-toothed grain bettle.

Infested granaries should be sprayed with pirimiphos-methyl 10 g a.i./litre *or* dust the floors with a mixture containing malathion (20 g a.i./kg) and gamma-HCH (5 g a.i./kg). For treatment of infested bags see A.L. 469.

Chapter 6
Pests and diseases of potatoes

6.1   Pests

*Angleshades moth (*Phlogophora meticulosa*). The caterpillars of this moth feed on many herbaceous plants. On potatoes they bite irregular holes in the leaflets and occasionally destroy the flowers. If chemical control were necessary against this pest or against caterpillars of other moth species, a larvicide such as trichlorphon could be used.

Ants (*Lasius flavus, L. niger*). Ants often shelter within tubers in holes made by other pests. They may cause superficial damage to tubers, their excretory materials producing small sunken pits in the skin. Chemical control is rarely necessary even in heavily infested fields or gardens.

Aphids—Plant lice, greenfly (A.L. 139, 278). Aphids cause damage (a) by direct injury as they suck plant juices (b) by increasing plant susceptibility to damage from anti-blight fungicides (see p. 128) and (c) by acting as vectors of certain potato virus diseases.

**Bulb and potato aphid (*Rhopalosiphoninus latysiphon*). Although of no importance as a vector of potato viruses, this species can kill shoots of sprouting seed tubers in the chitting house. In medium and heavy soils, particularly in dry summers, it colonizes roots and underground stems of the growing plants, which may wilt and die or have a reduced tuber yield. Chemical control in the chitting house is by gamma-HCH smoke *or* dimethoate spray at 30 g a.i./100 litres *or* formothion spray at 44 g a.i./100 litres using hand lances with long extensions *or* atomize malathion 12·5 g a.i./100 m³ *or* parathion smoke at manufacturer's rate. Sulfotep smoke formulation is effective but is not yet approved. No measure has yet been developed for chemical control in the growing crop.

***Peach-potato aphid (*Myzus persicae*), ***Potato aphid (*Macrosiphum euphorbiae*), ***Glasshouse-potato aphid (*Aulacorthum solani*) and ***Buckthorn-potato aphid (*Aphis nasturtii*). In the chitting house, all except *A. nasturtii* may be present and cause damage to the young sprouts; in addition *M. persicae* and, to a lesser extent, other aphid species may transmit potato viruses, the plants being very susceptible to infection at the chitting stage. For chemical control—see *R. latysiphon* above. In the growing crop, numbers of each species vary considerably between seasons. *M. euphorbiae, A. solani* and *A. nasturtii* are of little importance as field vectors of potato viruses, but they may cause physical damage to the plant foliage with consequent yield loss. *M. euphorbiae* may be associated with 'top roll' symptoms, occasionally seen in Majestic and other varieties. *M. persicae* is occasionally present in sufficient numbers to cause direct feeding damage, but is more important as the vector of leaf roll and rugose mosaic diseases.

Other aphid species, such as the leaf-curling plum aphid (*Brachycaudus helichrysi*) and the black bean aphid (*Aphis fabae*) may colonize potato foliage in midsummer. At least one (*B. helichrysi*) plays some part in the transmission of potato viruses.

Table 6.1 summarizes the chemical treatments available for aphid control in the field.

Thiometon is available as a 10 per cent dust as an alternative to the spray materials mentioned above. A maximum of 4 to 5 spray or dust applications is usually necessary of which the first 3 are generally most important; late spread of potato viruses occurs in some seasons and later spray applications are advocated to counter this spread. Unless aldicarb, disulfoton or phorate have been used at planting, the first spray or dust application should be of one of the systemic aphicides listed above.

These insecticidal treatments, especially of pre-planting granular materials, may effectively reduce the spread of leaf roll within the crop. They cannot prevent entry of this disease into the crop, nor the entry and spread of virus Y (rugose mosaic).

For ware crops, insecticidal cover may be unnecessary in years when aphid numbers are low. If required, use one of the above listed chemicals or malathion spray at 1·28 kg a.i./1000 litres/ha. Application should be made (in anti-blight sprays if compatible) before aphid numbers build up and in July to reduce the numbers of migrating *M. persicae*.

Aldicarb, demephion, demeton-S-methyl, disulfoton, oxydemeton-methyl, parathion, phorate, phosphamidon, sulfotep and thiometon are included in the Agriculture (Poisonous Substances) Regulations.

Cultural control is effected by growing crops intended for seed in as much isolation as possible, and not to prolong the life of the seed stock in areas where leaf roll and virus Y are prevalent. For first and second earlies, the benefits from saving seed (acclimatization etc.) may outweigh the risks of virus infection. Only healthy seed should be used at planting and this should be followed by careful roguing of virus-infected plants during the early part of the growing season. The build-up of aphid parasites and predators, including ladybirds and hoverfly larvae, when these are present may be encouraged by using chemicals having selective toxicity towards aphids. Burning off or mechanically destroying the haulm of seed crops should be done as early as possible.

Capsid (Mirid) bugs. The three main potato-infesting species are the common green capsid, *Lygocoris pabulinus,* the potato capsid, *Calocoris norvegicus* and the tarnished plant bug, *Lygus rugulipennis*. These, together with occasional feeders like *Dicyphus errans,* feed on many herbaceous and woody plants. Damage takes the form of extensive brown necrotic spots on potato foliage, this brown tissue later collapsing to leave holes. Young shoots and foliage may die or become distorted under heavy attacks. Damage is usually confined to plants on the headlands.

Table 6.1   Chemical control of potato aphids in the field

Chemical	Rate per ha	Application method and timing	Comments
1. Granules			
Aldicarb	See Table 6.2.	See Table 6.2.	Rates primarily designed for use against potato cyst eelworm. Supplement with late aphicidal spray (see **2.**) in July for crops grown where *M. persicae* is present and from which seed is to be saved.
Disulfoton	For earlies 0·85 kg a.i.; for maincrops 1·1 kg a.i. on mineral soils, 1·6 kg a.i. on peat soils.	Apply as a band to base of furrow at planting, using special granule applicator.	Also affects leafhoppers.
Phorate	For earlies 1·1 kg a.i.; for maincrops 1·7 kg a.i. on mineral soils, 2·25 kg a.i. on peat soils.	As disulfoton.	High rates reduce wireworm infestations; either rate affects capsids and leafhoppers.
2. Sprays			
Demephion	250 g a.i. in at least 225 litres LV.	Seed crops, apply at 80 per cent plant emergence and repeat at 14-day intervals until senescence or haulm destruction. Ware crops, apply when wingless aphid colonies first found, repeat as necessary.	Also controls leafhopper damage.

Demeton-S-methyl	245 g a.i. in at least 200 litres LV for ground application. 245 g a.i. in 20–50 litres for aerial application.	Timing as for demephion.	See demephion.
Dimethoate	340 g a.i. in 200–1100 litres for ground application. 340 g a.i. in at least 20 litres for aerial application.	Timing as for demephion.	Use w.p. formulation for addition to blight spray.
Ethoate-methyl	340 g a.i. in 200–1000 litres.	Timing as for demephion.	See demephion.
Formothion	415 g a.i. in 220–1100 litres.	Timing as for demephion.	See demephion.
Menazon	285 g a.i. in 200–1000 litres.	Timing as for demephion.	See demephion.
Oxydemeton-methyl	240 g a.i. in at least 200 litres LV for ground application. 240 g a.i. in 20–50 litres for aerial application.	Timing as for demephion.	See demephion.
Phosphamidon	225 g a.i. in 200–100 litres.	Timing as for demephion.	See demephion.
Thiometon	280 g a.i. in 200–1000 litres.	Timing as for demephion.	See demephion.
3. Dust			
Thiometon	280 g a.i.	Timing as for demephion spray—see 2 above.	See demephion, above.

Chemical control is seldom necessary and may be confined to treatment of headlands with dimethoate spray at 340 g a.i./200–1100 litres/ha *or* nicotine spray at 560 ml 98 per cent conc/1100 litres/ha. Phorate granules applied at or before planting to control aphids will also control capsid damage.

*Chafer grubs (A.L. 235). Grubs of chafer beetles may attack potatoes planted after old pasture. The chief offenders are the larvae of the cockchafer, *Melolontha melolontha* and the garden chafer, *Phyllopertha horticola*. In some areas grubs of the summer chafer, *Amphimallon solstitialis,* the rose chafer, *Cetonia aurata* and the brown chafer, *Serica brunnea* may be important.

Chemical control can be obtained with the chemicals listed below for use against cutworms. For cultural control, thoroughly cultivate and disc soil before ridging. Potatoes should not be grown immediately after old pasture in areas where chafer damage is frequently seen.

**Cutworms (A.L. 225). Cutworms are the caterpillars of a number of Noctuid moths and bite roots and stems of potatoes near ground level or more importantly, later tunnel into tubers and leave gaping holes in the skin and galleries in the flesh near the tuber surface. Attacks are more prevalent on lighter soils and in warm, dry summers. Cutworm damage is often ascribed to slugs.

Chemical control may be obtained with a spray or bran surface bait containing trichlorphon or chlorpyrifos or one of the materials listed for leatherjacket control (see p. 122). For cultural control, keep the land free from weeds as these encourage egg-laying and provide food for caterpillars.

Death's head hawk moth (*Acherontia atropos*). Larvae of this moth are occasionally found eating potato foliage, especially of garden and allotment crops.

Earwigs cause damage by biting neat holes in potato foliage; in severe attacks leaves on individual plants may be stripped and left ragged. Chemical control is unnecessary.

***Eelworms (Plant nematodes). Potato cyst eelworm (*Heterodera rostochiensis, H. pallida*) is an important pest in the major ware-growing areas and is present in most gardens and allotments. Advice on the frequency of planting potatoes on infested land is based at present on soil sampling services. In Northern Ireland cropping is controlled by legislation. Great Britain conforms with other member countries of EEC in not permitting seed potatoes to be offered for sale unless they have been grown on land declared free from potato cyst eelworm. Cropping is officially controlled on fields shown to be infested. The pest has important implications for the export of plant material, for many importing countries prohibit the entry of such material (e.g. bulbs, nursery stock) unless it was grown on land shown by soil sampling to be uninfested.

Potato cyst eelworm exists in a number of different forms, until recently designated 'biological races' or pathotypes, of which at least

Table 6.2   Pesticides for use against potato cyst eelworm (PCE)

Chemical and formulation	Rate per ha	Application method	Comments
Aldicarb (granules)	(a) For earlies 2·25 kg a.i. on 460 mm rows; for maincrops 3·35 kg a.i. on 760 mm rows or 2·8 kg a.i. on 910 mm rows. (b) 3·35 kg a.i. broadcast overall treatment.	(a) Apply in a 150 mm band to the furrow when planting, using a special granule applicator. (b) Apply to soil surface just before planting, using special granule applicator. Thoroughly incorporate into topsoil, using rotary implement, especially on peat soils.	All methods give early season control of potato aphids and some control of 'spraing' and should give an economic yield response against PCE up to moderately high infestation levels. PCE numbers may decline after cropping only when high broadcast rates are used.
Dazomet (98% prill)	220 kg on very light or medium soils only.	Apply to soil surface in late autumn or early winter preceding potato crop. Incorporate to a depth of 180—200 mm with rotary cultivator.	Plant potatoes under warm soil conditions in spring following treatment.
Dichloropropene (100% liquid)	(a) For earlies in one year rotations 145 litres; in three or more year rotations 305 litres. (b) For maincrop 270 litres as an overall soil injection to at least 200 mm below soil surface.	(a) Inject overall immediately after early potatoes are lifted. Use a light roller to seal soil surface immediately after application. (b) Inject in early autumn preceding planting maincrop potatoes. Soil temperature should be above 5°C and soil should be in a friable seedbed condition.	Do not use on heavy clay soils, and do not plant wheat for two years after fumigating soil at rates in excess of 135 litres/ha or any crop within six months of treatment. Hand injectors are available to treat small areas; special injection machinery is available as a contract service for large areas.
Oxamyl (granules)	(a) 3·9 kg a.i. on sandy soils with less than 5 per cent organic matter. (b) 5·6 kg a.i. on all other soil types.	Broadcast over well cultivated soil during seedbed preparation and incorporate to a depth of 92 mm, using a rotary implement, especially in peat soils.	PCE multiplication rate sometimes lowered after oxamyl treatment.

three have been identified in this country. Now two species (*H. rostochiensis, H. pallida*) have been recognized within this complex.

Resistant potato varieties have become commercially available for use on land infested with Pathotype A (*H. rostochiensis*), which predominates in some parts of the country.

Several nematicides are now obtainable; when used correctly they can give economically worthwhile yield responses and in conjunction with other measures such as correct rotations and resistant potato varieties their use can lead to a fall in eelworm numbers after cropping. They are listed—Table 6.2.

Potato tuber eelworm (*Ditylenchus destructor*) (A.L. 372) may be introduced within infested seed tubers. It persists in fields where field mint and creeping sowthistle are prevalent, these weeds being the major hosts. No effective chemical control measures are known.

Stem eelworm (*Ditylenchus dipsaci*) (A.L. 178) is of little importance to the potato crop in this country.

Migratory root-feeding nematodes include species of *Pratylenchus* shown to damage potatoes in south-west England and the Scilly Isles. The genus *Trichodorus* contains species which transmit tobacco rattle virus, one of the causes of 'spraing' in potatoes. The materials mentioned above for potato cyst eelworm control give useful control of *Trichodorus* nematodes, while a mixture of dichloropropane and dichloropropene is effective against migratory nematodes in south-west England when injected into soils at 455 kg mixture/ha on 3-year rotations and at 225 kg mixture/ha in 1-year rotations. Efficient weed control measures help to reduce the numbers of migratory eelworms.

*Flea beetle (A.L. 109). The potato flea beetle (*Psylliodes affinis*) feeds on potato leaves and its larvae mine the roots, but serious damage rarely occurs nowadays.

*Leafhoppers. Green leafhoppers (*Empoasca decipiens* and *E. flavescens*) and potato leafhoppers (*Eupteryx aurata* and *Typhlocyba jucunda*) frequently suck sap from potato foliage causing specking, browning or wilting of leaves. The damage is never serious and these insects do not appear to trasmit virus diseases in the UK. Phorate, *or* demeton-S-methyl, *or* oxydemeton-methyl *or* dimethoate, at rates used for potato aphids, also give control of leafhoppers.

*Leatherjackets. These are the grubs of crane-flies (daddy-longlegs) including *Tipula* spp. and *Nephrotoma maculata*. They are usually found in grassland, but will feed on roots and stems of many crops including potatoes, on which they are of little importance. Populations of up to 3 million larvae per hectare are of no economic consequence on potato crops. If necessary use a surface bait of 31 kg bran plus 1·1 kg Paris Green or 1·1 kg fenitrothion/ha. Methiocarb pellets at 220 g a.i./ha give some control of leatherjackets; this and Paris Green give partial control of slug and cutworm damage. Fenitrothion is of use against cutworms but not slugs. Note that Paris Green is a poison included in Part II of the 2nd Schedule of the Poisons List (1962).

Of the newer chemicals still in the experimental stage, chlorpyrifos has proved very effective for leatherjacket control.

*Millepedes (A.L. 150). The spotted millepede (*Blaniulus guttulatus*), the black millepede (*Cylindroiulus londinensis*) and the flat millepede (*Polydesmus angustus*) commonly shelter in holes in potato tubers, being attracted to tissues decaying as a result of attacks by other soil organisms. They can, however, act as primary feeders, scabbing the tuber surface and even tunnelling into the flesh.

Aldicarb, oxamyl or methiocarb, applied at rates suitable for potato cyst eelworm and slug control respectively, should give some control of millepede damage. Chlorpyrifos, an experimental material, may also be of use in this respect.

**Slugs (A.L. 115). Slug damage may be serious in some years in heavier soils, including clays, clay loams and medium to heavy silts. The garden slug (*Arion hortensis*) is the chief species affecting potatoes on a field scale, although the grey field slug (*Agriolimax reticulatus*) may also be responsible. The keeled slugs (*Milax gagates* and *M. budapestensis*) are occasionally important while the black slug (*Arion ater*) is often a serious pest in gardens and allotments. Although slugs are often secondary feeders, enlarging holes in tubers already bored by wireworms etc, they can also penetrate the tuber skin as primary feeders. Unlike cutworm damage, the entry hole is usually small and circular, while large pits are eaten deep into the flesh of the tuber. Damage, which begins before the tubers are mature, greatly increases in severity before lifting, and is worst in late autumn following warm, wet summer and autumn months. Varietal differences in susceptibility are marked, and seem to depend on relative differences in the attractiveness of tuber skin and flesh. Maris Piper, Ulster Glade and King Edward tubers are much more susceptible to slug damage than Stormont Enterprise, Pentland Crown and Majestic. Slugs may also bite ragged holes in the potato foliage lying over the ridged soil, but this seems to be of no economic significance.

Recent experimental work suggests that worthwhile reduction of tuber damage can be achieved by applying methiocarb pellets at 220 g a.i./ha or metaldehyde pellets (6 per cent) at 940 g a.i./ha over the ridges in late July and again in August. The precise time of treatment can be gauged by test baiting with methiocarb pellets under tile traps; when dead slugs are found under most traps in the field, slug activity is increasing and chemical control measures have their greatest effect. Further reductions in tuber damage have sometimes been obtained with an additional application made just after planting, but a large number of autumn applications do not provide much better control than two properly timed ones.

For cultural control, lift maincrop potatoes, especially of susceptible varieties, as soon as possible after tubers become mature. Avoid planting the most susceptible varieties (e.g. Maris Piper) on land known to be infested with large numbers of slugs.

Springtails. Both the garden springtail (*Bourletiella hortensis*) and *Onychiurus* spp. may cause slight holing of the young leaves lying close to the soil surface; the latter species are very common in fen soils. Chemical control is unnecessary.

**Swift moths (A.L. 160). Caterpillars of the ghost swift moth (*Hepialus humuli*) and the garden swift moth (*H. lupulina*) may feed readily on potato roots when the crop follows pasture or weed-infested land. Frequent cultivation after ploughing old pasture usually reduces the number of swift moth larvae to negligible proportions, so that chemical control measures are then unnecessary.

*Symphylids (A.L. 484). Damage by the glasshouse symphylid (*Scutigerella immaculata*) is occasionally seen in field crops in southern England and the Channel Isles. Feeding lesions on roots and root hairs may provide entry for secondary disease organisms.

Chemical control is rarely necessary. Diazinon or parathion spray on seed tubers in the furrow will serve this purpose; chlorpyrifos, an experimental material, is also active against symphylids.

***Wireworms (A.L. 199). Wireworms, the larvae of certain click beetles, are especially numerous in the first years after ploughing old turf, but populations as low as 75 000 per hectare, often found on arable fields, can damage the potato crop. Some soils habitually carry higher numbers of wireworms than others, but in recent years there has been a decline in numbers in all soils. Early attacks on seed tubers and sprouts are not usually of much consequence. Later in the season, holing of newly formed tubers affects quality and provides access for slugs, millepedes and other soil organisms; damage increases in late autumn in the few weeks prior to lifting maincrops. Earlies are not generally affected.

Chemical control is probably not justified for early-lifted crops, nor for maincrops which are planted 4 or more years after old turf or a long ley. Where chemical control is necessary, use aldrin dust at 2·4−3·0 kg a.i./ha on mineral soils or 4·7 kg a.i./ha on fen soils, broadcast and worked in before planting *or* aldrin spray at 2·5−4·5 kg a.i./340 litres water/ha on mineral soils (4·5 kg/ha should be used on fen soils) worked thoroughly into the soil before ridging, after ridging and before splitting back the ridges or after the attack on sets is noticed when the ridges are sprayed, harrowed and set up again. Phorate granules at 3·35 kg a.i./ha applied into the furrow along with seed tubers will control both wireworm damage and aphid attack.

Higher rates of aldrin as shown above should be used on peat soils. When potatoes follow grass leys or stubble, aldrin can be sprayed immediately before ploughing in the autumn preceding planting potatoes, as this affords aldrin longer time to exert its maximum effect.

Neither HCH nor gamma-HCH should be used for wireworm control in potatoes, and potatoes should not be planted on land treated with HCH or within 2 years of treating with gamma-HCH as off flavour may be imparted to tubers.

Phorate is included in the Agriculture (Poisonous Substances) Regulations.

Cultural control rests in growing resistant crops (e.g. peas, beans) after infested turf before taking the first potato crop. Grow early varieties where damge is anticipated, or lift maincrop varieties as soon as the tubers are mature and their skins are set. Cultivate the soil well before ridging.

6.2 Diseases

***Blight** (*Phytophthora infestans*) see A.L. 271. Blight can be reasonably well controlled by spraying, but only if the fungicide applications are suitably timed; to achieve this a knowledge of the course of development of an epidemic is needed.

The most important sources of blight each season are:

1. Imperfectly cleared clamp-sites and dumps of discarded potatoes. The latter are particularly dangerous, because they tend to occur in damp and sheltered sites such as ditches, where sporulation of the fungus is likely to take place earlier than in the open field and where the usually dense growth of potato shoots favours rapid build-up from the first affected plants to form an 'initial focus' (see below).

2. Slightly infected seed tubers of early varieties planted with the crop. Early varieties are in general more susceptible than maincrops, and the earlier development of the foliage encourages earlier spread of the disease. Furthermore, since the yield of earlies is seldom affected by blight it often happens that no measures are taken to check its spread, even though the earlies are growing alongside a susceptible maincrop.

3. Slightly infected seed tubers of susceptible maincrop varieties planted with the crop.

The relative risk from these sources undoubtedly varies from year to year and from place to place, but recent evidence suggests that infected seed tubers planted with the crop are not consistently of such primary importance as was previously thought, except in areas where early varieties predominate.

Slightly infected seed tubers are almost impossible to detect, but there is no excuse for permitting sources of type (1) to exist. Clamp sites should be kept clean throughout the period of riddling and any new growth of potato plants in the spring should be destroyed before the new crop emerges. The site should not be overplanted since this increases the difficulty of dealing with any regrowth which may occur. Dumps should be made in some easily accessible place where precautions can be taken to deal with haulm growth in the spring. Growth can be burnt off with desiccant herbicides as soon as it appears, but this requires very frequent applications if it is to be at all effective. A better plan is to prevent growth completely by the use of a persistent herbicide applied before the tubers have sprouted. Granules of chlorthiamid or dichlobenil may be applied at a rate of 100 g a.i./40 m² of dump surface and the site immediately covered with soil. Alternatively sodium chlorate may be used, preferably in a formulation containing sodium chloride or borax to reduce handling risks. Persistent herbicides should not be used on sites which it is intended to crop during the subsequent 12 months. An alternative method of dealing with small dumps is to spray them with water and cover them with plastic sheet held down at the edges with soil. Under these conditions the potatoes very soon rot away.

There are usually 3 stages in the development of an epidemic, though it should be realized that these so-called stages are simplifications of what is in reality a complicated process in which there are often no clear-cut distinctions between the various phases. The stages are:

(a) Focus-building—the spread by rain-splashed spores from an 'initial infector' plant (i.e. one which has grown from an infected tuber) to a group of neighbouring plants, the whole group forming an 'initial focus' of blight. The special case of the development of an initial focus on a dump has already been mentioned.

(b) Short-range spread—the spread by wind-borne spores from an initial focus on a dump into a nearby potato field, or from a focus in a crop to other parts of the field or into adjacent fields. In either case most of the spread will probably be confined to within a short distance of the initial focus, though a few infections may occur some hundreds of yards down-wind. The new infections develop into 'daughter foci', each of which will, in favourable conditions, grow and act as a centre for the further dispersal of the disease. This short-range spread seldom occurs before June except in south-west England and West Wales. If it is delayed until July or later this stage may not be easily recognizable, in the more favourable infection conditions of late summer blight may spread very quickly from the daughter foci so that they do not remain distinct for long. In such cases stage (b) merges quickly into stage (c).

(c) Long-range spread—in favourable weather the improved infection conditions and the vast number of spores now available in initial and daughter foci lead to widespread and often almost simultaneous infection over large areas. This may occur in June in south-west England and West Wales, but is unusual before July in the chief maincrop districts.

The first signs of this general spread—a scattering of infected plants in many crops—are usually seen 7–14 days after favourable weather conditions have occurred. This general appearance of blight is what is generally termed the 'outbreak' of the epidemic, though in fact it is a fairly late stage in its history. Given that foci are known to be present in an area, it is possible to forecast 'outbreaks' from a knowledge of when infection periods occurred. Various meteorological criteria have been used to define infection conditions. The most widely accepted have been those devised by Beaumont—48 hours during which the temperature does not fall below 10°C, nor the relative humidity of the air below 75 per cent. Slightly more accurate, and making more efficient use of modern weather reporting communications, is the 'Smith Period'—defined as 'two consecutive 24 hour periods (1300–1200) in which the minimum temperature is 10°C, or above, and in each of which there are at least 11 hours with the relative humidity above 89 per cent'. During the growing season information regarding the occurrence of Smith Periods and the spread of blight over the country is issued at weekly intervals by the MAFF via the ADAS, the press and the radio.

Protective spraying is only really effective when the amount of disease present is small. Once blight gets a firm hold on a field and begins to increase rapidly, further protective spraying is almost useless. Spraying should, therefore, be concentrated against the earliest stages of infection in a crop.

In some parts of northern England blight may develop so late that protective spraying is unnecessary except on very susceptible varieties. Even in the south not all areas are equally prone to the development of the disease. The following suggestions may be useful to growers, but they should be modified according to local experience.

In intensive potato growing areas, especially those where blight susceptible varieties predominate, the risk of being near an initial focus is great enough to warrant routine early spraying as an insurance against infection during stage (b) of the epidemic (above). Routine precautionary spraying of this kind should be started just before the crop meets across the rows. Only susceptible varieties such as King Edward and Bintje (and the second early, Craigs Royal) need be sprayed at this stage.

In less intensive areas it is usually safe to wait until a Smith period has occurred, susceptible varieties should then be sprayed within the next few days. In such areas less susceptible varieties such as Majestic need not be sprayed until blight has been seen in neighbouring susceptible varieties or, of course, in the crop itself.

The usual practice for later applications is to continue spraying at 10–14 day (or dusting at 7–10 day) intervals unless the weather turns really hot and dry.

An alternative method is to apply sprays (after the first) only when about 12·7 mm of rain has fallen in a running period of 5 days. This system has the advantage that spray deposits washed off by heavy rain will be replaced at once, and it prevents the wasteful applications of sprays in very dry weather. In humid weather, however, blight (once in a crop) can develop quickly even in the absence of appreciable actual rainfall; in such conditions a routine spray might have delayed in its progress considerably.

High volume (900–1300 litres/ha) and low volume (200–450 litres/ha) applications are about equally effective, and most modern materials can be used equally well in either way. Ultra-low volume spraying (20–60 litres/ha) from aircraft is also satisfactory, provided that the special formulations necessary for this work are used, that the field is suitably shaped and situated, and that the work is done by a reliable firm. In open country there is little difference in efficiency between fixed-wing aircraft and helicopters, but in difficult country and with irregular and badly obstructed fields helicopters are to be preferred.

Dusting is also satisfactory, provided that allowance is made for the lower tenacity of dusts. In wet or windy weather particularly, more applications, in the proportions of about 5 dusts to 3 sprays, have to be made. Moreover, it is safest to assume that anything like a heavy rain will have removed most of a dust deposit; it should then be replaced as soon as possible. Modern dusters enable this to be done quickly and easily and in some ways the more frequent applications necessary with dusts are an advantage in that there is less unprotected new growth than with the more widely-spaced sprays.

In view of the importance of correct timing, growers who normally have their spraying done by contract are well advised to have machines and materials available so that in emergency they can do the work themselves if the contractor is held up by bad weather.

The fungicides recommended for blight control are as follows:

Bordeaux mixture. This is made by mixing a solution of copper sulphate with a suspension of slaked lime. It is a good fungicide and has the valuable characteristic that the weathered remains of earlier deposits assist the adhesion of the later applications. For HV application the weight ratio of $CuSO_4 5H_2O : Ca(OH)_2 : water$ is $10 : 12 \cdot 5 : 1000$. Because of the great excess of lime, this mixture is only suitable for use in HV machines. To apply the same amount of copper/hectare through an ordinary LV farm spray it is necessary to use a formulation (e.g. $8 : 4 : 200$) in which the proportion of lime has been reduced.

One of the farmer's objections to Bordeaux mixture is the comparative difficulty of preparing the suspension, which must be freshly made. The dissolving of the copper sulphate, which is the time-consuming part of the operation, should therefore be done in advance, using wooden barrels or plastic containers.

Burgundy mixture. This differs from Bordeaux in containing washing soda or soda ash instead of lime. For HV mixtures the ratio of chemicals generally used is $8 : 10 : 800$. As with Bordeaux mixture, more concentrated mixes such as $8 : 5 : 200$ (where the second figure refers to soda ash—anhydrous sodium carbonate) can now be made.

Other copper compounds. Various inorganic compounds of copper—e.g. basic copper chloride (copper oxychloride), basic copper carbonate and cuprous oxide, are the essential ingredients of many formulated w.p.s. for blight control. They are usually compared on the basis of the quantity of metallic copper present. In most cases the recommended rate of application is 2–3 kg metallic copper/hectare though the so-called colloidal brands, which are more finely divided, are used at 1 kg metallic copper/hectare.

Dithiocarbamates and allied fungicides. These give much the same degree of control of blight as do the copper compounds but have the advantage of being practically non-phytotoxic. When first applied they are rather more active fungicidally than are copper compounds but they are slightly less stable and persistent. Rates of application of the main dithiocarbamate fungicides are tabulated below:

Fungicide	Usual formulation	Rate of application kg a.i./ha	kg product/ha
Cufraneb	80% w.p.	1·36	1·7
Mancozeb	80% w.p.	1·36	1·7
Maneb	80% w.p.	1·36	1·7
'Polyram'	80% w.p.	1·8	2·25
Propineb	70% w.p.	1·58	2·25
Zineb	70% w.p.	1·54	2·2

In addition to the 'straight' dithiocarbamates listed above, mixtures of maneb with zineb, maneb with zinc oxide, and propineb with manganese sulphate are also available, as are dithiocarbamate-complex fungicides containing manganese and zinc or manganese, zinc and iron. Most of the mixtures and dithiocarbamate-complex fungicides are so formulated as to require application at a rate of 1·7 kg/ha.

Some growers still prefer tank-mix preparation of dithiocarbamates in the field. For this purpose both soluble powder and liquid formulations of nabam are available. To obtain tank-mix zineb the nabam is mixed with zinc sulphate (1·75 kg nabam plus 1·25 kg hydrated zinc sulphate/1000 litres). For preparing zineb/maneb tank-mix a 28 per cent solution of nabam is sold for use with a prepared solution of zinc and manganese sulphates.

All dithiocarbamate fungicides can be irritating to the skin, eyes and nose.

Organotin fungicides. Fentin hydroxide and fentin acetate are rather more phytotoxic than the dithiocarbamates, though less so than the copper compounds. In some circumstances they give a measure of control of blight in the tubers as well as in the haulm. They are often used in spray programmes with the dithiocarbamates, the latter fungicides being used on the younger plants early in the season and then being replaced by organotins on the tougher, older plants when the need for tuber protection is more acute.

Fentin hydroxide is available in various wettable powder formulations containing 20 per cent to 50 per cent a.i. The 50 per cent w.p. is applied at 0·28 kg a.i./ha.

Fentin acetate is available only in mixtures with maneb. The proportion of tin to maneb varies from formulation to formulation and the rate of application varies accordingly—the higher the organotin content the lower the rate of application.

Fentin acetate is compatible with most organophosphorus insecticides. No organotin should be mixed with an oil based pesticide.

The Agriculture (Poisonous Substances) Regulations apply to the use of organotin compounds. All are dangerous to fish and livestock.

Captafol. As a wettable powder this fungicide was tried some years ago and abandoned because of the risk of dermatitis among operators. The chemical is now available in a liquid formulation which has greatly reduced this hazard. Captafol gives a good control of haulm blight and it has the advantages of being very persistent and non-phytotoxic. It is compatible with the insecticides menazon and demeton-S-methyl and may thus be applied with an aphicidal spray. The half-life of the chemical in the soil is 3–4 weeks and run-off from a routine spray programme is normally sufficient to ensure the constant presence of the chemical in the surface layers of the soil. By killing spores washed down from the haulm the fungicide in the soil gives increased protection against tuber blight. The liquid formulation of captafol contains 0·5 kg a.i./litre and should be applied at a rate of 3–3·5 litres/ha.

Chlorothalonil. This fungicide is active against potato blight though it has not yet received ACAS approval for use against the disease. The commercial product is a 75 per cent wettable powder. As a protectant it should be applied at 0·82 kg a.i./ha early in the season increasing to 1·14 kg a.i./ha as the haulm becomes more dense. The rate may be increased to 1·65 kg a.i./ha if the disease becomes established in the crop.

Haulm destruction. Haulm destruction reduces the risk of tuber blight by removing the source of infection (the infected but still living stems and leaves); it also facilitates lifting by destroying weed growth.

To achieve anything like complete control of tuber blight, haulm destruction must take place soon after blight is first seen in the field, but in practice this would often involve an unacceptably heavy loss of crop weight. On the other hand it is rarely worth destroying haulm already half dead with blight.

Factors to be considered before deciding to 'burn off' the haulm are: (1) the amount of blight on the leaves and stems, (2) the crop already formed; it is not worth risking an already good crop for the sake of a little extra weight, (3) the rate of bulking; it is useful to do weekly sample liftings to get some idea of this, (4) the nature of the soil and the state of the ridges; some soils seem to encourage tuber blight, either because they crack in the ridges or because they tend to retain water—most growers know the record of their fields in these respects, (5) the tuber susceptibility of the variety; it is rarely necessary to 'burn off' a variety as tuber resistant as Majestic for the control of tuber blight, though it may be desirable to do so to prevent 'second growth', or for the control of weeds. The crop should not be lifted for at least 10 days after the haulm is completely dead.

The materials in common use for haulm destruction are:

Sulphuric acid. The quickest kill, especially of stems, is obtained by spraying with concentrated (70 per cent) sulphuric acid (BOV) at the rate of 225 litres/ha. Acid-proof machines and protective clothing are necessary so that haulm destruction by this material is essentially a contractor's method.

Diquat. The proprietary formulation of this material is used at a rate of 5·6 litres/ha in 225–250 litres water. Only 1 application at this rate must be made to any one crop, and it must not be used unless the soil around the potato roots is thoroughly wet. It is particularly important to check this after a dry spell, even if a good deal of rain has fallen.

Dinoseb. Formulations containing about 10 per cent w/v dinoseb as an emulsifiable oil solution are used at the rate of about 9·1 litres in 450–1100 litres/ha. The higher volumes should be used if there is a heavy growth of haulm or weeds, but run-off should be avoided.

**Gangrene (*Phoma solanicola* f. *foveata*) (A.L. 545). Although the causal organism may be present in healthy tubers it normally causes rotting only in those which have been injured. Disease losses may be reduced if a 'curing period' (10 days at 55–60°F) is given after any process, such as lifting and riddling, which is likely to injure the tubers. The high temperatures accelerate wound healing and thereby check the advance of the pathogen.

In seed potatoes gangrene can be controlled to a considerable extent by dipping the tubers immediately after lifting in dilute solutions of certain organomercury compounds. Occasionally, however, this technique fails to give a satisfactory control of the disease—perhps because latent infections have been established before lifting. Methoxyethyl mercury chloride is the compound most generally used in this process. Unwashed tubers may be immersed for 0·5 to 1 minute in a solution containing 150 ppm of mercury, or the tubers may be washed before dipping and immersed for a longer period (up to 12 minutes) in a weaker solution (100 ppm of mercury). If unwashed tubers are used the solution must more frequently be changed as the soil from the potatoes reduces the activity of the chemical. Washing tubers before dipping them used to involve considerable risks of increasing the bacterial disease blackleg (*Erwinia carotovora* var. *atroseptica*—see A.L. 107). Recent improvements in technique have, however, greatly reduced this hazard.

Mercury dipping of seed potatoes is tending to be replaced by fumigation with 2-aminobutane—a fungicide which, unlike mercury, has ACAS approval for the control of gangrene and skin spot. Fumigation is carried out within 14 days of lifting in a gastight chamber fitted with a heated vaporizer and with a forced ventilation system to recirculate the air through the potatoes undergoing treatment. 280 ml of 99 per cent 2-aminobutane are used per tonne of tubers. Immature or badly skinned potatoes should not be treated since they may be damaged by the chemical. A Code of Practice for the fumigation of potato tubers with 2-aminobutane is issued by the Department of Agriculture and Fisheries for Scotland.

Because of the risks to operators mercury dipping and 2-aminobutane fumigation should be carried out only in premises subject to the Factories Act or in a process specifically cleared by Government Departments under the Pesticides Safety Precautions Scheme and, in the case of the fumigant, under licence from the National Research Development Corporation. Tubers which have received either treatment should on no account be used as food for humans or stock.

Good reductions in gangrene levels can also be achieved by treating with benzimidazole fungicides, of which the most widely used for this purpose is thiabendazole. No benzimidazole has yet received ACAS approval for use on potatoes, but they promise to be useful materials, especially since, unlike mercury and 2-aminobutane, they can be used on ware tubers.

Thiabendazole may be applied as a dip or as a dust, but both these methods of application present obvious difficulties. A more acceptable alternative is to apply the fungicide 40 g a.i./t potatoes as an ULV spray (34 g of the 60 per cent w.p./litre at 2 litres/t of potatoes) as the tubers pass over the grader. The material may also be introduced into potato stores as a thermal fog. A formulation available for this purpose contains 20 per cent of a mixture of thiabendazole and an organo-iodine complex (the latter component being a bactericide); 28 to 56 g of the liquid concentrate are applied per tonne of potatoes.

Thiabendazole treatments should be applied within 3 weeks of the harvesting of the crop.

Skin spot (*Oospora pustulans*) (A.L. 279). Skin spot can be controlled by mercury dipping or 2-aminobutane fumigation immediately after lifting (see under gangrene). Treatment with benzimidazole fungicides is also effective and has been found to reduce the level of the disease in the progeny of treated tubers.

Dry rot (*Fusarium* spp.) (A.L. 218). Dry rot can be controlled by the use of tecnazene (TNCB) dust applied at the rate of 135 g a.i./t at lifting. For clamped potatoes it is best to put on a layer of straw with a light covering of soil, adding the full winter cover as soon as practicable. Potatoes stored indoors should for preference be covered with straw. Tecnazene is a sprout depressant, and seed for first crops in early districts should be chitted in boxes in November. All other seed tubers should be aired for 6 weeks before planting. Dry rot can also be effectively controlled by organomercury disinfection immediately after lifting (see under gangrene).

Silver scurf (*Helminthosporium atrovirens*) (A.L. 279). This dease may be controlled by post-harvest organomercury dips and disease levels may be reduced by 2-aminobutane fumigation (see under gangrene). Benzimidazole fungicides are also active against the fungus, treatment of seed tubers has been shown to reduce disease levels in the progeny.

Black scurf and **stem canker** (*Rhizoctonia solani*) (A.L. 291). Organomercury dips (see under gangrene) provide a method of killing the fungus on seed tubers but they are not always completely effective as the thicker sclerotia may survive the treatment. Benzimidazole fungicides will also kill the fungus on tubers but, since infection may take place from the soil, treatment of the seed sometimes fails to reduce disease levels in the progeny.

***Powdery scab** (*Spongospora subterranea*) (A.L. 99). Levels of inoculum on seed tubers can be reduced by post-harvest mercury dipping (see under gangrene).

****Common scab** (*Streptomyces scabies*) (A.L. 5). Levels of the actinomycete present on seed tubers can be reduced by post-harvest mercury dipping but since the pathogen occurs so commonly in the soil such treatment has but little effect on disease levels in the progeny. Irrigation applied during tuberization offers the only practicable method of reducing disease levels.

***Blackleg** (*Erwinia atroseptica*) (A.L. 107). This disease is a common cause of losses in the field and it is most troublesome as a cause of storage rots—especially when wet loads of tubers have been placed in store. Bactericides such as dichlorophen and the organo-iodine complex mentioned above are available for the fogging of stored tubers, but chemical control of the disease is extremely difficult and if there is evidence that a large focus of rotting tubers is present in the bulk then the only safe course is to empty the store as quickly as possible.

Use of seed stocks derived from stem cuttings to avoid tuber-borne diseases. Potato stocks free of all tuber-borne diseases (including blackleg) can be produced by propagating tubers from rooted stem cuttings. This method is being used by the Department of Agriculture and Fisheries for Scotland for the production of high grade VTSC (= virus tested stem cutting) stocks for use in 'seed' production. Similar schemes are being initiated by the Ministry of Agriculture for Northern Ireland and by the Department of Agriculture and Fisheries for the Republic of Ireland. If strict hygienic measures are observed by 'seed' growers to reduce the chances of reinfection it is possible to obtain commercial 'seed' samples in which the levels of disease are very low. The use of systemic fungicides or fumigants in conjunction with VTSC material offers the hope of achieving a very high standard of health in 'seed' stocks.

One side effect of the use of healthier 'seed' stocks produced by these methods is that with some varieties (e.g. King Edward) such 'seed' produces a higher proportion of small tubers. Since overall yields are generally higher this seldom reduces the actual yield of ware, but it may present problems in the disposal of the small tubers and in a possible increase in the number of ground keepers.

***Virus Diseases.** The 2 main virus diseases of potatoes, Leaf Roll and Severe Mosaic (virus Y) are described in A.L. 139. Both are transmitted by aphids and in certain circumstances the spread of leaf roll can be controlled to some extent by the use of insecticides (see p. 117). Incidence of virus diseases in 'seed' stocks is kept at a low level by the statutory certification schemes.

Chapter 7
Pests and diseases of sugar beet, fodder beet and mangolds

(See MAFF Bull. 153, *Sugar beet cultivation*)

In 1975, 202 000 hectares of sugar beet were sown under contract with the British Sugar Corporation. Three-quarters of the crop follows cereals and a clause in the contract enforces a minimum rotation of one sugar beet crop in 3 years except after a ley of at least 3 years. Normally sugar beet is sown in late March and early April in 45–56 cm rows. Virtually all (99 per cent) of the seed is pelleted, mainly to help precision drilling; 86 per cent is monogerm and the remainder mostly polyploid multigerm. On 65 per cent of the area the seed is 'sown-to-stand', at a spacing of 12·5 cm or more, i.e. with the object of leaving a stand of plants from the seedlings that establish (usually 50–60 per cent of the seeds sown) without hand singling. If more seedlings emerge, or if herbicides are inefficient, hand work may be needed in the crop; conversely, if establishment is poor the plant population will be very irregular and below the optimum of 74 000/ha. The remaining 35 per cent of the area is sown at closer spacing and, on most of this, hand hoeing aims to remove only the excess of seedlings and leave the optimum number; this optimum is rarely reached, the average being about 62 000/ha. The crop is harvested from the end of September to about the end of December. Most roots are delivered to the sugar factories within a week or two on a permit system, but some may be stored in clamps for 8 weeks or more before delivery.

Mangolds (7,000 hectares) and fodder beet (less than a thousand hectares) are grown in a similar way to sugar beet but are stored in clamps or barns for use as stock feed mainly in spring and early summer. Stecklings for about 1000 hectares of sugar beet and mangold seed crop are sown either under cereal cover-crops in April, or in July–August in the open in areas where root crops are few. They are grown-on *in situ* or transplanted in October–March for harvesting the following summer.

7.1 Pests

(See MAFF Bull. 162, *Sugar beet pests*)

*Beet carrion beetle (*Aclypea opaca*). The beet carrion beetle has damaged beet rarely in recent years, but used to be an occasional, serious pest. Both the adults and larvae feed on the foliage and leave very characteristic blackened, ragged edges. The materials recommended for flea beetle will readily control this pest.

**Beet flea beetle (Mangold flea beetle) (*Chaetocnema concinna*) (A.L. 109). The beet flea beetle occurs in root crops, but the amount of damage caused differs considerably from year to year. Outbreaks are commonest in the drier parts of the country, especially in sheltered fields during a cold, dry spring. Diagnosis is sometimes confused by damage caused by birds, but whereas birds nip off part or all of the cotyledons, flea beetles chew irregular pits and holes in the leaves. The most serious damage is caused to seedlings in the cotyledon stage and insecticide must be applied promptly; use DDT at 560 g/ha, *or* gamma-HCH at 280–560 g/ha provided the land will not be cropped with potatoes or carrots within 18 months.

**Beef leaf miner (Mangold fly) (*Pegomya betae*) (A.L. 91). Attacks on sugar beet by the larvae of the beet fly vary considerably from district to district and year to year; there is a tendency for damage to be more on lighter soils, especially near the coast, and the pest is favoured by cool, moist conditions. White eggs are laid singly or in groups on the underside of the cotyledons and true leaves; the larvae hatch within a few days and burrow straight into the leaf from the underside of the eggs, producing at first linear and later blotch mines between the upper and lower leaf surface. There are two or three generations each year but only the first generation attack, during May and early June, is worth controlling, and only when the attack is severe. When the number of fresh eggs plus living larvae exceeds the square of the number of rough leaves, yield is likely to be diminished unless the pest is controlled. This is achieved by spraying, preferably at 200 or more litre/ha, with one of the following materials: dimethoate at 84 g a.i./ha, *or* formothion at 115 g a.i./ha *or* trichlorphon at 450 g a.i./ha are all suitable for any level of attack or stage of larvae, and all kill rapidly. Treatment should be when egg-laying seems complete and when the first mines are appearing.

The circumstances that lead to severe losses—poor, backward growth with large infestations—are not common in England. Probably no crop past the 8-leaf stage is worth spraying against beet leaf miner alone, but in years when aphids invade early, the two pests can be simultaneously checked by using dimethoate at 335 g a.i./ha which controls beet leaf miner excellently; green aphid control is usually adequate but an increased dosage is necessary if *Aphis fabae* are present. A mixture of 225 g trichlorphon and 300 g demeton-S-methyl per hectare controls both pests and virus yellows well. Trichlorphon should not be used without the addition of an aphicide when any aphids are present in areas where virus yellows becomes prevalent.

'Band-spraying' is suitable for sugar beet in May and early June if done carefully, and it avoids wasteful, and possibly harmful, use of insecticide.

Granules of aldicarb at 560 g a.i./ha *or* oxamyl at 900 g a.i./ha, applied in the seed furrow for control of Docking disorder and soil-inhabiting pests, controls the first generation larvae.

**Birds. Game birds, sparrows and skylarks defoliate seedlings sometimes so severely as to kill them; the damage is indistinguishable from that caused by some mammals. A single defoliation at the early cotyledon stage is the most damaging to subsequent sugar yield, but repeated defoliation is even more damaging at whatever stage it occurs. Spraying repellents, such as anthraquinone, is ineffective.

***Black bean aphid (*Aphis fabae*) (A.L. 54). The black bean aphid or blackfly is one of the most serious pests of sugar beet. Severe and widespread epidemics occur in some years, causing considerable damage to the seed and root crop. The aphid transmits beet yellow virus but the direct physical damage it causes when feeding on the plant is probably more important. The aphid overwinters in the egg stage, mainly on spindle (*Euonymus europaeus*), and migrates to beet in May and June. The primary infestation develops rapidly in hot dry weather, especially when predators of the aphid are few, and the plants lose yield as soon as there are more than 2 aphids per leaf. When left unchecked, infestations can blacken the crop and cause large losses of yield, especially on late-sown crops in drought.

Granules of aldicarb (3·5 g a.i./100 m of row), applied in the seed furrow to control other pests controls *A. fabae* until about mid-June; foliage treatment is then necessary, or earlier if aldicarb has not been used. Colonies of aphids must not be allowed to grow so large as to visibly damage the heart leaves of the plants. Only systemic insecticides are satisfactory and, if as sprays, then in not less than 450 litres water/ha. Suitable materials are, per hectare, demephion at 250 g a.i. *or* demeton-S-methyl at 245 g a.i. *or* dimethoate at 340 g a.i. *or* menazon at 420 g a.i. *or* oxydemeton-methyl at 240 g a.i. *or* phosphamidon at 280 g a.i. *or* pirimicarb at 140 g a.i. *or* thiometon at 280 g a.i. For larger plants the dosage rate recommended should be increased by at least 25 per cent, especially of dimethoate and menazon. The application of granules of disulfoton at 1·1 kg a.i./ha *or* phorate at 1·1 kg a.i./ha, with the granules concentrated in bands over the rows, controls black aphids for longer periods and is preferable to spraying when the plants are heavily infested or late in the season; alternatively, these amounts may be split into two applications and are then approximately equivalent to 2 sprays. Treatment with insecticide after the middle of July is unlikely to be worthwhile. For control of black aphids in seed crops, see p. 146.

*Capsids. Sugar beet can be damaged by several species of Miridae, especially the potato capsid (*Calocoris norvegicus*) and the tarnished plant bug (*Lygus rugulipennis*).

Damage by the potato capsid is confined to crop edges close to hedgerows, woods and orchards from where the nymphs migrate from woody winter hosts to feed on the beet. The active bugs are seen among the beet and injury shows as necrotic spots, puckering of the lamina, and general distortion and yellowing of the leaf, especially at the tip. Damage is rarely severe and, at most, only the field margins need treatment. Phorate granules, at the rate recommended for black bean aphids, control the pest; DDT and gamma-HCH are best avoided because they tend to increase aphid and virus damage. The tarnished plant bug migrates as an adult into the beet field very early in the season and feeds on the growing point of the young seedlings, producing blind distorted seedlings. These symptoms do not show for some time, too late for control measures. Preventive control might be worth while where attacks are expected, but they cannot be forecast; damage is controlled by aldicarb, applied in the seed furrow at the rate recommended for virus yellows control, p. 146.

*Chafer grubs (A.L. 235). Larvae of the cockchafer (*Melolontha melolontha*) and occasionally of the summer chafer (*Amphimallon solstitialis*) sometimes damage beet, especially in well-wooded areas where the soil is light. They feed entirely below the soil surface, at depths of as much as 30 cm. Control is impossible in the growing crop, and soil treatment before drilling impractical because of the difficulty of incorporating the insecticide deep enough and of anticipating damage.

*Cutworms (A.L. 225). The two main species are caterpillars of the turnip moth (*Agrotis segetum*) and the garden dart moth (*Euxoa nigricans*). Turnip moth caterpillars hatch in mid-summer, and feed just below soil level for the rest of the season. Damage is sometimes extensive (e.g. autumn 1959) but not severe enough to require control measures. The garden dart moth is local in the fenland regions. The caterpillars hatch in early spring and then feed until mid-June, sometimes killing many plants and leading to thin stands. Attacks cannot be forecast, so preventive measures are not usually possible. When damage is occurring apply DDT in the late afternoon or evening, preferably under moist conditions which favour surface feeding by the cutworms. Use a spray at not less than 1 kg a.i./ha in 225 litres water. The insecticide should be concentrated along the rows and worked into the soil by steerage hoeing. Alternatively, 280 g a.i. DDT, mixed thoroughly with 30 kg of bran and sufficient water to moisten, should be applied as a poison bait at 30–40 kg/ha during the late afternoon or evening; this is a convenient method for small areas but may not be as effective as dusting or spraying. Gamma-HCH, either as spray or bait, may well be an effective alternative to DDT, provided that 560 g a.i./ha is not exceeded and potatoes or carrots are not to be grown within 18 months.

***Eelworms. Beet cyst eelworm (*Heterodera schachtii*) (A.L. 233) is a potential menace to the beet industry but causes little actual yield loss. 'Beet sickness' (crop failure or severe damage due to root eelworm) occurs rarely in Great Britain, largely because crop rotation is controlled in two ways: (1) the contract made between the grower and the British Sugar Corporation specifies that beet may be grown only after a minimum of 2 years of non-host crops (i.e. crops other than *Cruciferae* and *Chenopodiaceae*) except following a ley of at least 3 years; (2) The

Beet Eelworm Order enforces a 3-year rotation of all susceptible crops within a scheduled area of the fens where beet eelworm is prevalent and a 4-year rotation at least in every field where beet eelworm is known to exist. Partial control is possible with aldicarb *or* oxamyl granules mixed into the soil but is not economic because of the large amount of pesticide that must be used.

Free-living eelworms (principally *Trichodorus* and *Longidorus* spp.) (A.L. 582) damage seedling root systems and stunt plant growth on light sandy soils, causing a trouble known as Docking disorder. Damage is usually patchily distributed in affected fields, being most severe on the sandiest areas, and is also characterized by nitrogen and magnesium deficiency symptoms in foliage, poor root development, frequently a fangy (forked) tap root, and occasional plants showing symptoms of tobacco rattle virus ('yellow blotch') or tomato blackring virus infection ('ringspot'). Rainfall, cultivations, herbicide and fertilizer usage, and previous cropping can affect the incidence of the trouble which, in the worst cases, halves or even quarters yield. A liberal application of nitrogen in the seedbed can obviate very severe yield depression, but control of nematode damage is necessary for the best yield and root shape. Dichloropropane-dichloropropene mixture at 3·3 ml/m row *or* dichloropropene at 2·2 ml/m row, injected 15–20 cm deep in the rows where the beet are to be drilled give adequate nematode control. Fumigants should not be applied into very wet soil which prevents their dispersing; the interval between fumigation and drilling should be about 10 days but this can be considerably reduced if soil conditions are favourable. Granular nematicides applied at very low rates in the furrow with the seed give equally good results; use aldicarb at 560 g a.i./ha *or* oxamyl at 900 g a.i./ha. Both, but especially aldicarb, can have beneficial side effects; they control other soil-inhabiting pests that may be present, also the seedling foliage pests such as beet leaf miner, tortrix moth, capsid bug (*Lygus*) and aphids up to 6 weeks after sowing.

The northern root-knot eelworm (*Meloidogyne hapla*) occasionally attacks beet and other crops such as potatoes and carrots on light sandy soil in East Anglia. Another species (*M. naasi*) occurs in Wales and the western counties of England where it is most often found on the roots of cereals and grasses but also attacks beet. No control measures can be recommended, or are known to be necessary.

Stem eelworm (*Ditylenchus dipsaci*) (A.L. 178) is endemic, but rarely causes severe damage. Seedlings are invaded by the eelworms, which cause galling, bloating and distortion of the stem or petioles and mid ribs, and sometimes death of the growing point. Obvious symptoms of infection are absent during the summer but reappear in the autumn as crown canker—a dry, corky canker in the region of the lower leaf scars—which develops rapidly and eventually invades the whole crown. There are many hosts for the stem eelworm, especially oats and onions, and beet should not immediately follow these crops if they were infested. Seedling damage can be avoided by using aldicarb at 560 g a.i./ha *or* oxamyl at 900 g a.i./ha.

*Leatherjackets (A.L. 179). Leatherjackets are usually a minor pest, occurring mainly in wet soils and immediately after grass or ley but damage was more extensive in spring, 1968. The larvae feed on the plants just below, at, or just above soil level. The commonest of several species is the larva of the marsh crane fly, *Tipula paludosa*. Control is by spraying gamma-HCH at 1·1 kg a.i./ha provided the land will not be cropped with potatoes or carrots within 18 months; or by using gamma-HCH in a bran bait at 250 g a.i./30 kg bran/ha, broadcasting in the late afternoon or evening; or by using fenitrothion (560 g a.i.) as a bait in the same way.

*Mammals. Wood mice (*Apodemus sylvaticus*) locate and dig out the 'seeds', and extract and eat the true seeds from the husk. Damage only occurs in the period between sowing and emergence of the radicle and was particularly prevalent in the dry springs of 1971, and especially 1974. Repellents on the seed are being tested but no recommendation can be made. Mice can be trapped, but poisoning is more practical; it must be done immediately damage begins. Use chlorophacinone, mixed with a suitable bait such as small cereal, and place in short lengths of pipe; three are needed per hectare and they should not be placed too near the hedgerows.

Mice, hares and rabbits defoliate seedling beet; the last two feed later on the petioles and crowns. Repellents have not been tested and damage is best minimized by local control of these pests.

**Millepedes (A.L. 150). Serious attacks by millepedes on seedling beet are sporadic; the area affected appears to be increasing and the pest has recurred in fields where damage was noted in the previous beet crop in the rotation. The spotted millepede, *Blaniulus guttulatus*, is the species most commonly responsible for damage, but the flat millepede, *Brachydesmus superus*, occurs more widely; the latter and several other species occasionally cause damage.

Seed treatment gives only very slight protection but soil treatment with gamma-HCH before drilling is effective; use 1·1 kg a.i./ha, worked lightly into the seedbed before sowing. Aldicarb 560 g a.i./ha *or* oxamyl 900 g a.i./ha limit damage; because they are systemic, they also control some pests of the seedling foliage.

There is no treatment that can be recommended to control the pest when it has started to attack the seedlings. Some growers have obtained partial control by applying gamma-HCH at 1·1 kg a.i./ha in as large a quantity of water as possible and preferably concentrated in a band along the rows; inter-row cultivation should follow treatments to help mix the insecticide into the surface soil. Potatoes or carrots should not be grown within 2 years because of the risk of gamma-HCH taint.

***Peach-potato aphid (*Myzus persicae*). This green aphid is the most important vector of sugar beet viruses; its biology and control are described under Virus Yellows (p. 144).

Pygmy beetle (Pygmy mangold beetle) (*Atomaria linearis*). Small, blackened pits in the hypocotyl of young beet plants are the characteristic form of damage by this pest. Seedlings attacked when small may be killed, either directly or because fungi invade through the wounds, but once the stem starts to thicken this pest does little harm. The beetles often also feed on the young heart leaves; when these later expand they are very tattered. The pest migrates in fine weather in April, and later, from the previous year's beet fields to the new ones; its damage is most common in intensive beet-growing areas. Beet is very rarely grown after beet because of the risk of multiplying beet cyst eelworm (see p. 138); it is permitted after a ley of at least 3 years' duration but control of pygmy beetle damage is then almost invariably necessary.

Seed treatment with dieldrin gives very little protection, but the replacements for dieldrin now being introduced give at least partial control. Soil treatment with gamma-HCH is more effective; use 1 kg a.i./ha and work lightly into the seedbed before drilling. Damage to the leaves is rarely severe enough to warrant treatment but may be prevented by spraying with gamma-HCH at 500 g a.i./ha in sufficient water to give slight run-off; potatoes or carrots should not be grown within 18 months. As for millepedes above, granules of aldicarb at 560 g a.i./ha *or* oxamyl at 900 g a.i./ha are effective alternatives when applied in the seed furrow.

Rosy rustic moth (*Hydraecia micacea*). The caterpillars of this moth burrow inside the swelling root of beet plants from late May onwards and may kill them. Damage is usually negligible since so few plants are attacked.

Sand weevil (*Philopedon plagiatus*). Damage caused by sand weevil is confined to sandy soils and is most prevalent in the Breckland regions of Norfolk and Suffolk. The adults feed on the foliage from late April to early June and can be controlled by the insecticides recommended against beet flea beetle (p. 136). Damage when the beet are in the cotyledon stage can be serious and prompt treatment is essential.

Slugs (A.L. 115). Injury by slugs, usually the grey field slug (*Agriolimax reticulatus*), and sometimes species of *Arion*, is recognized only rarely, but is probably fairly common on the heavier soils. Seedlings can be seriously damaged, either above or below ground, and prompt action with metaldehyde bait is necessary; 560 g of metaldehyde in 30 kg of bran is a standard bait for up to 1 ha and should be moistened before distributing it uniformly, or a proprietary slug bait of metaldehyde and pelleted carrier may be used. Methiocarb controls slugs even more effectively than metaldehyde and is less affected by rainfall; the 4 per cent pellets should be used at a rate equivalent to 220 g a.i./ha.

Springtails. The soil-inhabiting *Onychiurus* spp. have recently been recognized as a serious pest of the germinating seed radicle and seedling roots. They are active in the seedbed, even at the low soil temperatures experienced in March; later in the season they go deeper as the soil surface layers dry out and become warmer. Damage is most prevalent on the heavier soils, especially those with ample organic matter. Gamma-HCH soil treatment 1·1 kg a.i./ha before sowing gives some additional protection to that from the seed treatment; seed-furrow treatment with aldicarb (560 g a.i./ha) *or* oxamyl (900 g a.i./ha) is the most effective.

Symphylids cause damage similar to that by springtails, but are more localized in distribution. The same treatments can control damage but are inconsistent in effect.

Thrips (*Thrips angusticeps*). Injury by thrips is not common. They overwinter in the soil as brachyapterous (virtually wingless) adults and are found on the seedlings in April and May, mainly feeding on the still-curled heart leaves; when these expand they are elongated and even straplike, roughened, with irregular and partially reddened or blackened margins and tips. Small silvery lesions on the leaf surfaces are also usually evident. Control by spraying with DDT at 1·1 kg a.i./ha in at least 340 litres of water.

Tortrix moth caterpillars, usually *Cnephasia interjectana*, bind parts of a leaf or leaves together and feed on the leaf surface within. Only in some years are numbers sufficient for damage to be noticeable and even then is of minor significance. Trichlorphon (450 g a.i./ha) kills them rapidly but, if aphids are present, it may increase subsequent yellows incidence because it also kills aphid predators. Aldicarb (560 g a.i./ha) in the seed furrow controls the damage.

****Wireworms** (*Agriotes* spp.) (A.L. 199). Despite the steadily diminishing seed rate per hectare, and the greatly increased use of genetical monogerm seed, wireworms are not at present a serious pest.

Seed treatment with gamma-HCH in combination with fungicide was introduced in 1948 and dieldrin seed treatment was later introduced also, as an alternative. In 1961 the British Sugar Corporation decided to supply only treated seed to growers and dieldrin became the preferred material because of gamma-HCH's occasional phytotoxicity. Until recently dieldrin 40 per cent dressing (200 g a.i./100 kg of seed) has been used for all seed, whether pelleted or not. In 1975, 99 per cent of the seed was pelleted; 75 per cent was treated with dieldrin and 25 per cent with methiocarb (200 g a.i./ha). Dieldrin will be replaced completely by the best alternative material as soon as possible. Where the wireworm population exceeds 500 000 per hectare additional treatment of the soil with gamma-HCH is needed; apply 560 g a.i./ha and work into the seedbed. Increase the rate to 840 g a.i./ha on fen soil. If using extremely low seed rates so as to achieve a final population of 62 000 to 100 000 plants per hectare with no thinning or singling, i.e. 'planting-to-stand', soil treatment will be needed where there are only 250 000 wireworms per hectare, especially for early drillings. Control of wireworms damaging the seedlings in April or May is very difficult; a tentative recommendation is to apply gamma-HCH as suggested for millepedes.

7.2   Diseases

(See MAFF Bull. 142, *Sugar beet diseases*)

***Black leg. Of the several fungi that can cause black leg, *Pleospora bjoerlingii* is the most prevalent and important. These fungi attack at different stages of the growth of the seedlings: germinating seeds may be killed below ground, the young seedlings may 'damp-off' soon after they come above ground, or they may survive the early infection and then die from stem girdling later on. These fungi rarely cause complete crop failure, but lead to thin and gappy stands of plants. The disease is controlled by seed disinfection; all seed is treated by the seed merchants with ethylmercury phosphate (EMP). The seed is soaked in a 40 mg/litre water solution of EMP (72.3 per cent mercury), i.e. 29 mg mercury/litre, for 20 minutes, drained, dried and regraded.

In some countries, seed is treated with organic fungicides, such as thiram, as a protection against soil organisms, but this appears not to be necessary in Great Britain. The soil fungus *Aphanomyces cochlioides* has recently been detected more frequently than in the past, damaging late-sown crops when the soil remains wet. It is widespread in soil, but crops are normally sown when the soil is too cold for the fungus to attack the seedlings. It sometimes infects the fine roots of sugar beet later in the season.

*Cercospora leaf spot (*Cercospora beticola*) is a major disease in warmer climates, but is seldom seen in Great Britain.

***Docking disorder (see under Eelworms, p. 139).

**Downy mildew (*Peronospora farinosa*). This disease is prevalent in some years in areas where a cycle of infection is maintained between seed crops and root crops. Contracts for growing seed crops specify minimum distance between stecklings, seed crops and root crops, and this separation helps to check the spread of the disease.

Stecklings sown in the summer and early autumn are particularly susceptible. Spraying with fungicide decreases but seldom more than halves the incidence of the disease. Materials used are copper oxychloride (equivalent to 2·24 kg copper/ha) *or* maneb (1·34 kg a.i./ha) *or* zineb (1·8 kg a.i./ha). Spraying should start shortly after seedling emergence and be continued at fortnightly intervals; the fungicide can be incorporated in the sprays of organophosphorus insecticides used against aphids.

Although the disease occurs in root crops regularly, the area severely affected is usually small and routine spraying of crops is therefore unnecessary. Varieties differ in susceptibility (see NIAB Farmer's Leaflet No. 5). Plants infected in June may have their root yield more than halved but infection in September has small effect on yield. Infection in July depresses sugar content from 16 per cent to 11 per cent, for instance, but earlier or later infection depresses it less. The purity of the root juice is greatly decreased by the disease and this adversely affects sugar extraction in the factory.

*Powdery mildew (*Erysiphe* spp.) is common on foliage in the late summer of hot. dry years when it may reduce sugar yield by up to 10%. In recent experiments sprays of benomyl or fentin hydroxide controlled the disease and increased sugar yield. The disease is sometimes troublesome on experimental plants in the glasshouse and is controlled by dusting the foliage with flowers of sulphur, spraying with dinocap, or treating the seed with ethirimol.

*Ramularia leaf spot (*Ramularia beticola*) occurs most years in the south-west where it defoliates the crop in a wet summer. In trials, sprays of fentin acetate (560 g a.i./ha) *or* zineb (1·8 kg a.i./ha) *or* copper oxychloride (equivalent to 2·24 kg copper/ha) in late summer gave some control and increased yield considerably; the problem is a local one in an area where few beet crops are grown and no recent experiments have been made.

Some sugar beet seed crops in Gloucestershire, Lincolnshire and Oxfordshire are defoliated in July by *Ramularia*. Sprays of fentin hydroxide at 670 g a.i./ha have delayed defoliation in trials but their effect on yield or quality of seed has been erratic.

*Rust (*Uromyces betae*) shows as reddish brown pustules on the leaves of beet and may become numerous in late summer killing some of the older leaves, but usually too late to cause appreciable loss of root yield, so no control measures are taken.

*Silvering (*Corynebacterium betae*). See under Beetroot (p. 317).

*Violet root rot (*Helicobasidium purpureum*). This is common in sugar beet but rarely causes serious loss. The roots of affected plants have a purplish fungus growth on the surface below which the root tissues decay. The fungus survives in the soil as resting sclerotia and grows on the roots of many crops and weeds. Soil sterilization with chemicals has not proved practical and control is by crop rotation, generous nitrogen fertilizer dressings, keeping the land free from weeds and deep and thorough cultivation.

***Virus yellows is the disease caused by beet yellows virus (BYV) and beet mild yellowing virus (BMYV), infection with either of which turns the leaves yellow and makes them more susceptible to fungal infection. The disease depresses yields and quality of sugar beet, mangolds and red beet, to an extent depending on how early in its development the plant is infected. The viruses are spread to healthy plants by aphids that have previously fed on an infected plant. BYV persists in aphids for a few hours but, once infected with BMYV, an aphid remains infective for most of its life. The viruses are carried into young root crops in spring or early summer by winged aphids that infect a few, usually widely scattered, plants. Aphids then spread the disease from these centres of infection often producing well-defined circular patches of yellow plants. These patches may increase gradually in size or, where spread is not checked, the whole crop may rapidly become yellow.

The main field vector is *Myzus persicae,* but other aphids may spread the viruses, e.g. *Aphis fabae* can also spread BYV but not BMYV.

Various species of green aphids, such as the potato aphid (*Macrosiphum euphorbiae*), are common on beet in the spring, and some are vectors of yellows viruses, but their main significance is to indicate that heavy infestations of *M. persicae* are likely to follow soon.

Control measures are aimed against clamped mangolds, 'cleaner-loader' sites, seed crops and other host plants from which the viruses spread in spring and also against aphids spreading the disease within the root crop. Insecticides are used extensively to achieve both these objectives. Early sowing, a regularly spaced, dense population of plants and quick growth tend to decrease the incidence of yellows. Control of weeds is important, for some common weeds, such as groundsel (*Senecio vulgaris*) and shepherd's purse (*Capsella bursa pastoris*) may become infected and survive the winter to act as a source of yellows viruses in the following spring. Varieties better able to tolerate yellows viruses are now available commercially (see NIAB Farmers' Leaflet No. 5).

On average, about one-quarter of the few remaining mangold clamps in mid-April are infested with aphids, and nearly all of these contain the mangold clamp aphid, *Rhopalosiphoninus staphyleae*, but about one-fifth contain *M. persicae*. Many of the mangolds are infected with yellows viruses. Winged aphids leave the clamps from mid-April onwards and carry the viruses acquired from the mangold shoots to young root crops. Their effects are often obvious later in the season by the greater incidence of yellows around clamp sites than elsewhere, but some winged aphids doubtless fly considerable distances and infect occasional plants over a wide area.

Aphid infestation in clamped mangolds can be controlled by methyl bromide fumigation but this must be done by specialist contractors. Infestation is best avoided firstly by cutting away all leaves and leaf stalks when the mangolds are harvested, without cutting into the flesh of the root; secondly, by delaying covering the clamps until the roots have cooled, and removing the covering in March. Where practical, all roots should be fed to stock before the end of April. The mangold area in Britain has declined and mangold clamps are now a less important source of virus yellows than formerly. Their place has been taken by beet remnants left in the field during machine harvesting, or on soil dumps at cleaner-loader sites; every effort should be made to destroy these overwintering beet before April.

Contracts between the British Sugar Corporation and the seed merchants, and between merchants and growers of sugar beet and mangold seed, specify measures to be taken to control pests and diseases of seed crops. The stecklings are inspected in the autumn and only those certified as having less than 1 per cent of plants with yellows and as reasonably free from downy mildew are grown in the second year to produce seed.

Stecklings grown under cover crops, especially those sown in April under cereal, are unlikely to need protection with an insecticide until the cover crop is harvested; if there is any risk of aphids infecting the seedling beets, treating the seed with menazon 80 per cent dust at 40 g a.i./kg seed is recommended. Stecklings grown in open beds for transplanting and direct sown stecklings without cover or grown on *in situ* must be protected with systemic insecticides from the time seedlings emerge until late October, and stecklings grown under cover need protection once the cover crop is harvested. The standard recommendation is frequent spraying, at least 3 times, with liquid formulations of demephion, demeton-S-methyl, dimethoate, menazon, oxydemeton-methyl or pirimicarb, at the rates listed on p. 137 for control of black bean aphid, in not less than 300 litres water/ha. Or any two spray applications except the first may be replaced by one application of disulfoton *or* phorate granules at 1·1 kg a.i./ha, or a split application; they should be concentrated in a band along the row so that the plants retain as many granules as possible. Adequate and timely protection of the very young seedlings is difficult by foliar application, since sprays will not persist and granules are not retained, but can be achieved by treating the seed with menazon as above. This treatment should control aphid infestation until at least the first pair of true leaves have expanded; protection afterwards can be by applying liquid or granule formulations, as detailed above (but note the warning below of o.p. resistance in *M. persicae.*). Such treatment to protect the stecklings from aphids and yellows is obligatory, being specified in the contract between the British Sugar Corporation and the seed merchants.

Aphid infestations of the seed crop need checking to prevent winged ones leaving the crops and spreading the viruses to root crops and, if the infestation is heavy, to prevent damage to the seed crop by the aphids' feeding. This is achieved by spraying with any of the chemicals listed above, using the same quantities but in at least 500 litres/ha, just before the crop grows too tall for tractor-mounted sprayers. Band-application to the foliage of disulfoton *or* phorate granules at a minimum of 1·68 kg a.i./ha is likely to give good and lasting control of aphids when applied just before the plants exceed 50 cm in height. Infestations that develop only after the crop is too tall to be sprayed by tractor-drawn machines can be sprayed from aircraft.

Removing the sources from which yellows viruses spread in spring will diminish the initial infection in root crops but inevitably some infective aphids will eventually introduce and spread the disease. This spread can be checked by well-timed treatment with systemic insecticides. The time at which aphids invade root crops differs greatly in different years and different districts so seed treatment with menazon cannot be recommended as a routine, especially as its incorporation into the seed pellet runs the risk of decreased seedling numbers.

Aldicarb granules, applied in the seed furrow, control aphids until mid-June, partially control yellows and increase sugar yield; use 3·5 g a.i./100 m of row (840 g a.i./ha on 460 mm row width). When aldicarb is used for control of Docking disorder, or seedling pests (*q.v.*) at 2·5 a.i./100 m of row, satisfactory control of aphids will be given until at least the beginning of June. In 1975 trials, thiofanox controlled aphids better and more persistently than aldicarb at the above rates.

Timing of foliage treatments is important and the British Sugar Corporation warn growers by postcard when the time seems opportune for treatment to start. The issue of warning cards is decided on the basis of local and general information about the development of aphid populations. In the areas where yellows is commonly prevalent an average population of about 0·25 wingless green aphids/plant justifies the warning cards, but the prevalence of yellows sources, aphid activity, stage of the crops, weather, etc., are also taken into consideration. Growers are advised to examine their own crops regularly and treat if green aphids occur on more than 1 in 4 plants on average. In early seasons aphids may infest the crop in the seedling stage. Spraying with any of the materials listed for control on stecklings (p. 146) checks the infestation and the insecticide remains effective for a week or more. Band spraying is effective with a suitable machine provided it is done carefully enough. A second spray 2–3 weeks later may pay, but a third one rarely does. In all cases a liberal volume of water is essential. Granules of disulfoton (1·1 kg a.i./ha) or phorate (1·1 kg a.i./ha) applied to the foliage kill aphids for longer, control yellows more effectively and increase yield more than do any spray materials. The granule applicator can be mounted on the steerage hoe, thus combining insecticide treatment with inter-row cultivation; the granules should be directed, in a band the width of the plants, so that as many as possible are retained by the foliage. The granules should be applied on receiving the British Sugar Corporation's spray warning or when green aphids are found on the crop and the dosage can be split between two applications. Granules should not be used on plants before singling, especially during dry weather.

Peach-potato aphids resistant to some organo-phosphorus aphicides occur locally throughout the sugar-beet growing areas. If control is unsatisfactory following thorough application of an organophosphorus material at the correct dosage rate and further control is necessary, do not use the same material again; instead, use a carbamate spray.

Chapter 8
Pest and disease control in grass and fodder crops

Pests

Grass and fodder crops grown for conservation or grazing are subject to attacks by a wide range of pests; the damage caused is of little or no economic significance because of the high plant population density. Even where pest damage is noticeable, expensive chemical control measures cannot usually be justified for crops of relatively low cash value. Under exceptional circumstances (e.g. drought) a fodder crop may be vitally important to the livestock farmer and chemical control measures must then be used to reduce pest damage. Chemical methods of control may also be necessary when the pest is the vector of an important virus disease.

A suitable period of time, depending on the insecticide used, should elapse between application and allowing animals to graze treated crops.

Herbage and fodder seed crops are also affected by pests attacking the inflorescence and developing seed. As the cash value of seed crops is often high, chemical control measures are more often justifiable on economic grounds.

New recommendations were introduced for the use of DDT on agricultural crops in October 1971. This pesticide should not be used on permanent or temporary grassland for grazing, silage or hay, although, its use on new leys is acceptable where necessary. DDT should not be used on brassica seed crops or beans except for the control of pea and bean weevil on seedling crops.

Pollinating insects, especially bees, are of prime importance to some brassica and clover seed crops and also to field beans. Chemicals which are toxic to bees should not be used on crops in open flower, and their use (if unavoidable) should be restricted to dull, cloudy days and application made in early morning or late evening. Local beekeepers should be given adequate warning before such chemicals are used, so that hives can be closed temporarily or removed from the vicinity of the area to be treated.

Information on the relative toxicity of pesticides to bees is given in the current list of *Approved products for farmers and growers* of the Agricultural Chemicals Approval Scheme.

8.1   Cereals

RYE

***Frit fly** (*Oscinella frit*)—see under cereal pests (p. 111).
*Stem eelworm** (*Ditylenchus dipsaci*)—see under cereal pests (p. 112).

MAIZE

***Frit fly** (*Oscinella frit*). Heavy attacks on young seedlings cause stunting and distortion of the plants, leaves become twisted and ragged. Attacked plants product multiple tillers and poor cobs. Lighter attacks result in small holes in the expanded leaves. Control measures are not often justified but attacks can be prevented using phorate at 1·7 kg a.i./ha drilled with a suitable applicator close to, but not in direct contact with, the seed *or* chlorfenvinphos at 2·25 kg a.i./ha as a 150 mm band along the rows at 50 per cent crop emergence. Phorate should not be used on soils with over 10 per cent organic matter.

*Cereal leaf aphid** (*Rhopalosiphum maidis*)—causes little direct damage but may transmit virus disease. The chemical control methods are as for grass aphids below.

8.2   Grasses

Pests common to several grasses include:

Antler moth (*Cerapteryx graminis*). Outbreaks are very infrequent but the larvae occasionally reach epidemic proportions in upland areas. Chemical control measures are unlikely to be economically worthwhile.

***Aphids.** The bird-cherry aphid, the grain aphid, the grass aphid, the rose-grain aphid and the apple-grass aphid, described on p. 137, may cause serious damage, particularly to grass seed crops in warm, summer months following mild winters.

Aphids can be controlled with LV sprays (at least 225 litres/ha) of dimethoate 335 g a.i./ha *or* demeton-S-methyl 245 g a.i./ha *or* formothion 420 g a.i./ha *or* oxydemeton-methyl 240 g a.i./ha.

Demeton-S-methyl and oxydemeton-methyl are included in the Agriculture (Poisonous Substances) Regulations.

Burning the seed crop stubbles after harvest effects a degree of cultural control.

*Cereal leaf beetle** (*Lema melanopa*). The adults and larvae of this beetle eat longitudinal strips from the leaves of grasses and cereals and transmit a virus disease of cocksfoot (see p. 164). Control measures are unnecessary.

Chafer beetles (A.L. 235). Grubs of garden chafer and other species destroy grass roots; the turf is then easily rolled back. Fine leaved grasses are often preferentially attacked. Cultivation and reseeding of infested pasture gives effective control. Gamma-HCH soil treatment has been used to control chafers on the Continent but little work has been done in this country.

Common leaf weevil (*Phyllobius pyri*). The larvae feed on grass roots chiefly on light, sandy soils. Perennial ryegrass and fescues are chiefly affected, cocksfoot less so. Adult feeding on grass foliage is of little consequence.

Rolling of the sward repeatedly with heavy rollers when damage is first seen can give some control.

*Common rustic moth** (*Mesapamea secalis*). The caterpillars live and feed within the central tillers of grasses. Control measures are not advocated.

Frit fly (*Oscinella frit*) (A.L. 110). The larvae destroy the central shoots of ryegrass, fescues and bents. They may affect both establishment and development of a young ley after an arable crop or direct reseeding, causing bare patches or allowing coarse grasses to thrive at the expense of sown grass species.

*Grass and cereal flies.** A number of flies lay their eggs on various grasses and the larvae destroy the central shoots. They include *Opomyza florum*, *O. germinationis*, *Cetema* spp., *Geomyza* spp. and *Meromyza saltatrix*. Control measures in grassland have not been advocated, but work is in progress to determine the importance of stem borers in grassland.

*Grass and cereal mite** (*Siteroptes (Pediculopsis) graminum*). This mite is associated with a fungus in causing 'silver top' of herbage seed crops, especially of timothy and cocksfoot. The burning of infested stubbles immediately after harvesting the seed crop serves as a control measure.

*Grass thrips.** *Aptinothrips rufus*, *A. stylifer* and other species, notably *Limothrips cerealium*, feed in both adult and nymphal stages on seed heads of many grasses. The effect on yield is insufficient to warrant control measures.

Leatherjackets (A.L. 179). Old grassland and newly established swards may be severely affected, with bare patches appearing in spring. Less severe attacks may delay the first flush of vegetative growth in spring, a matter of some consequence where 'early bite' is required for livestock.

Spray with DDT 1·05 kg a.i./1100 litres/ha or gamma-HCH 1·12 kg a.i. at least 225 litres/ha, harrowing the soil after treatment and before direct re-seeding or sowing the nurse crop. Do not use DDT on old grassland. For attacks in young leys, surface baits of bran plus insecticide (see p. 102) may give better results.

*Marbled minor moth** (*Procus strigilis*). Though the caterpillars bore up the central shoots of several grasses, they are of no practical importance.

*Oat spiral mite** (*Steneotarsonemus spirifex*). The feeding of this mite on grasses may result in a twisted rachis, failure of the ear to emerge, blind spikelets and reduced grain. Control measures are not warranted.

*Rustic shoulder knot moth (*Apamea sordens*). The caterpillars feed within central shoots of many grasses. Control measures are not necessary.

**Slugs (A.L. 115). On heavy soils particularly in wet seasons, the grey field slug, *Agriolimax reticulatus,* may completely destroy newly established swards or herbage seed crops. For control measures, see under cereal pests (p. 103).

*Swift moths (A.L. 160). The caterpillars occasionally cause serious damage to meadows and newly established swards; for control see under vegetable pests (p. 313), but do not use DDT on established grassland.

*Wireworms (A.L. 199). These pests cause little trouble to established swards but newly sown seedlings may be destroyed. Chemical control is based on insecticidal seed treatments or sprays—see under cereal pests (p. 103).

COCKSFOOT

*Cocksfoot aphid (*Hyalopteroides humilis*). Although often present in large numbers, this aphid causes little damage by its direct feeding; it may transmit cocksfoot streak virus. Control is rarely necessary but see under grass aphids above.

*Cocksfoot gall midges. White larvae or dark brown puparia of *Mayetiola dactylidis* live within swollen stem bases, resulting in a rotting of leaf and stem tissue. Chemical control is unwarranted, for attacks are usually controlled by a high degree of parasitism. Larvae of *Contarinia dactylidis* feed on developing seed within the glumes, control measures are not required in this instance nor in the case of *Dasyneura dactylidis, Sitodiplosis dactylidis* or *Stenodiplosis geniculati,* the larvae of which also feed within cocksfoot seed heads.

*Cocksfoot moth (*Glyphipteryx cramerella*). The larvae feed on the seed heads and pupate in the stem. Chemical control measures are unnecessary. Burning the stubble after harvesting seed crops gives some control.

*Red-legged earth mite (*Penthaleus major*). This causes silvering and die-back of foliage, with patches of thin tufts of grass appearing in late autumn; the pest is very rare.

FESCUES

*Gall midges. The yellowish larvae of *Contarinia festucae* live in large numbers within flower heads while reddish larvae of *Dasyneura festucae* live singly within florets. Neither species is of economic importance.

MEADOW FOXTAIL

***Gall midges.** Larvae of the three species *Contarinia merceri, Dasyneura alopecuri* and *Stenodiplosis geniculati* feed within the developing seed heads and seriously affect yield.

Control for *D. alopecuri* and *S. geniculati* whose larvae remain within infested seed consists in dry-heating the seed to 59–60°C for 35 minutes.

MEADOW GRASSES (*Poa* spp.)

*Gall midges.** Larvae of *Mayetiola schoberi* live within rotting tillers of *Poa pratensis*. Larvae of *M. joannisi* similarly infest shoots of *Poa annua*. Chemical control has not been advocated against these or larvae of *M. poae* and *Caulomyia radifica*, also found within the stems. Larvae of *Contarinia poae* and *Sitodiplosis cambriensis* feed in the flowers of *Poa pratensis* and *P. trivialis*, thus preventing seed formation. Attacks are of no economic significance.

RYEGRASSES

*Gall midges.** Yellow larvae of *Contarinia lolii* feed on ovules but are of no practical importance to the seed crop.

TIMOTHY

*Timothy fly** (*Amaurosoma* spp.). The yellowish larvae feed on the developing flower head before it emerges from the leaf sheath. Although damaged heads are conspicuous within the seed crop, loss of seed in most years is slight. Sprays of DDT, dimethoate or parathion have given variable results because of the need for precise timing when applying sprays.

*Timothy tortrix moth** (*Amelia paleana*). This moth is sometimes prevalent in low-lying marshy areas. The caterpillars web together the leaves and feed on leaf tissue and seed heads. Control measures are unnecessary.

8.3   Herbage legumes (excluding field beans)

Pests common to several leguminous crops include:

Aphids. Large numbers of the black bean aphid (*Aphis fabae*) (A.L. 54), pea aphid (*Acyrthosiphon pisum*), vetch aphid (*Megoura viciae*) or of *Aphis craccivora* may be present on several legume plant species and affect plant growth and vigour; in addition they may transmit certain virus diseases of leguminous crops. Their chemical control depends on methods given for aphids on field beans (see below).

*Cabbage thrips (*Thrips angusticeps*) feed on a wide variety of plants including lucerne and clovers, the foliage becoming speckled and growth retarded. Chemical control is rarely necessary; if so use DDT spray at 1·05 kg a.i. 1100 litres/ha.

*Capsid bugs. A number of species feed on the foliage and may destroy the flowers and developing seeds, but attacks are of no economic significance.

*Clover leaf weevils (*Hypera variabilis, H. nigrirostris*). The adult weevils and larvae feed openly on foliage, buds and seed of clover, lucerne, sainfoin, trefoil and vetches; in some years they affect seed production of these crops. Badly attacked crops should be cut as late as possible to remove pupating larvae with the hay.

For chemical control, apply DDT spray at 1·05 kg a.i. not less than 225 litres/ha 10–14 days after the first cut and before flower buds form.

**Clover seed weevils (*Apion* spp.). Larvae feed on developing ovules and seeds, particularly of red clover. Adult feeding on the foliage is less important. For chemical control spray with DDT at 1·05 kg a.i. not less than 225 litres/ha. Apply the sprays to the headlands in early May and 10–14 days after the first cut when flower buds begin to form. On white clover use DDT 1·05 kg a.i. 225 litres/ha *or* malathion 1·1 kg a.i. 450 litres/ha where white flowers are just beginning to appear. Sprays should not be applied once flowering is well advanced because of the risk to pollinating insects.

*Leafhoppers (*Empoasca* spp.). Speckling of the foliage may result from adult and nymphal feeding. A number of other species transmit clover phyllody virus and other closely related viruses. Special chemical controls are unnecessary, the insects being killed by insecticides applied against other pests.

**Leatherjackets (*Tipula* spp.) (A.L. 179). Legume seedlings and established plants, especially of clovers and lucerne, may be seriously affected. For control measures, see under cereal pests (p. 102).

**Pea and bean weevils (*Sitona* spp.) (A.L. 61). Adults feed on the foliage of a wide range of leguminous crops, the larvae feeding within the root nodules. Attacks are rarely important, but for control measures see under vegetable pests (p. 332).

**Slugs (A.L. 115). The rasping of seedling legumes may occasionally be serious, but damage to the foliage of mature plants is of less consequence. The chief species involved is the grey field slug, *Agriolimax reticulatus*. Control measures are given under cereal pests (p. 102).

**Stem eelworm (*Ditylenchus dipsaci*) (A.L. 409). A number of distinct races, often with overlapping host ranges, infest legume crops and may cause complete failures. Resistant varieties should be grown and over-cropping with susceptible legumes should be avoided.

Chemical control for clovers and lucerne is possible by the methyl bromide fumigation of seed, carried out by the seed merchant in special chambers. As this operation requires careful control and methyl bromide is extremely poisonous, it should not be used by the farmer.

Pests of individual legume crops:

Gall midges. Larvae of the following gall midges infest the leaflets, leaf axils, florets or pods of legumes. Only four are of economic importance, namely, *Dasyneura leguminicola, D. gentneri, D. viciae* and *Contarinia medicaginis*. Chemical control may be effected by DDT sprays at 1·05 kg a.i. 1100 litres/ha, applied before the flower heads are formed; optimum timing of the spray is often difficult. For cultural control, the crop should be cut early for hay or silage.

The clover flower midge (*Dasyneura gentneri*) is sometimes serious on red and white clover; the clover seed midge (*D. leguminicola*) may be serious on red clover, and the clover leaf midge (*D. trifolii*) may be found on both red and white clover. Less common are the lucerne flower midge (*Contarinia medicaginis*), the lucerne leaf midge (*Jaapiella medicaginis*) and the red clover gall gnat (*Campylomyza ormerodi*), the latter being found on red clover. Other midge pests are *Dasyneura ignorata*, on lucerne, and the sainfoin flower midge (*Contarinia onobrychidis*) and the sainfoin leaf midge (*Bremiola onobrychidis*), both on sainfoin. The trefoil flower midge (*Contarinia loti*) is rare but the vetch leaf midge (*Dasyneura viciae*) may be serious on vetches.

8.4   Field beans

***Black bean aphid (*Aphis fabae*) (A.L. 54). Although both winter-sown and spring-sown field beans are attacked by this aphid, the damage caused to winter-sown beans is rarely important. In southern counties of England, spring-sown beans are frequently attacked and whole field preventive treatments, applied as a routine measure at early flowering before aphid colonies build up, and when wheel damage is minimal, are often worthwhile. Use disulfoton granules at 1·1 kg a.i./ha *or* phorate granules at 1·1 kg a.i./ha *or* LV sprays, at least 225 litres/ha, of demephion 250 g a.i./ha *or* demeton-S-methyl at 245 g a.i./ha *or* menazon at 300 g a.i./ha *or* thiometon at 275 g a.i./ha.

In Midland, Eastern and Northern counties, spring-sown field beans are less frequently attacked and routine measures are not justified. In these areas, and in the south on fields where preventive treatments have not been used, eradicant treatment may be required in years when heavy attacks develop. Use LV sprays, at least 225 litres/ha, of demephion at 250 g a.i./ha *or* demeton-S-methyl at 245 g a.i./ha *or* disulfoton granules at 0·85 kg a.i./ha *or* oxydemeton-methyl at 240 g a.i./ha *or* phorate granules at 1·1 kg a.i./ha *or* pirimicarb at 140 g a.i./ha *or* thiometon at 275 g a.i./ha. As eradicant treatments are likely to be applied when the crop is in full flower, the risk of killing bees is greater and granule treatments are to be preferred as these are practically non-toxic to bees.

The sprays should only be applied in the evening after honey bees have ceased flying. Pirimicarb is less toxic to bees than the other sprays. Narrow-wheeled high-clearance applicators are preferable for eradicant treatments to reduce the amount of crop damage.

A number of other aphicides including malathion and nicotine are also effective. Dichlorvos, dimethoate, formothion, mevinphos and phosphamidon give a good control of black bean aphid but are more toxic to bees than the materials recommended above.

In most Midland counties, routine preventive treatments of demephion, demeton-S-methyl, disulfoton, menazon, oxydemeton-methyl, phorate, thiometon, *or* a mixture of menazon and pirimicarb at the rates given above, applied to the headlands of fields at early flowering will often give effective control. These headland sprays may have to be followed with whole field treatments in years when aphid attacks are heavy.

*Green aphids. The vetch aphid (*Megoura viciae*) and pea aphid (*Acyrthosiphon pisum*) colonize field beans, but are usually important only in their role as vectors of bean viruses. The preventive chemical control measures applied against black bean aphid (*q.v.*) will also control green aphid infestations.

**Pea and bean weevils (*Sitona* spp.) (A.L. 61). Although leaf margins are often notched by adult weevil feeding, the damage is of little consequence. Control measures if necessary can be obtained under vegetable pests (p. 332). When used as a soil treatment with late sown field beans (sown after mid-April) phorate granules will also give control of weevils.

8.5   Cruciferous crops

**APHIDS

Mealy cabbage aphid (*Brevicoryne brassicae*) (A.L. 269). This aphid colonizes most cruciferous crops and may cause serious leaf-curl especially in swedes. Late infestations on brassica seed crops building up after the pods have filled are unlikely to be important but, where heavy attacks develop during flowering, control measures may be justified to present pod loss; in these crops chemical control (see under vegetable pests, p. 318) during July and August can be effected only by aerial means.

Peach-potato aphid (*Myzus persicae*) overwinters on many brassica crops and may damage the foliage. It transmits certain virus diseases to healthy plants. For chemical control, see under potato pests (p. 116).

***BEETLES

Blossom (=Pollen) beetles (*Meligethes* spp.). A serious pest of brassica seed crops, especially of spring-sown crops. Both adults and larvae damage flower buds and thus reduce the number of pods set.

The following pesticides will control blossom beetle: azinphos-methyl at 460 g a.i./ha (2 sprays) or 310 g a.i./ha (3 sprays) *or* azinphos-methyl/demeton-S-methyl sulphone 420 + 130 g a.i./ha (2 sprays) or 280 + 80 g a.i./ha (3 sprays) *or* gamma-BHC at 280–560 g a.i./ha *or* malathion at 1·26 kg a.i./ha *or* phosalone 460 g a.i./ha. Endosulfan at 490 g a.i./ha can also be used to control pollen beetle on oil-seed rape and mustard.

On winter rape a spray should be applied before flowering when more than 20 beetles per plant are present.

On spring rape two sprays may be necessary, one at early bud if more than 3 beetles per plant are present, and the second at yellow bud stage if the crop becomes reinfested.

Brown mustard should be sprayed as soon as the first buds are visible, regardless of the number of beetles present, followed by a second spray at yellow bud. White mustard is less susceptible to attack and sprays are only required when heavy infestations occur.

To reduce the risks of off-flavour to root crops for human consumption avoid growing these within two years of applying gamma-HCH sprays.

Damage to tall crops may be reduced by using high-clearance machines fitted with large-diameter land wheels with narrow treads. For very tall seed crops, ground spraying is impracticable. The spray boom should be adjusted to clear the crop by 300–450 mm and low pressures with fan-type nozzles should be used.

Malathion, gamma-HCH and azinphos-methyl are very toxic to bees and should not be used on flowering crops, and in any case it should not be necessary to treat against blossom beetles after the crop has begun flowering. Treatments should be applied in early morning or late evening to reduce hazards to bees.

Early drilling for both overwintered and spring-sown seed crops ensures a vigorous plant able to compensate partly for loss of buds.

Cabbage stem flea beetle (*Psylliodes chryocephala*). The larvae tunnel within leaf stalks and stems of overwintering brassica plants and may seriously affect the vigour of the young plants. Control may be obtained with a gamma-HCH spray at 280 g a.i./ha applied in October–November. If more than one larval mine occurs per 75 mm of plant height, spraying on coleseed may be worthwhile. If gamma-HCH is used do not plant potatoes or carrots within 2 years, so reducing the risk of off-flavour.

Overwintering brassica crops should be grown as far as possible from the site of the previous year's crops.

*Chafer beetles (A.L. 235). The grubs occasionally attack cruciferous crops planted in newly ploughed grassland; for control, see under potato pests (p. 120).

***Flea beetles (*Phyllotreta* spp.) (A.L. 109). The cotyledons and stems of fodder and seed crops are holed and often destroyed by adult beetles during April and May. Damage is accentuated in dry seasons on spring-sown crops. Larval feeding at the roots is unimportant. For control use a combined gamma-HCH/thiram (75 per cent/10 per cent) seed treatment at 15–60 g/kg of seed. If the attack is prolonged or where attacks develop on untreated crops use, at not less than 225 litres/ha, a DDT spray at 875 g a.i./ha (do not use DDT on seed crops), *or* carbaryl 850 g a.i./ha (for late attacks), *or* gamma-HCH as a dust broadcast at 500 g a.i./ha or a spray at 280 g a.i./ha. Because of the risk of imparting off-flavour, potatoes or carrots should not be planted within 2 years in fields treated with gamma-HCH.

Sow spring crops as early as possible on a fine tilth, and provide adequate fertilizer in the seedbed.

*Mustard beetle (*Phaedon cochleariae*). Both adults and larvae occasionally affect seed crops of mustard, turnips, rape and swedes. The second generation attack in August–September is of chief importance. For chemical control, use gamma-HCH dust at 110 g a.i./ha. Apply in May and repeat later if necessary.

As gamma-HCH may impart off-flavour to potatoes and carrots, these should not be grown within 2 years of treatment.

For cultural control, burn stubbles of seed crops, and remove any cruciferous weeds; clean hedge bottoms and dykes bordering infested fields.

***Seed weevil (*Ceutorhynchus assimilis*) is an important pest of brassica seed crops except white mustard. The adults lay eggs in the young pods and each grub eats several seeds before boring exit holes in the pod wall during June and July.

The following pesticides will control seed weevil: gamma-HCH at 280 to 560 g a.i./ha *or* malathion at 1·26 kg a.i./ha *or* azinphos-methyl at 460 g a.i./ha (2 sprays) or 310 g a.i./ha (3 sprays) *or* azinphos-methyl + demeton-S-methyl sulphone at 420 + 130 g a.i./ha (2 sprays) or 280 + 80 g a.i./ha (3 sprays).

Quite low populations of adult weevils can result in high levels of pod infestation in both winter and spring sown brassica seed crops (excluding white mustard) and the pest can cause considerable losses in areas where seed crops are grown intensively. Control measures are usually justified when one or more adult weevils per plant are found on the crop. Brown mustard crops should be sprayed as a routine.

Adult weevils usually begin to invade winter-sown crops at early flowering and sprays may be necessary at this stage but the hazard to bees from pesticides is then greatly increased. Phosalone at 460 g a.i./ha has a low toxicity to bees, and if sprays have to be applied during the flowering stage, it is less likely to cause losses than the other pesticides. Sprays should be applied in the late evening or early morning. Endosulfan at 490 g a.i./ha is also effective and is less toxic to bees but does not kill weevils at temperatures below about 18°C. On spring-sown crops sprays at the late yellow bud stage may also be effective.

While self-pollinated types of rape and trowse mustard do not require bees for pollination; seed yields of other brassica crops, such as Brussels sprouts and most *B. oleracea* are improved when bees are present.

***Stem weevil** (*Ceutorhynchus quadridens*) is an important pest of brassica seed crops, especially of spring-sown ones; older hollow-stemmed varieties of white mustard may be less affected than modern short-stemmed ones. The larvae tunnel through leaf-stalks and into stems during May and June, affecting plant vigour and seed yield. Adult weevils feed on the leaves without serious effect.

A partial control is effected (*a*) by the combined gamma-HCH/thiram seed dressing used against flea beetles (see above) using a high rate of 30–60 g of seed dressing/kg of seed, *or* (*b*) by the spray programme directed primarily against blossom beetles and seed weevil.

Turnip gall weevil (*Ceutorhynchus pleurostigma*) (A.L. 196). The larvae live within rounded root galls, and root decay following attack may be important in the seedling stage of mustard and other brassica seed crops. Control is not normally required for fodder or seed crops.

Wireworms (A.L. 199). Many brassica crops are highly resistant to wireworm attack and can be safely grown after ploughing infested grassland without using chemical control measures.

**BUTTERFLIES AND MOTHS

Caterpillars of the following insects feed on a wide range of cruciferous fodder and seed crops, though attacks are rarely serious enough to warrant chemical control (see under vegetable pests, Chapter 11).

Beet webworm (*Loxostege sticticalis*), **Cabbage moth (*Mamestra brassicae*) (A.L. 69), **Cutworms (A.L. 225), **Diamond-back moth (*Plutella xylostella*) (A.L. 195), Garden pebble moth (*Mesographe forficalis*), **Green-veined white butterfly (*Pieris napi*), Large white butterfly (*Pieris brassicae*), Small white butterfly (*Pieris rapae*), all described in A.L. 69, and Swift moths (*Hepialis* spp.) (A.L. 160).

**FLIES

Cabbage root fly (*Erioischia brassicae*) (A.L. 18). This fly is locally troublesome on fodder and seed crops, but special control measures are not normally required (see also under vegetable pests, p. 319).

GALL MIDGES

Of the species whose larvae infest cruciferous plants, the following are of some importance

**Brassica pod midge (*Dasyneura brassicae*) lays its eggs in pods damaged by feeding or oviposition of seed weevil (*q.v.*) or other agencies. The midge larvae cause attacked pods to ripen prematurely and shed their seed during July and August. Overwintered seed crops are worst affected. The measures recommended for the control of the seed weevil (p. 157) effectively control midge attacks.

**Swede midge (*Contarinia nasturtii*) attacks cruciferous fodder crops especially in the south-western and northern counties of the UK. Its yellow-white larvae are also occasionally found in seed crops.

LEAF MINERS

*Cabbage leaf miner (*Phytomyza rufipes*). The larvae tunnel within leaf midribs and stems of rape and other brassica plants. Chemical control measures are rarely if ever necessary. The larvae of *Scaptomyza apicalis* form blotch mines in cruciferous plants, but control measures are unnecessary.

*Leatherjackets (*Tipula* spp.) (A.L. 179). The larvae may cut off the stems of plants but are only occasionally troublesome in fodder or seed crops, and then at the seedling stage. If control measures are required, see under potato pests (p. 122).

*Turnip root fly (*Erioischia floralis*). Larval damage is occasionally important on turnip and swede fodder crops in Scotland and northern England. The control measures for cabbage root fly (p. 319) are effective.

Diseases

The problems of disease in grass and fodder crops are of a different nature to those of most other crops. These crops may be grown in mixed swards or in pure stands, and with the grasses and legumes, compensation by unaffected plants will often obscure the effects of the disease. Only occasionally will conditions be especially favourable to the disease and severe loss be recognized; in fact the extent of damage caused by these diseases is largely unknown. Control is primarily by management, and very rarely are direct control techniques involving fungicides economically justified, and then usually as seed treatments.

8.6   Cereals

Rye is subject to most of the cereal diseases (see Chapter 5) but, when grown for early bite or ensilage, is little affected. No control measures are normally adopted apart from the use of an organomercury seed dressing (containing equivalent of 1–2 per cent mercury), on rye to control **Stripe smut** (*Urocystis occulta*) at the rate of 2·25 g mercury/100 kg of seed.

Maize

Damping off (*Pythium* spp.). The use of a seed treatment containing thiram at 112 g a.i./100 kg of seed is officially approved. Manufacturers also recommend seed treatment by machine with a liquid preparation containing 30 per cent drazoxolon at the rate of 150 g a.i./100 kg of seed.

Smut (*Ustilago maydis*) can be troublesome on land which has been cropped with maize for a number of years. It is considered useful to treat maize seed with an organomercury seed dressing (1–2 per cent mercury) at the rate of 225 g/100 kg to prevent the introduction of this disease. This dressing will also control pre-emergence damping off.

****Stem rot and Stalk break** (*Fusarium* spp.). This disease attacks the roots and lower parts of the stem, leading to lodging, rendering harvest difficult. Some cultivars are more susceptible than others and should be avoided. The disease builds up in the soil, and whilst an organomercury seed treatment will control disease borne on the seed, it will not prevent the plants picking it up from the soil.

Take-all (*Gaeumannomyces graminis*). Maize is susceptible to root infection by the take-all fungus, and infected maize root fragments remaining in the soil will infect wheat after 6 months.

8.7   Grasses

The grower may be confronted by any of the following diseases:

Ryegrass. **Mildew (*Erysiphe graminis*), **foot rot and leaf spot (*Drechslera siccans*), **net blotch (*D. dictyoides, D. catenaria*), *leaf blotch (*Rhynchosporium orthosporum, R. secalis*), *leaf fleck (*Mastigosporium album*), **crown rust (*Puccinia coronata*), **ergot (*Claviceps purpurea*), root rot (*Fusarium culmorum*), root rot (*Ligniera junci*) and **blind seed disease (*Gloeotinia temulenta*).

Cocksfoot. **Mildew (*Erysiphe gramminis*), **leaf fleck (*Mastigosporium rubricosum*), *black leaf spot (*Phyllachora graminis*), *leaf blotch (*Rhynchosporium orthosporum, R. secalis*), *leaf spot (*Drechslera siccans*), *leaf streak (*D. phlei*), net blotch (*D. catenaria*), **halo spot (*Selenophoma donacis*), **yellow rust (*Puccinia striiformis*), *stem rust (*P. graminis*), *rust (*Uromyces dactylidis*), **ergot (*Claviceps purpurea*), **choke (*Epichloe typhina*), take-all (*Gaeumannomyces graminis*) and *yellow slime (*Corynebacterium rathayi*).

Fescues. **Mildew (*Erysiphe gramminis*), *leaf spot (*Drechslera siccans, D. dictyoides*), *black leaf spot (*Phyllachora sylvatica*), **crown rust (*Puccinia coronata festucae*), *stem rust (*P. graminis*), *rust

(*Uromyces dactylidis*), **ergot (*Claviceps purpurea*), *choke (*Epichloe typhina*), *leaf spot (*Spermospora lolii*), *halo spot (*Pseudomonas coronafaciens*).

Timothy. **Leaf spot (*Mastigosporium cylindricum*), *leaf spot (*Cladosporium phlei*), *leaf streak (*Drechslera phlei*), *leaf blotch (*Rhynchosporium secalis*), *halo spot (*Selenophoma donacis*), *blotch and char spot (*Septogloeum oxysporum*), **stem rust (*Puccinia graminis phlei-pratensis*), **ergot (*Claviceps purpurea*) and *choke (*Epichloe typhina*).

The multiplicity of the foliage diseases is evident from the lists given under the different hosts, but they can be considered under the headings of Mildew, Rust and various Leaf Spotting diseases.

****Mildew** (*Erysiphe graminis*) is destructive under certain conditions, particularly after a dry period, and gives a white powdery appearance to the upper sides of leaves. The fungus exists in many strains and attacks most of the grasses, though there appears to be no record of it attacking timothy. Potash manures will increase resistance, and the disease is worst under conditions of high nitrogen.

****Rusts.** Ryegrass can be severely attacked by crown rust (*Puccinia coronata*) in warm summers, and the rust can be particularly damaging to the aftermath, rendering it unpalatable to stock. It has been noted that leys under intensive production receiving high applications (400 + units) of nitrogen are less severely affected. Good control of this disease has been obtained by fungicidal sprays using maneb + nickel sulphate, and the use of systemic rust fungicides is being investigated. Work is proceeding on the breeding of resistant cultivars. Timothy is frequently attacked by a stem rust, *Puccinia graminis phlei-pratensis,* which can be destructive to the aftermath in hay meadows, affects seed production and can also be responsible for the failure of newly-sown leys.

Cocksfoot is attacked by four rusts. The most destructive is yellow rust (*Puccinia striiformis*) which forms yellow stripes on leaves and sheaths and develops inside the glumes where it can severely affect the yield and quality of the seed. This rust is common on wheat and barley, but the cocksfoot strain does not attack cereals. The timothy stem rust can also cause damage to cocksfoot and another variety of stem rust, *P. graminis avenae* from oats, has been recorded in Scotland on cocksfoot. A fourth rust, *Uromyces dactylidis*, is common on cocksfoot from midsummer onwards.

Fescues are attacked by a specialized form of crown rust, *P. coronata festucae*, but in Wales the commonest rust on these grasses is a form of the stem rust of wheat, *P. graminis*.

LEAF SPOTTING DISEASES

Leaf fleck (*Mastigosporium rubricosum*) can cause quite severe damage to cocksfoot in the autumn and early spring months. Another species, *M.*

cylindricum, is recorded on timothy but does not cause much damage. *M. album* has recently been recorded on perennial ryegrass.

The leaf blotch disease of barley, *Rhynchosporium secalis,* has been found attacking ryegrass, cocksfoot and timothy, and isolations from Italian ryegrass have been shown to infect barley. Another species, *R. orthosporum,* causes trouble in cocksfoot and Italian ryegrass.

Selenophoma donacis causes a halo spot on timothy, but is more damaging to cocksfoot, particularly seed crops, which are attacked at heading time in wet seasons, and yields are reduced.

Drechslera siccans causes a leaf spot of ryegrass and occurs less frequently on fescues and cocksfoot. This disease was the predominant one in a recent national survey on perennial ryegrass. A related species *D. dictyoides* is also found on ryegrass, cocksfoot and timothy. Another leaf attacking fungus is *Phyllachora graminis* which produces shiny black stromata on the leaves of cocksfoot in winter, and a related species *P. sylvatica* is quite commonly recorded on fescues. Timothy is also attacked by two other fungi, *Cladosporium phlei* causing a leaf spot, and *Septogloeum oxysporum,* responsible for a blotch and charspot.

INFLORESCENCE DISEASES

**Ergot (*Claviceps purpurea*) (A.L. 548), in which the grass seed is replaced by a fungal sclerotium, attacks most grasses. The ergots can cause gangrene and lameness in cattle or sheep, and are sometimes blamed for causing abortions, but it is very probable that abortion is seldom, if ever, a true symptom of ergot poisoning. It is most common in old pastures, and where attacks have been serious, these areas should be mown before coming into flower. When such areas are reseeded, the ground should be deeply ploughed, so that the ergots are buried more than 250 mm deep and will be unable to produce their fruiting bodies above the ground. Seed should not be taken from affected crops to avoid disseminating ergots with the seed. The fungus exists in several strains and the form on ryegrass, timothy and couch is only slightly infectious to wheat, whilst the strain on black grass and meadow fescue is highly infectious. Work is proceeding on the suppression of apothecial formation by benzimidazole fungicides.

**Choke (*Epichloe typhina*). The inflorescence is converted into a fungal stroma, first white then turning yellow, and results in complete blindness and no flowers or seed are produced. Most of the grasses are attacked, though there appears to be no record for ryegrass. It is most important on cocksfoot seed stands which may be rendered unproductive after the third harvest year. It has been recorded on timothy seed stands but not to the same extent.

**Blind seed disease of ryegrass (*Gloeotinia temulenta*) is usually recognized by poor germination figures for the seed. The latter is rotted internally though this may not be evident from outward appearance. The disease has been found throughout the country, but particularly in the wetter parts, and can be serious in Northern Ireland. Control consists mainly in not taking seed from affected crops, as infected seed sown normally will produce the fruiting bodies of the fungus, which by means of aerial spores infect the flowering heads of the new crop. Burning the straw and stubble after harvesting seed will assist control and work is proceeding using benzimidazole fungicide for the suppression of apothecial formation. Valuable nuclear stocks can be cleared of infection by immersing the seed for 20 minutes in a 0·25 per cent solution of an organomercury compound (containing the equivalent of 1 per cent Hg) held at 50°C.

*Yellow slime disease of cocksfoot (*Corynebacterium rathayi*). This disease has been recorded a few times in this country, and is sometimes found in imported seed. If infected seed is sown, bare and stunted patches are caused, with failure to tiller; finally the leaves and panicles are coated with a yellow slime, and the seed becomes infected. Disinfection of the seed is not normally practised for, under the certification scheme, seed is only taken from healthy crops.

ROOT DISEASES

Take-all disease (*Gaeumannomyces graminis*) of cereals is known to affect grasses but is rarely damaging. Grasses, however, carry over take-all infection to a subsequent cereal crop; in this connection, it is the stoloniferous weedy grasses which are the most important and the main herbage grasses are not considered to play a serious role.

Root rot (*Fusarium culmorum*). This fungus has frequently been isolated from the roots of perennial ryegrass which have shown loss of stand. In many instances cultural and management factors have been at fault. Pathogenicity tests to ryegrass have been negative.

The fungus (*Ligniera junci*) has been found in rotting roots of Italian ryegrass in young leys showing premature loss of stand.

SEEDLING DISEASES

Damping-off fungi (*Pythium* spp. *Fusarium* spp. and species of *Drechslera*) can cause pre-emergence and seedling losses. For this reason manufacturers recommend for the direct seeding of ley grasses, the use of a liquid preparation containing drazoxolon applied by machine at the rate of 75 g a.i./50 kg for dense seeds or 100 g/50 kg for less dense seeds.

*Virus diseases. Barley yellow dwarf virus exists in several strains and can be found attacking most of the grasses which are tolerant to this virus. A recent random survey of S.24 Perennial Ryegrass fields showed 90 per cent to be infected, mostly without symptoms, and this must serve as a reservoir of infection available to infect cereals. The virus is transmitted by several species of aphids, and is of the semi-persistent type and may therefore be controlled by insecticides though these are not normally used.

Another aphid-borne virus is cocksfoot streak, which is common in leys. Cocksfoot is also susceptible to cocksfoot mottle virus, which can cause serious losses in seed crops. This disease is transmitted by both adults and larvae of the cereal leaf beetle, *Lema melanopa*, and can also be transmitted by mechanical contact.

Ryegrass can be extensively infected with ryegrass mosaic virus, which is transmitted by a mite *Abacus hystrix*, commonly present between the ribs of the leaves on the upper surface. The virus exists in strains of differing virulence, the mild strains producing a mottling and streaking of the leaves, whilst severe strains cause a dark brown leaf necrosis. The disease is severe and widespread in the south of the country affecting both seed crops and leys for conservation and grazing. Infected swards can lose up to 30 per cent of their production. Surveys have indicated that Italian ryegrass is more susceptible than perennial, crops sown in the spring (usually under a cereal) are more severely affected than crops direct sown in the autumn. Early grazing, especially by sheep in the autumn, reduces the mite population and so virus infection in the following year. Research is in progress on controlling the mites by acaricidal sprays, which at present would not be economic, and on the selection and breeding of tolerant cultivars.

8.8   Leguminous crops

Of fungi causing disease in these crops, the following merit mention:

Clover. *Leaf spot (Ascochyta trifolii), *ring spot (Pleospora herbarum), **leaf spot (Pseudopeziza trifolii), *leaf spot (Stagonospora meliloti), *leaf spot (Cercospora zebrina), *mid-vein spot (Mycosphaerella carinthiaca), *black blotch (Cymadothea trifolii), **mildew (Erysiphe trifolii), *downy mildew (Peronospora trifoliorum), *rust (Uromyces trifolii), *white clover rust (U. nerviphilus), **scorch (Kabatiella caulivora), *white clover burn (Leptosphaerulina trifolii), *black stem (Ascochyta imperfecta), ***clover rot (Sclerotinia trifoliorum), *violet root rot (Helicobasidium purpureum), *root rot (Corticium solani), *grey mould (Botrytis cinerea), *Anther mould (Botrytis anthophila), *wilt (Verticillium dahliae) and *bacterial leaf spot (Pseudomonas syringae).

Trefoil. *Leaf spot (Pseudopeziza medicaginis), *ring spot (Pleospora herbarum), *downy mildew (Peronospora trifoliorum), *rust (Uromyces pisi), **clover rot (Sclerotinia trifoliorum), *Violet root rot (Helicobasidium purpureum), **black stem (Ascochyta imperfecta).

Lucerne. **Leaf spot (Pseudopeziza medicaginis), *ring spot (Pleospora herbarum), *burn (Leptosphaerulina trifolii), *downy mildew (Peronospora trifoliorum), *rust (Uromyces pisi), **anthracnose

(*Colletotrichum trifolii*), **black stem (*Ascochyta imperfecta*), **clover rot (*Sclerotinia trifoliorum*), **crown wart (*Physoderma alfalfae*), *violet root rot (*Helicobasidium purpureum*), *root rot (*Corticium solani*) and ***wilt (*Verticillium albo-atrum*).

Sainfoin. *Leaf spot (*Asococchyta orobi*), *leaf spot (*Ramularia onobry chidis*), *leaf spot (*Septoria orobina*), *ring spot (*Pleospora herbarum*), **mildew (*Erysiphe trifolii*), **clover rot (*Sclerotinia trifoliorum*), *grey mould (*Botrytis cinerea*), *rust (*Uromyces pisi*) and *wilt (*Verticillium dahliae*).

Tares or Vetches. *Chocolate spot (*Botrytis fabae*), **downy mildew (*Peronospora viciae*), *rust (*Uromyces fabae*) and *clover rot (*Sclerotinia trifoliorum*).

The extent of loss caused by the many foliage diseases is largely unknown and in many instances obscured by compensation. No fungicidal spraying is normally practised. The commonest leaf spot disease is caused by two species of *Pseudopeziza* which are present in most crops of red clover and lucerne, and can attack White Clover, trefoil and other clovers. They can be quite destructive in wet autumns. In lucerne, trials have shown that an earlier second cutting can reduce leaf and protein losses and retard disease outbreak by 1 week, whilst still allowing sufficient growth after the first cut. ***Black blotch** (*Cymadothea trifolii*) may be found causing extensive infection in both red and white clover in late summer, but seldom causes serious damage. ****Mildew** (*Erysiphe trifolii*) may cause extensive infection in red clover, and sainfoin is sometimes severely attacked. Other leaf diseases including downy mildew, rusts and various leaf spotting fungi occur, but are seldom important.

*Burn (*Leptosphaerulina trifolii*) causes a fine spotting on white clover leaves, and reduces yield and quality. The crude protein content is reduced, and the disease has been implicated in raising the leaf oestrogen content, with possible adverse effects on animal health and reproduction. A strain of this fungus specific to lucerne has been recorded.

STEM DISEASES

**Black stem (*Ascochyta imperfecta*) is very widely distributed in Britain in these crops. It is seed-borne and attacks clovers, trefoil and lucerne, causing cankering and death of shoots. It is most severe in the early months and may affect the first cut of lucerne. Organomercury seed treatments are known to be effective against the disease, but are not normally used in this country. The disease has been shown to be completely controlled on trefoil seed by soaking the seed in 0·2 per cent thiram solution held at 30°C for 24 hours.

**Red clover scorch (*Kabatiella caulivora*) may cause the death of leaves and a shrivelling of flowers by forming lesions which girdle leaf and flower stalks. Temporary severe loss of foliage, and in a seed crop, a total failure of seed production may occur. The disease is seedborne and eradication of the fungus from the seed coat can be achieved with captan or organomercury seed treatments.

Anthracnose (*Colletrotrichum trifolii*) causes wilting and death of stems by forming a girdling lesion at the base of the shoots of lucerne, and can cause temporary loss of foliage.

***Clover rot** (*Sclerotinia trifoliorum*) (A.L. 266) can be serious, particularly in red clover when a short rotation is followed. Infection occurs in October from spores produced by the germinating resting bodies (sclerotia) in the ground. In mild seasons of high humidity the disease spreads killing large patches of plants, and severe loss of stand may be evident at the end of the winter. Red clover is the most susceptible, but there are new cultivars becoming available which are very much more resistant than the old Broad red. Trefoil was once thought to be resistant, but is sometimes more severely attacked than red clover in the same field, and it is thought there may be a special strain which attacks trefoil. The disease can be very damaging to young lucerne in its first year, killing many plants, but established lucerne crops are generally resistant, the disease causing temporary loss of crop before the first cut, and the plants are not killed. Susceptible crops should not be grown more than once in 8 years, and care should be taken to secure seed from uncontaminated sources, as the sclerotia can be disseminated with the seed. Grazing the crops off in the autumn is beneficial, as it reduces the amount of foliage available for infection, and reduces conditions of high humidity within the crop. Work is proceeding on the suppression of apothecial formation by benzimidazole fungicides.

Crown wart of lucerne (*Physoderma alfalfae*) produces characteristic warty galls at the crown of the plant, and may lead to wilting in hot weather and loss of yield. This disease is invariably associated with poor drainage, and is often found in fields which have been 'poached' by winter grazing of cattle. Control is best effected by attending to drainage and lengthening the rotation.

***Verticillium wilt** (*Verticillium albo-atrum*) has caused most trouble in lucerne in recent years and can infect the ground for a considerable, and as yet undetermined, number of years. The symptoms are a yellowing and wilting of the leaves and stems, followed by a shrivelling of the whole plant from the base up. Symptoms appear after the first cut and increase throughout the season. Regrowth may be stunted with shortened internodes, and eventually the whole plant may be killed and lucerne stands rendered quite worthless in their third year. A recent survey showed that in one-third of infected fields, the disease had most probably been introduced with the seed. Experiments in the UK have shown that the use of a (50 per cent) thiram seed dressing at the rate of 450 g a.i./100 kg of seed appears promising. The dressing may be applied by the merchant prior to bacterial inoculation by the farmer before sowing. Although field information is lacking, it is recommended that all lucerne seed should be treated before sowing in unaffected land, to prevent the possible introduction of this troublesome disease, and this is now officially approved. Resistant cultivars are now becoming available.

VIRUS DISEASES

**Phyllody of clover, in which the inflorescence is converted to a green vegetative and sterile condition, can cause considerable losses in white clover seed crops. It is transmitted by leafhoppers (see p. 153) and can also be carried to strawberries in which it produces the troublesome green petal disease. This disease has now been shown to be caused by a mycoplasma and not by a true virus.

A recent survey of white clover in permanent leys revealed the presence of 7 different viruses whose relative significance has not been determined.

8.9   Field beans (*Vicia faba*)

The following are the more important diseases:
***Stem Rot (*Ascochyta fabae*), ***Chocolate spot (*Botrytis fabae*), *Leaf Spot (*Cercospora zonata*), *Downy Mildew (*Peronospora viciae*), *Net Blotch (*Pleospora herbarum*), *Rust (*Uromyces fabae*), **Sclerotinia Rot (*Sclerotinia trifoliorum* v *fabae*), *Root Rot (*Phytophthora megasperma*, *Corticium solani* and *Fusarium* spp.), **Damping off (*Pythium* spp.).

*****Stem rot** (*Ascochyta fabae*). The increasing popularity of beans as a break crop has brought this disease into prominence, and it can reduce yields to less than 2·5 t/ha. The disease is seed-borne, and the use of a benomyl plus thiram seed treatment at the rate of 72 g benomyl + 72 g thiram/100 kg of seed is officially recommended. Ideally only seed free from disease should be sown, and farmers can have their seed tested for the presence of the disease by the Official Seed Testing Centre, Cambridge. Volunteer plants from previous infected crops in adjoining fields also carry over the disease, and these should be eliminated.

*****Chocolate spot** (*Botrytis fabae*). In wet seasons favourable to this disease, winter-sown beans can be very seriously affected. Spring-sown beans are not so severely attacked. Infection is carried over on debris of previous crops in adjoining fields, on self-sown seeds and occasionally on seed. An inadequate supply of potash may aggravate the incidence of the disease. Manufacturers recommend the application of benomyl at 0·55 kg a.i. in not less than 230 litres of water/ha. If infection occurs before the early pod stage 2 applications at half strength with a 3 week interval are advised. An approved non-ionic wetting agent should be added to the spray tank. Alternatively spray with carbendazim 0·55—0·66 kg a.i. in at least 200 litres/ha *or* thiophanate-methyl 1·1 kg a.i. in 330—660 litres/ha.

***Damping off** (*Pythium* spp.). The use of seed treatments of captan *or* thiram at 112 kg a.i./100 hg of seed *or* a mixture of 72 g benomyl + 72 g thiram/100 kg are officially approved to combat this trouble and seed decay. Manufacturers also recommend farmers to use seed treated by the merchant with drazoxolon at the rate of 50 g a.i./100 kg of seed.

Sclerotinia rot (*S. trifoliorum* v *fabae*). This is a specialized form of the clover rot fungus, and follows too close cropping with beans, and the failure to eradicate ground keepers. The common form of clover rot has been found also on beans which should not be included in the break to control clover rot in red clover.

Virus diseases. Field beans are subject to several virus diseases which can seriously affect growth and yield, if substantial infection develops before or during flowering. The two most prevalent diseases, broad bean true mosaic and broad bean stain viruses are both carried in the seed and are mainly spread by the clover seed weevil *Apion varax* and to some extent by the bean weevil *Sitona lineata*. Insecticidal control measures (see p. 149) are feasible, but are not generally adopted for economic reasons. Seed should not be taken from affected crops. Two other viruses, bean leaf roll and bean yellow mosaic are prevalent in the south, and are transmitted by aphids. Leaf roll is not seed borne, and yellow mosaic seldom unless the parent crop is infected unusually early. Aphicides applied against *Aphis fabae* may help to reduce the spread of these viruses. Late sowing of spring beans, especially adjacent to winter beans, favours early infection.

8.10   Brassica crops

The following are the more important diseases:

Kale. **Dark leaf spot (*Alternaria brassicicola* and *A. brassicae*), *leaf spot (*Ascochyta brassicae*), *grey mould (*Botrytis cinerea*), *wire stem (*Corticium solani*), *white blister (*Cystopus candidus*), *Mildew (*Erysiphe cruciferarum*), *light leaf spot (*Cylindrosporium concentricum*), *downy mildew (*Peronospora parasitica*), *ring spot (*Mycosphaerella brassicicola*), *canker (*Leptosphaeria maculans*), **root rot (*Phytophthora megasperma*), **club root (*Plasmodiophora brassicae*), *sclerotinia rot (*Sclerotinia sclerotiorum*), *black rot (*Xanthomonas campestris*) and **soft rot (*Pectobacterium carotovorum*).

Rape. *Dark leaf spot (*Alternaria brassicae*), *damping off (*Phytophthora cryptogea*), *damping off (*Corticium solani*), *damping off (*Pythium* spp.), *mildew (*Erysiphe cruciferarum*), *downy mildew (*Peronospora parasitica*) and **club root (*Plasmodiophora brassicae*).

Turnip and Swede. *Leaf spot (*Alternaria brassicae*), *leaf spot (*Ascochyta brassicae*), *root rot (*Botrytis cinerea*), *white spot (*Cercosporella brassicae*), *white blister (*Cystopus candidus*), *mildew

(*Erysiphe cruciferarum*), **downy mildew (*Peronospora parasitica*), *light leaf spot (*Cylindrosporum concentricum*), ***club root (*Plasmodiophora brassicae*), **dry rot and canker (*Leptosphaeria maculans*), *sclerotinia rot (*Sclerotinia sclerotiorum*), *violet root rot (*Helicobasidium purpureum*), *damping off (*Pythium* spp. and *Corticium solani*), *scab (*Streptomyces scabies*), *soft rot (*Pectobacterium carotovorum*) and *black rot (*Xanthomonas campestris*).

The many leaf-attacking diseases are generally considered to be of little importance, and no protective spraying is normally practised. The use of zineb at 1·3 kg a.i./1000 litres has given some control of **Downy mildew** (*Peronospora parasitica*) in field swedes, where this disease has been found troublesome in seedling crops.

Mildew (*Erysiphe cruciferarum*) gives a white powdery appearance to the leaves, attacks all these crops, but is more prevalent on turnips and swedes, the latter suffering more severely. Spraying is not economic, and there is some evidence that late sowing may reduce the trouble, though yields may be considerably affected.

***Clubroot** (*Plasmodiophora brassicae*) (A.L. 276) will attack all these crops, but is rarely seen on kale. Swedes and turnips may be seriously affected. Fungicidal alleviative measures are practised on vegetable crops, but not normally on fodder crops. The use of lime to produce an alkaline reaction in the soil, improvement in drainage, and in some instances, e.g. swedes, the use of resistant varieties are the cultural methods practised with these crops. An interval of 8 years between susceptible crops is normally sufficient for the disease to have died down sufficiently, unless there have been susceptible weeds to carry over the trouble.

A root and foot rot caused by *Phytophthora megasperma* is one of the most destructive diseases of kale, usually under conditions of poor drainage, which is the predisposing factor. Correction of the drainage is the best means of control.

Swedes and turnips may be attacked by a dry rot of the roots caused by *Leptosphaeria maculans*, which splits the roots and may rot them completely. The trouble may be picked up from the soil or introduced with the seed. It is best to adopt a rotation of 6 years between crops to allow the soil infection to die down. This fungus may also cause a serious stem canker in kale and other brassicae crops. It is seed-borne, and seed can be effectively decontaminated by hot water treatment, consisting of immersing the seed in hot water at 50°C for 25 minutes, a practice usual for vegetable crops, but not normally for the fodder crops. The recently introduced warm water thiram soak treatment (see p. 314) has been shown to be completely effective for *Leptosphaeria maculans*, but does not achieve complete control with *Alternaria* leaf spot and is not effective against bacterial black rot (*Xanthomonas campestris*). The latter rot is controlled by the hot water treatment.

New hybrids of turnip crossed with radish are becoming increasingly used as 'catch crops' after early harvested cereals. The present varieties are very susceptible to leaf spot, *Alternaria brassicae*, which appears to be contracted from other brassica crops. Resistant varieties are urgently required.

Chapter 9
Pests and diseases
of fruit and hops*

Low volume spraying

Chemical doses per hectare for low volume spray applications are commonly quoted in this chapter as about 22·5 to 28·0 times the amount per 100 litres for a high volume spray. The general basis for such figures is the thesis that a low volume spray should apply about the same amount of chemical per hectare as would be applied by a high volume spray by automatic machines, and that 2250–2800 litres/ha embraces common or 'average' rates. Most commercial recommendations quote low volume dosage rates as 22·5 times the dosage/100 litres, but in some cases this is suggested for 'medium' sized trees and a 28 times dosage rate suggested for 'large' trees. The effectiveness of doses arrived at in this way has been verified, or in some cases modified, by large orchard scale trials of simple design, and by experience. A few experiments have shown that with good spray distribution smaller amounts of toxicant can in fact give adequate control of certain pests, but in practice there is a tendency to advocate more than the minimum dose, on the grounds that the excess may help to compensate for inadequacies of spraying; the chemical deposit may later be redistributed by rain, or if it remains patchy, the more concentrated the toxicant in the patches the more likely they are to affect mobile insects.

It is unfortunately true to say that deposit requirements for most pests and diseases, in terms of size of droplets, their distribution and density on the tree, and concentration of toxicant, for each of the chemicals in use, are not known with any precision. Until these are known the optimum quality and quantity of mist spray for any given plant of fruit trees or bushes cannot be closely defined, nor criteria offered by which a grower can decide for himself how much to use. Because of this lack of information it has been difficult to present doses per hectare in this chapter without some inconsistencies, and manufacturers' recommendations should always be studied, especially as these are liable to be modified with increasing experience. Some inconsistencies are also likely to occur now that recommended HV and LV rates have been metricated.

While experiments have shown that with a good standard of spraying, and appropriate doses of toxicant, good control of common pests and diseases on apple trees can be obtained when the same amount of insecticide or fungicide is applied in low, medium or high volumes of carrier, they have also shown that the degree of control tends to improve

*(See also, 'A Fruit Grower's Guide to the use of Chemical Sprays', MAFF, ADAS, 1976)

as the volume is increased. This is because of a more uniform distribution of spray deposit, and is particularly noticeable with summer sprays against red spider mite, codling moth and apple mildew. It should be noted that if the spray volume is increased to obtain better cover, coalescence of droplets begins to occur to a marked extent on the foliage and fruit in the volume range of about 900–1350 litres/ha. If a compound spray of several insecticides and fungicides is being applied this can mean an increased risk of phytotoxicity. Unless improved spraying techniques or more frequent applications at shorter intervals show that less toxicant per hectare than that recommended for low volume spraying can be effective, it would be safer, at least with some mixtures of insecticides and fungicides, to increase the rate to a full high volume spray.

9.1   Apple

9.1A   Pests

Pests usually calling for annual routine sprays are aphids, winter moth and apple sucker (at green cluster); apple sawfly (at petal fall); fruit tree red spider mite (at petal fall or fruitlet stages); and codling and tortrix moths (at fruitlet stages). For most of these pests insecticides can be included with routine applications of fungicides for scab and mildew. Sprays may not be necessary every year for winter moth, except where orchards adjoin woodland, nor are special sprays needed for apple sucker which is kept in check by insecticides commonly used against aphids. Of the other pests mentioned below many are controlled by insecticides regularly used against the above pests, while others require treatment only on special occasions when they occur. A few, such as fruitlet mining tortrix, may need annual sprays in certain localities.

The Advisory Committee on Pesticides and Other Toxic Chemicals, in 'Further Review of Certain Persistent Organochlorine Pesticides Used in Great Britain', published in 1969, recommended that certain uses of these pesticides should cease as soon as could be arranged. The uses on fruit crops which were phased out by the end of 1971 were DDT post-blossom on top fruit; DDT on raspberries, loganberries, blackberries, gooseberries and red currants; DDT for aphid control on any crop; TDE (Rhothane) on raspberries and loganberries, and post-blossom on apples and pears; endrin on apples; and aldrin on strawberries. In other cases where the use of these pesticides is still permitted and less harmful alternatives are available, preference should be given to the latter. Work to find alternatives to the more persistent pesticides continues, and both endrin and TDE have now been removed from the published list of Approved Products for Farmers and Growers.

***Apple aphids** (A.L. 106). Apart from woolly aphid, which is described separately, there are 4 principal species: apple-grass aphid

(*Rhopalosiphum insertum*), green apple aphid (*Aphis pomi*), rosy apple aphid (*Dysaphis plantaginea*) and rosy leaf-curling or red-leaf aphid (*D. devecta*). These begin hatching from winter eggs on apple trees at about the time of bud break of Cox's Orange Pippin or Bramley's Seedling. Hatching is virtually complete by the green cluster stage of these cultivars, but with the rosy leaf-curling and green apple aphids it may be a little later.

Apple-grass aphid causes slight curl of rosette leaves and departs soon after petal fall for grasses, especially annual meadow grass. Heavy infestations are more likely to follow summers with sufficient rainfall to maintain continuous growth of the grass. An indication of the degree of infestation to be expected in the following spring can be obtained by counting the small, wingless, yellow-green aphids (the egg-laying females) on the undersides of the leaves in late October. An average of one or fewer aphids per leaf from a sample of 20 leaves from each of about 8 trees across the orchard indicates a light infestation.

Rosy apple aphid causes severe leaf curl, and the leaves may turn yellow, but never red; it disperses to plantains in June and July although some colonies may persist on apple into August. The majority of the fruitlets stay on infested trusses but remain small and distorted in shape.

Rosy leaf-curling aphid, a local pest tending to appear unless controlled on the same trees year after year, causes severe leaf curl with conspicuous red areas; it lays its winter eggs deep in crevices under the bark in June. It is restricted to apple and spreads very slowly from tree to tree.

Green apple aphid is less common in the spring in orchards than the apple-grass aphid. In late May, June and July it disperses to other apple and related trees such as pear, hawthorn and rowan. It infests mainly young extension growth and is more a pest of young trees which become re-infested in the summer.

Either (1) winter or delayed dormancy washes as ovicides, or (2) spring sprays against the newly-hatched aphids, give good control. Spring sprays, which lend themselves more readily to low volume applications, are the more widely used.

(1) Winter washes. Except with rosy leaf-curling aphid these are effective if a good cover of the eggs is obtained; HV sprays are normally recommended as a good cover is difficult at LV. Washes are: tar oil, 5 litres (e.c.) or 5–6 litres (stock emulsion)/100 litres (HV), or tar-petroleum oil, 10 litres (e.c. or stock emulsion)/100 litres (HV), when buds dormant (December-February); or DNOC-petroleum oil, at 0·1 per cent DNOC and 3 per cent oil (=4·5 litres conc.)/100 litres, at the delayed dormant stage when the buds begin to break (March).

The tar oil spray also kills scale insects, hibernating caterpillars of small ermine moth, eggs of apple sucker and some of those of winter moths. With very hard or brackish waters stock emulsion types are preferable to e.c. types.

DNOC-petroleum oil also controls scale insects, and eggs of winter

Table 9.1   Apple—green cluster stage, spring sprays for aphid control

Compounds	Rate HV g a.i./100 litres	LV g a.i./ha
Azinphos-methyl + demeton-S-methyl sulphone	25 + 7·5	560 + 170
Chlorpyrifos	21	475
Demephion	7·5	170
Demeton-S-methyl	7·5*	160*
Dichlorvos	—	1050†
Dimethoate	15	335
Dioxathion	60	1320
Fenitrothion‡	31	700
Formothion	18·5	415
Malathion	56	1260
Mevinphos	6–9§	140–210§
Oxydemeton-methyl	7·4	160
Phosphamidon	30	675
Thiometon	9	210
Vamidothion‖	20	—

* Use at 22 g a.i. (HV), 490 g a.i. (LV), on Bramley's Seedling.

† Apply in at least 225 litres water. HV sprays give the best results if applied under calm weather conditions and air temperature of 15°C or over.

‡ Mixtures of fenitrothion with binapacryl, dodine, or with more than one other insecticide or fungicide, may be incompatible or cause phytotoxicity—consult product label.

§ Use the higher rate in cool weather; if caterpillar control is also required use 12 g a.i. (HV) or 280 g a.i. (LV).

‖ Maximum 2 applications per season.

moths, apple capsid (at 7·5 per cent, see p. 176), and gives some control of winter eggs of fruit tree red spider mite.

Do not apply winter washes in windy or frosty weather, or when the trees are wet, and do not use DNOC-petroleum oil after the buds have reached the breaking stage. Some apple cultivars, e.g. Beauty of Bath, Gladstone, Cox's Orange Pippin and Lane's Prince Albert, are more susceptible to oil damage to buds and should be sprayed first.

(2) Spring sprays. The most reliable control is given by the systemic or penetrant organophosphorus insecticides (Table 9.1) applied at the time when Cox or Bramley have reached the green cluster stage, when egg hatch is virtually complete.

Other treatments include: nicotine at 50 ml (95–98 per cent)/100 litres (HV) or 560–1700 ml/ha (LV) which can be effective if the air temperature is 15°C or over, but is non-persistent; gamma-HCH (25 g a.i./100 litres), which is not so effective as the systemic organophosphorus compounds and, for aphid control, is best applied at late bud burst (which with added DDT is too early for a combined control of other pests such as caterpillars and apple sucker).

Azinphos-methyl, carbaryl or phosalone at green cluster will reduce aphid numbers, and where used against winter moth caterpillars (see p. 190) will probably give adequate control in years when infestations are light.

For rosy leaf-curling aphid vamidothion appears to be the most effective and 1 spray at green cluster or petal fall may virtually eliminate the pest. Good cover is important.

With young trees subject to infestation by green apple aphid in the summer, spray in late May and again later as required.

To reduce hazards to bees, do not spray open blossoms; cut down flowering weeds beneath the trees before spraying. Avoid contamination of ponds and streams. Because of the risk of causing taint, do not allow gamma-HCH to fall or drift onto currants at any time, or onto other soft fruit between flowering and picking, or onto cauliflowers before curd forming, or onto beetroot, onion, carrot and peas. Always observe the minimum intervals between application and harvest; consult product label.

*Apple blossom weevil (*Anthonomus pomorum*) (A.L. 28). Adult weevils winter under loose bark, in leaf litter, etc., emerging in early spring when they feed on young apple foliage. From bud burst onwards the female bores holes into blossom buds, laying one egg in each blossom, and the larva feeds on the stamens and base of the flower. The petals are prevented from expanding and turn brown ('capped blossom'). At one time a serious pest, but now rare in commercial orchards owing to extensive use of DDT. Pear, quince and medlar may also be attacked.

Annual sprays against this pest are not necessary. Where required, control may be obtained by azinphos-methyl, chlorpyrifos or fenitrothion applied at green cluster against caterpillars Table 9.8, p. 190 though for the best results with heavy infestations these insecticides should be applied at bud burst. DDT is effective at bud break or bud burst at 75 g a.i./100 litres (HV); low volume sprays are less effective. Gamma-HCH (560 g a.i./ha) gives good control though is less effective than DDT.

To reduce hazards to bees, and the tainting of crops by gamma-HCH observe the precautions given under apple aphids (p. 175).

*Apple capsid (*Plesiocoris rugicollis*) (A.L. 154). This pest virtually disappeared when DDT became widely used. If capsid attacks occur it is important to discover whether they are due to this species or the common green capsid (see p. 182), which is a serious pest on some farms, as control measures are not the same for the two species. Apple capsid overwinters as eggs inserted in the bark, which have started hatching by green cluster of Cox or Bramley. The young bugs puncture leaves, shoots and fruit. Reddish spots form on the leaves which may become distorted in shape; shoots are scarred and stunted, and rough russeted areas with scattered pits and pimples appear on the fruit. The bugs mature and lay their winter eggs from mid-June to about mid-July.

Table 9.2 Apple—green cluster stage, spring sprays for apple capsid control

Compound	Rate HV g a.i./100 litres	LV g a.i./ha
Azinphos-methyl	16·5	375
Azinphos-methyl + demeton-S-methyl sulphone	25 + 7·5	560 + 170
Carbaryl	75	1700
Chlorpyrifos	21	475
DDT	100	2240
Fenitrothion*	31	700

* Mixtures of fenitrothion with binapacryl, dodine, or with more than 1 other insecticide or fungicide, may be incompatible or cause phytotoxicity—consult product label.

The eggs may be killed by petroleum oil washes, containing not less than 5 per cent petroleum oil on dilution, if thoroughly applied, but spring sprays against the newly hatched bugs are more efficient. For winter washes, use DNOC-petroleum oil, 7·5 per cent (HV), at bud break (March) *or* tar-petroleum oil, 10 per cent (HV), at delayed dormant bud (before end February), *or* a winter petroleum oil, 7·5–10 per cent (HV), at bud burst (late February or early March). The precautions against oil damage given under apple aphids (p. 174) should be observed.

For spring sprays, any one of the following (Table 9.2) is effective at the green cluster stage.

Dichlorvos, 1050 g a.i./ha has given good control in trials; still weather, air temperatures of 15°C or over, and HV sprays are required for the best results. Nicotine, 50 ml (95–98 per cent)/100 litres (HV), can also be effective but requires temperatures of 15°C or over, and being non-persistent may require a second spray. Experiments suggest that the organophosphorus compounds are the most effective.

To reduce hazards to bees observe the precautions given under apple aphids, p. 175.

*Apple fruit miner (*Argyresthia conjugella*). These moths emerge in June and lay eggs on the fruit, and the caterpillars form a maze of small tunnels in the flesh. When full grown they leave the fruit, spin a cocoon on the trunk or among dead leaves, entering the pupal stage in October. The insect also occurs on rowan or mountain ash.

Measures used against codling moth (Table 9.4, p. 182), should also control this pest.

*Apple fruit rhynchites (*Caenorhinus aequatus*). This is a small brownish red weevil occurring on the trees from May to July, but now scarce as a pest, probably due to the widespread use of insecticides. Cylindrical holes are drilled in the side of the fruit, resembling those made by pushing in a pencil point.

If required, spray at pink bud with DDT at 100 g a.i./100 litres (HV) and to avoid danger to bees, do not spray open blossoms; also, cut down flowering weeds before spraying. No work has been done on alternatives to DDT but azinphos-methyl, carbaryl, chlorpyrifos, fenitrothion, malathion or phosalone, at dosage rates for winter moth control Table 9.8, p. 190, would probably be effective.

*Apple leaf midge (*Dasyneura mali*). Although generally a minor pest, several conspicuous infestations have occurred in recent years. Eggs are laid in the unopened or partially uncurled leaves and, as with pear leaf midge, the leaf margins become tightly rolled inwards. Affected leaves turn reddish and finally black, falling from the tree. The larvae are bright pink and feed within the rolled leaves, most of them dropping on to the ground when full-fed to pupate. There are three overlapping generations in the year, from May to August. Miller's Seedling seems to be particularly susceptible.

There is little information on control, although carbaryl *or* vamidothion, applied at petal fall or when curled leaves are seen, appear to be effective. Summer sprays of azinphos-methyl or demeton-S-methyl have little effect.

***Apple sawfly (*Hoplocampa testudinea*) (A.L. 13). Adult sawflies emerge about the time when mid-season cultivars are flowering; they are active in warm sunny weather and are only attracted to trees in blossom. Eggs are laid, usually one per flower, in a slit-like cut just below the calyx. Hatching normally begins 4 or 5 days after 80 per cent petal fall and is complete within 14–15 days. The larva mines under the skin, then penetrates to the core. It later leaves to bore straight into another fruitlet, making a large entry hole at which sticky frass accumulates. When fully fed the larva drops to the ground and builds a cocoon in the soil. Adults emerge the following spring, but some not until the second spring.

Worcester Pearmain, Charles Ross, James Grieve, Ellison's Orange and Early Victoria are particularly susceptible, while culinary cultivars, except Edward VII, are generally lightly attacked.

One of the most effective insecticides is gamma-HCH, 12·5 g a.i./100 litres (HV) or 280 g a.i./ha (LV). Spray within 7 days after 80 per cent petal fall, using Worcester Pearmain or Bramley's Seedling as a general guide for timing. Late cultivars, e.g. King Edward VII, should be sprayed separately. With heavy infestations repeat this spray for at least 2 seasons.

In regularly sprayed orchards infestations are kept in check by organophosphorus pesticides used at petal fall for red spider, e.g. azinphos-methyl + demeton-S-methyl sulphone, demeton-S-methyl, dimethoate, dioxathion, oxydemeton-methyl, phosphamidon, triazophos and vamidothion (see Table 9.6, p. 185); tetrachlorvinphos, 75 g a.i./100 litres (HV), is also effective. Nicotine 50 ml (95–98 per cent)/100 litres (HV), gives variable results.

IFH—7

To reduce hazards to bees, and the tainting of crops by gamma-HCH, observe the precautions given under apple aphids (p. 175). See pp. 29–63. Observe the minimum intervals between last application of insecticides and harvest; consult product label.

Apple sucker (*Psylla mali*) (A.L. 96). This insect overwinters as eggs on the bark, and these hatch over a fairly long period which may extend through April into May. The young suckers feed on the leaves and flowers, and blossom trusses may turn brown as if killed by frost. The insects remain on the trees and eggs are laid in the autumn.

Either winter washes against the eggs, or spring sprays against the newly hatched suckers, may be used for control. For winter washes use either tar oil, 5 per cent (HV), *or* tar-petroleum oil, 10 per cent (HV), when the buds are dormant (December–February). The precautions to be taken are those given under apple aphids, p. 172.

Apple sucker is no longer a common pest in regularly sprayed orchards where it is probably kept in check by the organophosphorus pesticides used against aphids, sawfly and red spider mite. The insecticides listed in Table 9.3 are known to be effective at green cluster.

Among the more effective for heavy infestations are dimethoate, fenitrothion, formothion, mevinphos and gamma-HCH.

Table 9.3   Apple—green cluster stage, spring sprays for apple sucker control

Compound	Rate HV g a.i./100 litres	LV g a.i./ha
Azinphos-methyl	16·5	375
Azinphos-methyl + demeton-S-methyl sulphone	25 + 7·5	560 + 170
Chlorpyrifos	21	475
Dichlorvos	—	1050*
Dimethoate	30	680
Fenitrothion†	31	700
Formothion	37	830
Gamma-HCH	25	560
Malathion	56	1260
Mevinphos	6–9‡	140–210‡
Phosphamidon	30	675
Tetrachlorvinphos	75	—
Vamidothion§	40	900–1120

* Apply in at least 225 litres water. HV sprays give the best results if applied under calm weather conditions and air temperature of 15°C or over.
† Mixtures of fenitrothion with binapacryl, dodine, or with more than 1 other insecticide or fungicide, may be incompatible or cause phytotoxicity—consult product label.
‡ Use the higher rate in cool weather or if caterpillar control is also required.
§ Maximum 2 applications per season.

Carbaryl and phosalone at rates recommended for pre-blossom caterpillar control (p. 190) will reduce infestations of apple sucker, but in trials the degree of control has been variable. Nicotine, 50 ml (95–98 per cent)/100 litres (HV) or 560–700 ml/ha (LV), can also be effective but requires temperatures of 15°C and over for best results and is not persistent. DDT is not effective.

To reduce hazards to bees, and the tainting of crops by gamma-BHC, observe the precautions given under apple aphids (p. 175).

*Apple twig cutter (*Rhynchites coeruleus*). This is a small, deep blue weevil appearing in May and feeding on foliage, laying eggs in early June in the young shoots, which are then cut off just below the oviposition point. These shoots wither and drop off or remain hanging on the tree. A local pest, more important on young trees, it may be controlled by DDT, 100 g a.i./100 litres (HV) or 2·24–2·8 g a.i./ha (LV), applied at pink bud.

The following alternatives to DDT should prove effective: azinphos-methyl, carbaryl, chlorpyrifos, fenitrothion, or phosalone, at dosage rates as for winter moth control (Table 9.8, p. 190), *or* malathion or phosphamidon as for aphid control (Table 9.1, p. 174).

To reduce hazard to bees, do not spray open blossoms and cut down flowering weeds beneath the trees before spraying.

*Bark beetles and shot-hole borers. The shot-hole borers (*Anisandrus dispar* and *Xyleborus saxeseni*) bore into the heart wood of the trunk and main branches. The females excavate galleries in which the eggs are laid, and the grubs feed largely on fungi which grow on the tunnel walls. Grubs may be found at all times of the year, while adults are more common from January to June. The bark beetles (*Scolytus rugulosus* and *S. mali*) tunnel under the bark where the grubs excavate a maze of galleries. As a rule only trees unhealthy due to root disorders or disease are attacked.

There are no standard recommendations for the control of these pests but, in addition to cultural measures to increase the health of the trees, the following uses of insecticide have been suggested: tar oil, 10 per cent (HV), applied to the trunks in March or April to drive out the beetles, taking care not to spray any foliage or buds, *or* DDT, 500 g a.i./100 litres (HV), sprayed or painted on to the trunk and main branches in April, to kill beetles after they emerge, taking care not to wet any foliage.

*Brown leaf weevil (*Phyllobius oblongus*)—see Leaf weevils, p. 186.

*Bryobia mite, Apple and Pear Bryobia (*Bryobia rubrioculus*). The adults of this mite can be distinguished from those of fruit tree red spider by the greater length of their front legs, which are twice as long as the second pair, and by the absence of long hairs on the body. It overwinters as eggs on the bark of the trunk and main branches. The eggs begin hatching in March or April, about a fortnight earlier than those of fruit tree red spider.

The mites are active on the leaves in warm, sunny weather, retiring to crevices in the bark when air temperature drops below about 13°C. They also return to the bark for moulting, and whitish clusters of cast skins are another aid to the recognition of their presence. There are 3 generations during the summer, the period of egg to adult lasting some 6 weeks. Summer eggs are also laid on the bark, and the winter eggs are laid from about the third week of July onwards. The species also occurs on pear. Infestations are now uncommon.

The acaricides listed in Table 9.6, p. 185, for the control of fruit tree red spider mite are probably all as satisfactory against bryobia mite. Where heavy infestations have arisen, spray at green cluster with an acaricide effective against active stages; by choosing a suitable organophosphorus compound, aphids can be controlled at the same time. Follow with a second spray at petal fall. In subsequent years the pest should be kept under control by normal fruit tree red spider mite spray programmes.

For best results sprays should be applied in warm conditions (air temperature near 15°C if possible) when the mites are active on the leaves. On culinary cultivars lime sulphur at green cluster can be effective.

Observe the minimum intervals between last application and harvest.

Some control of the winter eggs may be obtained by winter washes such as: DNOC-petroleum oil, 4·5 per cent (HV), when the buds are beginning to break (March), or winter petroleum oil, 4·5 per cent (HV), at bud burst. The precautions given under apple aphids (p. 174) should be observed to reduce hazard of oil damage.

*Chafer beetles (A.L. 235). Adult cockchafer (*Melolontha melolontha*) and garden chafer (*Phyllopertha horticola*) beetles, which may be abundant in May and June in some seasons, occasionally bite into fruitlets and cause damage resembling that by caterpillars. The large white larvae, which live in the soil, may also attack the roots of nursery trees. Adults may be controlled when seen in May to early June by spraying with carbaryl, 1·87 kg a.i./ha, but only if fruit thinning is required. See under strawberry on p. 236 for the control of larvae.

*Cherry-bark tortrix (*Enarmonia formosana*)—see under Cherry, p. 210.

*Clay-coloured weevil (*Otiorrhynchus singularis*). Adult weevils, which are wingless, emerge from hibernation in late spring and early summer, sheltering under clods of soil, etc. by day, and ascending plants at night to feed. They eat holes in leaves and gnaw the bark from woody stems, including those of fruit trees, currant and gooseberry, cane fruits and hops. Damage is more likely to be serious on young trees and bushes. Eggs are laid in the soil in summer and the white grubs feed on the roots of various plants until the following spring. Populations sometimes build up in currant plantations, and these, together with woodland which shelters hibernating beetles, may be sources of infestation.

The adult weevils are not easy to kill with insecticide. The most effective is DDT at high dosage rates, e.g. 20 or 25 per cent w.p. used as a dust applied to the soil and lower parts of the plant at about 1·4–1·75 g a.i./plant, or a spray of 4480 g a.i. in at least 3370 litres/ha. Use a w.p. to avoid possible crop damage by e.c. solvents. There is little information on alternative insecticides, but carbaryl at 200 g a.i./100 litres (HV) is suggested.

Clouded drab moth (*Orthosia incerta*). Damage by caterpillars of this common species occurs locally. Adults appear in the spring, laying eggs in April and early May.

The caterpillars feed in the spring and early summer, sometimes excavating cavities, which may be deep, in the fruitlets.

Where required, a petal fall spray with one of the organophosphorus insecticides recommended for use against winter moth caterpillars (Table 9.8, p. 190) is suggested.

***Codling moth** (*Cydia pomonella*) (A.L. 42). In an average season these moths emerge from late May or early June until early August, with the main flight from late June to mid-July. In some years a small second generation occurs in late August or September. Eggs are generally laid in the evening, on leaves and fruit, and in most numbers when temperatures exceed about $15.5°C$. The caterpillars, which hatch in 10–14 days, bore into the fruit, feeding for the first few days in a cavity just beneath the skin, then tunnel to the core. The entry hole is small and covered with dry frass, in contrast with the large hole with a mass of wet frass typical of apple sawfly. About half of the early-hatched caterpillars enter by the calyx. When fully fed, in about 4 weeks, the larva leaves the fruit and spins a cocoon under loose bark or other shelter. Second generation moths arise only from larvae which have spun up by early August. Later caterpillars overwinter in their cocoons to produce moths the following summer.

Since egg hatch extends over at least 2 months, the ideal control would be to maintain an insecticide deposit effective against eggs or young larvae throughout this time, and even with the most persistent materials this would mean at least 3 sprays. Usually 2 sprays are given to destroy all except the latest-hatched larvae, and on most commercial farms, where codling moth infestations have for some years been light, this proves adequate. The second spray should follow 2 or 3 weeks later according to the material used (see below). The first should be applied just before the earliest eggs hatch, which in southern and eastern England is about 5 weeks after petal fall of Cox, i.e. mid-June in average seasons. Guidance on spray timing in any particular season is usually available from ADAS entomologists who operate light traps in the main fruit growing areas. Spray timing may also be based on catches of adult males in pheromone-baited sticky traps. These traps can also give some indication of the likely levels of codling moth infestations in orchards.

Of the various recommended insecticides (Table 9.4, p. 182) lead arsenate is a stomach poison, the remainder kill mainly by contact and have effect on the moths as well as larvae and/or eggs. High volume sprays of at least 1125 litres/ha are generally more effective than low volume sprays.

Table 9.4   Apple and pear—sprays for codling moth control

Compound	No. sprays	Timing weeks apart	Rate HV g a.i./100 litres	LV g a.i./ha
Azinphos-methyl	2	3	33	750
	3	2	22	460
Carbaryl*	2	3	75	1700
Chlorpyrifos	2†	2	42	945
Dioxathion	2	3	60	1320
Fenitrothion‡	2†	2	45	1050
Lead arsenate	(see text)		300	6710
Malathion§	—	—	112	—
Phosalone	2†	2	31	700
Phosphamidon	2	3	30	675
Tetrachlorvinphos	2	3	75	—
Triazophos	2	3	40	900

* Use between late pink bud and 3 weeks after petal fall may lead to thinning. Not recommended for pears.

† A third spray, 2 weeks after the second, is recommended for maximum control of heavy infestations or if tortrix moths are important pests (see p. 188).

‡ Mixtures of fenitrothion with binapacryl, dodine, or with more than 1 other insecticide or fungicide, may be incompatible or cause phytotoxicity—consult product label.

§ Less persistent than other insecticides but can be useful if a late spray against 2nd generation attack is required near picking time for early cultivars.

The performance of lead arsenate can be improved by applying an additional spray with spreader at petal fall before the calyx closes, but it is not as effective as the other compounds. It should not be used with lime sulphur or when air temperatures are above 21–24°C.

The insecticides recommended, with the exception of lead arsenate, have adverse effects on predators of fruit tree red spider mite. Azinphos-methyl, chlorpyrifos, dioxathion, phosalone and triazophos are also acaricides, but resistance by this mite to certain organophosphorus compounds is common, so it is generally necessary to combine codling moth control with suitable acaricides (see under fruit tree red spider mite, p. 184).

Moths from caterpillars overwintering in packing sheds and box stacks may be responsible for heavy attacks on trees nearby. Keep doors closed as much as possible.

On a small scale, corrugated paper bands can be tied round the tree trunks in mid-July to trap caterpillars, which are then readily attacked by birds, or can be removed and destroyed after the crop is picked.

**Common green capsid (*Lygocoris pabulinus*). Damage by this pest to apples and pears has become more common in recent years. Eggs are laid in the shoots in autumn and hatch in the following spring, over a period corresponding to that of early pink bud to petal fall of Bramley's Seedling. The young bugs puncture both fruitlets and young shoots.

Table 9.5 Apple and pear—sprays at petal fall, for common green capsid control

Compound	Rate HV g a.i./100 litres	LV g a.i./ha
Chlorpyrifos	42	945
Dimethoate	30	680
Fenitrothion*	31	700
Formothion	37	830
Malathion	112	2530
Triazophos	40	900

* Mixtures of fenitrothion with binapacryl, dodine, or with more than one other insecticide or fungicide, may be incompatible or cause phytotoxicity—consult product label.

When adult, they leave the trees for a second generation on herbaceous plants, and adults of this generation return to fruit trees, as well as bush fruit, in the autumn to lay winter eggs. It is difficult to predict the risk of attack in a particular orchard, but the pest does seem to recur in certain orchards.

The standard preventive measure of spraying with DDT, 100 g a.i./100 litres (HV) or 2240 g a.i./ha (LV), has not always given adequate control in recent years. The most effective materials are listed in Table 9.5, p. 183, and should be applied immediately at petal fall (included with green cluster sprays they will only give partial control). Most of the other organophosphorus insecticides used on fruit have proved to be insufficiently effective against this capsid.

*Common leaf weevil (Phyllobius pyri)—see under leaf weevils, p. 186.

*Dock sawfly (Ametastegia glabrata). There are 2 and sometimes 3 generations of this sawfly in the year, those of the last brood laying eggs in August or September. The light green caterpillars feed on docks and fat-hen. They hibernate in hollow stems or suitable crevices, and sometimes tunnel into apples in search of winter quarters. They may also injure young trees by tunnelling into the pith of branches, entering at pruning cuts.

Insecticides used against codling moth should also control sawfly caterpillars on weeds under the trees. Where keeping the ground weed-free is impracticable, spraying the weeds with carbaryl (1·7 kg a.i./ha) or derris (2 g rotenone/100 litres HV) or malathion (112 g a.i./100 litres HV) in mid-July and again in early August should be effective.

*Earwig (Forficula auricularia). Earwigs lay their eggs in the soil from December to March, and a second period of egg laying occurs in May and June. Earwigs shelter by day in dark crevices, ascending plants at night to feed. Numbers appear to be highest from July to September, and in some orchards in grass, damage to fruit occurs in the form of deep rounded cavities with small entry holes, and also soiling by frass where the insects shelter between the stalk and the fruit or a touching leaf.

Where earwigs are a pest, apply a coarse spray to the base of the trunk and surrounding vegetation with gamma-HCH, 25 g a.i./100 litres, in late July. A spray of carbaryl, 200 g a.i./100 litres, or trichlorphon, 120 g a.i./100 litres is also effective.

***Fruit tree red spider mite** (*Panonychus ulmi*) (A.L. 10). This species overwinters as bright red spherical eggs on the bark. Hatching normally begins at the pink bud stage of Cox's Orange Pippin or Bramley's Seedling, in late April or early May; about half the eggs have hatched by petal fall, the remainder hatching in the next 3–4 weeks, i.e. up to about mid-June. Surveys have shown that there are strains of the mite with different hatching periods, the time of peak hatch of these ranging from 'normal' to up to 3–4 weeks later.

The mites feed on the undersides of the leaves, causing a minute speckling, and in heavy attacks the leaves become dull green and then bronzed. Summer eggs are laid mainly on the undersides of the leaves, and 4 or 5 generations occur during a warm summer. The period egg to adult normally takes about 4 weeks (egg stage 2–3 weeks, young stages 2 weeks). Breeding ends in the first half of September, when daylight is reduced to about 14 hours, and winter eggs are laid. Winter eggs may be laid earlier on severely bronzed trees.

For control in the summer, 2 sprays applied 3 weeks apart are usually required. With heavy infestations—clusters of winter eggs evident on the bark—the first should be at petal fall or as soon afterwards as peak hatching is apparent, using an acaricide effective against the active stages. Where winter eggs are few this spray can be delayed until mid-June, when an acaricide active against the summer eggs is an advantage.

Repeated use of one acaricide is liable to result in the selection of a resistant strain of the mite, and in experiments this has occurred after 4 or 5 years of regular use of the same group of chemicals. Resistance to organophosphorus compounds is now widespread in south-east and eastern England and resistance to tetradifon, and other materials, is becoming increasingly common. A careful watch should be kept on the degree of control being obtained in orchards where the same material has been used for more than 3 years.

Recommended acaricides are listed in Table 9.6, p. 185, arranged according to their chemical grouping; the resistance of mite of the chemicals in one group does not appear to confer resistance to those in the other groups. The mite stages against which the chemicals are effective are also shown.

Binapacryl and dinocap, used regularly against powdery mildew, will control fruit tree red spider mite, at least until the end of July, provided the season does not start with a heavy infestation of winter eggs. To ensure full control after the mildew programme has ended, it may sometimes be necessary to include a specific acaricide with the last

Table 9.6   Apple—sprays for fruit tree red spider mite control

Compound	Rate HV g a.i./100 litres	LV g a.i./ha
A—Binapacryl (active stages, partially ovicidal)		
(as regular powdery mildew sprays)	See Table 9.10	p. 195
Dinocap (active stages, partially ovicidal)		
(as regular powdery mildew sprays)	See Table 9.10	p. 195
B—Propineb (suppressant) (as regular scab sprays		
on culinary apples)	105	2350
C—Mancozeb + zineb (suppressant)		
(as regular scab sprays)	See p. 198	See p. 198
D—Dicofol (all stages)	40	1120
E—Tetradifon (eggs and young stages)	12·5	280
F—Quinomethionate (active stages)	250 in at least 1500 litres/ha	
G—Cyhexatin (active stages)	18·7	420
H—*Organophosphorus compounds* (active stages)		
Azinphos-methyl	33	750
Azinphos-methyl + demeton-S-		
methyl sulphone	25 + 7·5	560 + 170
Chlorpyrifos	42	945
Demeton-S-methyl	22	490
Dichlorvos	—	1050*
Dimethoate	22·5	500
Dioxathion	60	1320
Malathion	112	2530
Oxydemeton-methyl	22	480
Phosalone	31	702
Phosphamidon	30	675
Triazophos	30	675
Vamidothion†	40	900

* Apply in at least 225 litres water. HV sprays give the best results if applied in calm weather and an air temperature of 15°C or over.
† Maximum 2 applications per season.

mildew spray, when dicofol, which is effective against all stages, would be suitable. Resistance to binapacryl has been found in some orchards where this chemical has been used regularly for at least 4 years, and such resistance renders the mites less susceptible to dinocap.

Propineb reduces the fertility of the mites; used regularly for scab control on culinary apples, it will also hold fruit tree red spider mite populations in check, at least until the scab spray programme is terminated.

Mancozeb with zineb, regularly used for control of scab, also has a suppressing effect on mite populations.

Quinomethionate is recommended at 80 per cent hatch of winter eggs, with a second spray 10–14 days later, and a third if necessary. It should not be applied in less than 1500 litres/ha, or at a higher dosage rate than

250 g a.i./ha. It should not be mixed with pesticides other than captan or HCH dispersible powder.

Tetradifon is slow acting, since although it kills eggs and young mite, its effect on adult females is to prevent them from laying fertile eggs. Where it is used in the spring it is therefore recommended that it should be applied as soon as the first adults are seen, usually late in the blossom period or soon after petal fall, with a repeat spray in early July.

Malathion is less persistent than the other organophosphorus compounds and gives only a short-term effect. Dichlorvos is non-persistent and is useful if a spray near harvest is required as the pre-harvest interval is 24 hours.

Cyhexatin kills young stages by contact: though not ovicidal, it also kills newly-hatched larvae crawling over spray deposits and has a slow effect on the adults. Two sprays 3 weeks apart are applied, the first when about 25 per cent of the winter eggs have hatched if infestations are heavy, or when hatching is nearly complete with light infestations.

For use in winter, washes containing 3 per cent petroleum oil when diluted give some control of winter eggs with thorough HV spraying, though they are less efficient than summer sprays. Use either DNOC-petroleum oil, 4·5 per cent (HV) when the buds are breaking in March, or winter petroleum oil, 4·5 per cent, at bud burst. The precautions given under apple aphids (p. 174) should be taken to reduce the hazard of oil damage.

*Leafhoppers. Two common species are *Erythroneura alneti* and *Typhlocyba froggatti*, which overwinter as eggs in young shoots; those of *T. froggatti* hatch between mid-April and early May. This species has 2 generations a year, eggs of the second generation being laid in or near the midrib of the leaf from June onwards, and adults occur from about mid-June into November. *E. alneti* has 1 generation, its eggs hatching in May, and adults occur from late June into October. *T. rosae* overwinters on rose and adults may fly to apple for the second generation, returning to rose in the autumn. Several other species occur in small numbers.

The insecticides listed in Table 9.7, p. 187, applied as HV sprays in late June and again 3 weeks later, have given good results in trials. LV sprays are probably as effective. Leafhopper infestations rarely require special sprays and are probably kept under control by appropriate insecticides used against codling moth.

*Leaf weevils (*Phyllobius* spp.). The brown leaf weevil (*P. oblongus*), common leaf weevil (*P. pyri*) and silver-green leaf weevil (*P. agrentatus*) occasionally damage leaves and flowers. The larvae feed at the roots of grasses and adults emerge about the blossom time of apples, sometimes appearing in large numbers but often disappearing after a week or so to feed elsewhere. Damage is only likely to be important on very young trees. No experiments have been made with alternatives to DDT, but it is probable that azinphos-methyl, chlorpyrifos, fenitrothion or malathion would be effective.

Table 9.7   Apple—late June sprays for leafhopper control*

Compound	Rate HV g a.i./100 litres
Azinphos-methyl	41
Carbaryl	75
Demephion	22·5
Demeton-S-methyl	22
Dimethoate	30
Formothion	37
Malathion	112
Oxydemeton-methyl	22

* Repeat spray 3 weeks later.

To reduce hazard to bees, do not spray open blossom and cut down flowering weeds beneath the trees before spraying.

*March moth (*Alsophila aescularia*) (A.L. 11)—see winter moths, p. 189.

*Mottled umber moth (*Erannis defoliaria*) (A.L. 11)—see winter moths, p. 189.

*Mussel scale (*Mytilococcus ulmi*) (A.L. 36). The adult is about 3 mm long, shaped like a mussel shell, and lies flat on the bark. Eggs are laid beneath the scale in late summer, and these hatch in May, the young insects settling in a suitable place and gradually developing the waxy covering. Winter washes thoroughly applied are very effective: tar oil, 5 per cent (HV) when the buds are dormant (December–February), or DNOC-petroleum oil, 4·5 per cent (HV), at bud break (March), or winter petroleum oil, 4·5 per cent (HV), at bud burst. The precautions to be taken in the use of winter washes are given under apple aphids (see p. 174).

If control is required out of the dormant season spray with malathion, 187 g a.i./100 litres (HV), in late May or early June. This spray is directed against the young stages or 'crawlers' and good cover of the wood and undersides of the leaves is essential. With heavy infestations a second spray 14 days later may be necessary. This pest has become uncommon in regularly sprayed orchards, probably being kept in check by insecticides used in spring and summer against other insects.

*Oystershell scale (*Quadraspidiotus ostreaeformis*). The scale is brown or grey, variable in shape and resembles an oyster shell. It lives on the bark, and the eggs which are laid beneath the parent scale hatch in June. The pest is now uncommon in sprayed orchards, but if present may be controlled by the methods given above under mussel scale.

**Red bud borer (*Thomasiniana oculiperda*). Midges appear from June to August. Eggs are laid in the cuts made when budding nursery stock and the red maggots feed under the bark. The tying material should be coated liberally with petroleum jelly immediately after tying in.

*Small ermine moths (*Yponomeuta* spp.) (A.L. 40). Moths appear in late July and in August, and lay their eggs in flat scale-like masses on the twigs. Caterpillars hatch in September but remain under the egg scales during the winter, crawling out in the spring when they first mine the young leaves; later, leaves are spun together to form a 'nest'. By late May and in June these 'nests' are large and conspicuous, with numerous caterpillars and defoliation of the enveloped twigs. The caterpillars, about 12 mm long when full grown, are grey with black spots. They build cocoons for pupation in July. *Y. padella* feeds on plum, hawthorn and related trees; *Y. malinellus* on apple. Moths infesting spindle and willow are different species.

Overwintering caterpillars can be destroyed by winter washes of 5 per cent tar oil when buds are dormant in December–February, or 4 per cent DNOC-petroleum oil when the buds are breaking in March.

If control in the spring is required, lead arsenate powder at 200–300 g/100 litres may be used. Alternatively, any of the organo-phosphorus alternatives to DDT for use against winter moth caterpillars (Table 9.8, p. 190) are probably effective. Spraying at green cluster is, however, too early and should be delayed until pink bud or petal fall, when all the small ermine caterpillars have emerged from hibernation but before protective 'nests' have been formed.

*Silver-green leaf weevil (*Phyllobius argentatus*)—see under leaf weevils, p. 186.

***Tortrix caterpillars. Some 16 species can be found on apple trees, but the larvae of only 3 are known to cause damage to the fruit. These are the fruit tree tortrix (*Archips podana*), the summer fruit tortrix (*Adoxophyes orana*) and the fruitlet mining tortrix (*Pammene rhediella*).

The fruit tree tortrix hibernates as young caterpillars in cocoons fixed to twigs or buds. These emerge over a fairly long period from late March to green cluster or pink bud stages, first boring into fruit buds and then feeding on the young leaves, which are frequently spun together. Pupation occurs between leaves spun together in late May to early June. Moths begin emerging 1–2 weeks later than codling moth and they occur until mid-August, with greatest numbers in late June and the first half of July. Scale-like eggs, in flat green batches, are laid on the leaves. Caterpillars hatch, feeding at first on the leaves, then on the fruit, eating out deep irregular areas under the protection of a leaf attached with silk. These caterpillars go into hibernation in the autumn but some, especially in fine summers, mature earlier and produce moths from late August to October. These lay eggs and the caterpillars may cause further damage to the apples before hibernating, though the most important damage is usually caused by caterpillars from the moths of the first brood. Caterpillars taken in apple stores continue to feed on the fruit.

The summer fruit tortrix occurs mainly in north and east Kent, and has two complete generations a year. It overwinters as young caterpillars under leaf fragments fastened with silk to crevices in the bark, etc., and these emerge in late March or early April to feed on the leaves. The caterpillars pupate and produce a first brood of moths which occur over a similar period to that of the codling moth. These lay eggs and the caterpillars feed on the leaves and also remove extensive areas of skin from the fruits. A second brood of moths occurs from late July to September, lays eggs, and the caterpillars feed a little before hibernating. This second brood is usually much larger than the first.

The fruitlet mining tortrix occurs locally in many of the fruit growing areas, the moths emerging in May and laying eggs on the undersides of leaves. The caterpillars attack the fruitlets, usually where 2 are touching, forming groups of round, black-rimmed holes. They are fully fed by early July and go into hibernation in cocoons under loose bark.

The bud moth (*Spilonota ocellana*) is a common species with a life cycle similar to that of the fruit tree tortrix, and the caterpillars emerging in the spring bore into the blossom buds.

Control of the fruit tree tortrix and summer fruit tortrix, to prevent damage to the fruit, is best obtained by the sprays recommended for codling moth (Table 9.4, p. 182), and the most effective insecticides are azinphos-methyl, carbaryl, chlorpyrifos, dioxathion, fenitrothion, phosalone, tetrachlorvinphos and triazophos. The first spray can be applied about a week later than the date recommended for codling moth in areas where the latter is not an important pest, followed by 1 or 2 further sprays at 2-week intervals, according to the severity of attacks. If control in the spring is required (and also of bud moth) use one of the insecticides mentioned above at green cluster.

A spray with one of the above-mentioned insecticides at petal fall should be effective against fruitlet mining tortrix, although experimental evidence in this country on alternatives to DDT is lacking.

To reduce hazard to bees, do not spray open blossom and cut down flowering weeds beneath the trees before spraying.

Wasps (A.L. 451). The 2 commonest species are *Vespula vulgaris* and *V. germanica*. They may cause damage to ripening fruit. The best control measure is systematic nest destruction. DDT or derris dusts can be placed in or at the entrance hole, ideally when the wasps are most active or in the late evening when workers are returning to the nest, so that the dust is stirred up and carried inside. About a tablespoonful of dust is applied per nest, using a spoon tied to a long cane. Alternatively, carbon tetrachloride can be used, 140–285 ml being injected into the entrance at dusk, the hole then being plugged with soil. Nests in brick or stone work can also be treated by painting chlordane concentrate round the entrance hole.

Winter moths (A.L. 11). The caterpillars of 3 species commonly

attack foliage, flowers and fruitlets, but only the winter moth
(*Operophtera brumata*) occurs in sufficient numbers to be a pest; the
March moth (*Alsophila aescularia*) and the mottled umber moth (*Erannis
defoliaria*) usually occur only in small numbers. The females of all 3
species have vestigial wings and cannot fly. Adults of the winter moth
appear in October to December, those of the mottled umber October to
December, but occasionally in January and February, and the March
moth in March. Eggs are laid on the bark, those of the March moth in
bands around twigs. Caterpillars hatch from about bud break or burst to
green cluster, and when fully fed drop to the ground and pupate in the
soil. Young winter moth larvae may be blown from one tree to another,
and from woods or hedgerows into neighbouring fruit trees.

Partial control can be obtained by winter washes; spring sprays are
more effective.

(1) Winter washes. The DNOC-petroleum oil (4·5 per cent) applied
HV in March at bud break for control of eggs of aphids, apple capsid and
red spider, or winter petroleum oil (4·5 per cent) applied at bud burst for
apple capsid and red spider, will also destroy eggs of winter moth and
mottled umber moth. For precautions against oil damage to buds, see
under apple aphids, p. 174.

(2) Spring sprays. Apply one of the treatments listed in Table 9.8,
p. 190, at the green cluster stage.

Table 9.8   Apple—spring sprays for winter moth control

Compound	Rate HV g a.i./100 litres	LV g a.i./ha
Azinphos-methyl	16·5	375
Azinphos-methyl + demeton-S-methyl sulphone	25 + 7·5	560 + 170
Carbaryl	75	1700
Chlorpyrifos	21	475
DDT	50	1120
Dichlorvos	—	1050*
Dioxathion	60	1320
Fenitrothion	31	700
Lead arsenate, powders	200	4480
Mevinphos	12·5	280
Phosalone	31	700
Tetrachlorvinphos	75	—

* Apply in at least 225 litres water. HV sprays give the best results if applied in calm
weather and at an air temperature of 15°C or over.
† Mixtures of fenitrothion with binapacryl, dodine, or with more than 1 other insecticide or
fungicide, may be incompatible or cause phytotoxicity—consult product label.

Table 9.9   Apple—sprays for woolly aphid control

Compound	Rate HV g a.i./100 litres	LV g a.i./ha
Chlorpyrifos	42	945
Demephion	22·5	—
Demeton-S-methyl	22	490
Dimethoate	30	680
Formothion	37	830
Malathion	112	2530
Menazon	50	—
Mevinphos	6–9*	140–210*
Oxydemeton-methyl	22	480
Phosphamidon	30	675
Vamidothion†	40	900–1125

* Use the higher rate in cold weather.
† Maximum 2 applications per season.

Preference should be given to the organophosphorus or carbamate insecticides; in experiments these have given at least as good control as that by DDT. Good spray cover is important and, especially in cold springs, spraying at late green cluster or early pink bud may give better results than earlier applications.

Woolly aphid or **American blight** (*Eriosoma lanigerum*) (A.L. 187). This aphid passes its whole life cycle on the tree. It overwinters as young aphids, devoid of 'wool', sheltering in cracks or under loose bark. These become active in March or April, secreting the typical waxy 'wool', and breeding colonies are present by May. Some winged forms occur in July but are not usually an important source of new infestations. Breeding continues through the summer. Eggs are laid in September, but are usually sterile, and the adults die as winter approaches.

The insecticides listed in Table 9.9, p. 191, are recommended, preferably as drenching HV sprays (especially if infestations are heavy) in late June. Vamidothion appears to be the most effective.

9.1B   Diseases

Blossom wilt (*Sclerotinia laxa* f. *mali*) (A.L. 155). Spores of the fungus are carried to the blossom where they germinate causing blossom wilt. The mycelium may continue along the flower stalk into the spur and even into the spur-bearing branch where it causes dieback. In moist weather, spore pustules develop on the flowers shortly after infection to continue the spread. Spurs and killed wood release spores the following spring, so renewing the disease cycle.

Where lime sulphur is used preblossom for scab control (p. 199), sources of infection are likely to be reduced. Routine spraying with tar oil winter washes in the late dormant period (see p. 173) is also effective. Sprays of benomyl 560 g a.i./ha in not less than 220 litres water at first open flower and again 7 days later will give good control. Infected trusses, cankers and dead wood should be removed and burnt. This can best be done in spring or early summer when infected parts can be recognized by the wilted blossoms. The varieties Lord Derby, Cox's Orange Pippin and James Grieve are very susceptible. Bramley's Seedling is resistant.

***Canker** (*Nectria galligena*) (A.L. 100). *N. galligena* is responsible for most cankers on apple trees, but it is not always easy to distinguish from cankers caused by *Gloeosporium* spp. and also by other agencies including mechanical damage and woolly aphid. Common canker is usually seen as sunken zones of bark around bud, leaf scar, base of small dead side shoot, or an open wound. Small branches are often encircled but on larger branches the canker is restricted to one side. Fruits are occasionally infected. Two types of spores are produced: summer spores ooze from cankers as white pustules; winter spores develop on the canker in red pear-shaped receptacles (perithecia), sometimes mistaken for red spider eggs; the spores may be shot out of the perithecia at any time of the year, particularly in winter.

Spores can infect only through breaks in the bark layer, including pruning or other wounds, leaf scars during autumn, woolly aphid, or scab injury. All badly infected small shoots should be removed and the cankered pockets in large boughs pared away. Large cut surfaces whether from healthy or diseased zones should be protected immediately with soft grafting wax or a special preparation for tree sealing. On shoots not more than half girdled a proprietary paint containing 3 per cent mercuric oxide or 30 per cent 2-phenylphenol may be used during the dormant period instead of cutting out. Treatment should not extend beyond areas of diseased wood. Paints containing 2-phenylphenol should only be applied when the trees are dormant, and should not be applied to maiden trees.

Attention to control of woolly aphid and to scab will prevent entry through injuries made by these. Spray programmes containing benomyl, carbendazim or thiophanate-methyl for the control of scab (see Table 9.12, p. 197) will help to suppress the spread of canker in the orchard. Where disease is severe spray with Bordeaux mixture applied (1) just before leaf fall begins, (2) at about 50 per cent leaf fall and (3) in the spring when the buds start to swell. The spray should contain the equivalent of 250 g copper/100 litres (HV) or 5·6 kg copper/ha (LV). Dipping or drenching the fruit in benomyl, 0·5 g a.i./litre, immediately before storing will reduce the incidence of fruit infection.

The apple varieties Cox's Orange Pippin, James Grieve and Worcester Pearmain are particularly liable to canker, whereas Bramley's Seedling,

Lane's Prince Albert and Newton Wonder are resistant. Environmental and cultural conditions influence liability to canker (see A.L. 100).

Collar Rot (*Phytophthora cactorum* and *P. syringae*). Infected areas of bark occur at, or just above, soil level. They are best seen in spring and autumn, often with cracks at margin. Small oily or watersoaked patches may occur on the infected area and also an ooze of reddish brown droplets. If extension of the area ceases, the bark shrinks and appears smooth and shrunken and cracks away clearly from the surrounding healthy bark.

Severe outbreaks have been recorded mainly in mature orchards of Cox's Orange Pippin and less severe in several other commercial varieties. The common commercially used Malling and Malling-Merton rootstocks are mostly resistant.

Rotted pockets of bark, together with about 5 cm of the surrounding healthy bark, should be cut out and the chippings burnt. The wound should be covered with Bordeaux mixed slurry made from Bordeaux mixture w.p. and water to protect from further infection.

The disease is soil-borne and all trees in the vicinity should be protected as follows: (1) clear trunk bases of debris and weeds, keeping free with herbicides; (2) remove soil from graft unions where feasible; (3) avoid mechanical injury to base of tree; (4) remove fallen fruit; (5) paint trunk bases with slurry made from Bordeaux w.p. and water. (Undetected infection will often appear reddish brown after slurry has dried.)

In subsequent years, tree bases and 30 cm soil width around should be protected by spraying with copper sulphate (pentahydrate), 2 kg/100 litres, or copper oxychloride w.p. or liquid formulation at 500 g copper/100 litres. These sprays should be applied in early March for *P. cactorum* and in early February and mid September for *P. syringae*.

Fireblight (*Erwinia amylovora*) (A.L. 571). First recorded in 1957 on pears, this disease has spread rapidly and in 1969 infected a large number of apple trees in Kent and Suffolk. As far as is known, all varieties of apples are susceptible. Infection of apples is usually through the shoots. Initially the tip of the infected shoot wilts and droops, and at this stage golden droplets of bacterial ooze are often seen on the affected stem. As the bacteria progress down the affected shoot the leaves and stem become brown. The disease does not appear to spread as rapidly in apple tissues as it does in pear, and scaffold branches or trunks of apple trees are rarely affected.

Most of the infection that has been seen in apple orchards can be related to spread down-wind from infected hawthorn hedges during a storm.

Precautions recommended for pear (see p. 206) should be taken where the disease is known to exist.

***Gloeosporium rot** (Bitter rot) of apple (*Gloeosporium album, G. perennans* and *G. fructigenum*). Most of the rotting of apples stored until

January or later is due to one or more species of *Gloeosporium*. These fungi exist on small cankers which may cause dieback of shoots or may be insignificant. From the cankers, spores are produced all the year round, but especially in autumn, when pruning cuts and other wounds may be infected. During wet periods the spores are washed to the fruit surfaces where after entry of the lenticel chambers they remain dormant until the fruit reaches a stage of maturity in store which permits further penetration by the fungus.

The variety Cox's Orange Pippin is particularly liable to the disease.

This disease can only be reduced in severity by spraying with captan at 100 g a.i./100 litres (HV) or at 2·8 kg a.i./ha (LV) at about mid-July and then at 2–3 week intervals until orchard conditions make further applications impracticable. Alternatively spray with benomyl (0·56 kg a.i./ha) in not less than 110 litres/ha *or* with carbendazim (50–60 g a.i./100 litres) *or* with thiabendazole (55 g a.i./100 litres HV or 1·38 kg a.i./ha LV) *or* with thiophanate-methyl (1·1 kg a.i./ha) in mid-July, mid-August and 1 or 2 weeks before harvest.

Dipping or drenching fruit immediately before storage with benomyl, (50 g a.i./100 litres), *or* with carbendazim (50–60 g a.i./100 litres) *or* thiabendazole (120 g a.i./100 litres) *or* thiophanate-methyl (100 g a.i./100 litres) will further reduce the level of fruit infection.

Pruning should be delayed until January at least. Cankers, dead shoots and mummified apples should be removed and burnt.

***Powdery mildew** (*Podosphaera leucotricha*) (A.L. 205). The fungus overwinters in buds and when an infected spur bud breaks the emerging growth appears white and mealy due to the presence of a large number of spores. Diseased blossoms and leaves wither and drop from the tree. A terminal bud in which the fungus has overwintered often produces a 'silvered' shoot with mildewed leaves. Primary outbreaks can cause secondary infections of unexpanded leaves, of new shoots, of growing points, and also of young buds. The latter begin the cycle again during the following year.

The cutting-out of diseased parts is strongly recommended (A.L. 205) but, where it is not practicable, spray with DNOC-petroleum oil, at 0·1 per cent DNOC and 2·9 per cent oil (4·5 per cent e.c.)/100 litres (HV) once in December or January during the dormant period but not later than bud swelling.

The following sprays (Table 9.10, p. 195) should be applied during the growing season.

The trees, especially the growing tips, must be thoroughly wetted. For mildew control, applications at high volume are considered more effective than those at low volume. Several apple varieties are sensitive to sulphur (see apple scab, Table 9.12, p. 197).

Specific apple replant disease. In some orchards, retarded development of trees planted on or next to grubbed trees of the same type

Table 9.10   Apple—sprays for powdery mildew control

Chemical	Rate	Time of application	Remarks
Benomyl w.p.	560 g a.i./ha	At 14 day intervals until extension growth ceases—see also recommendations for scab control (Table 9.12)	Apply in not less than 220 litres/ha in early season and in not less than 560 litres/ha in full leaf
Binapacryl w.p. or col.	560 g a.i./ha 840 g a.i./ha 120 g a.i./ha	From blossom time at 5–7 day intervals From blossom time at 8–10 day intervals From blossom time at 11–14 day intervals	Do not use where trees are underplanted with soft fruit
Carbendazim w.p.	0·55–0·66 kg a.i./ha (LV) 27·5—30·0 g a.i./100 litres (HV)	See recommendations for scab (Table 9.12)	
Dinocap w.p. or e.c.	12·5 g a.i./100 litres (HV) 0·28–0·35 kg a.i./ha (LV) 19 g a.i./100 litres (HV) 0·42–0·52 kg a.i./ha (LV) 25 g a.i./100 litres	From pink bud at 5–7 day intervals From pink bud at 5–7 day intervals From pink bud at 8–10 day intervals From pink bud at 8–10 day intervals From pink bud at 11–14 day intervals	
Lime sulphur	1 litre/100 litres	At early pink bud	May cause leaf damage
Wettable sulphur	450 g a.i./100 litres (HV) 8·5 kg a.i./ha (LV)	From blossom at 10–14 day intervals From blossom at 10–14 day intervals	May cause damage to some varieties
Thiophanate-methyl	1·1 kg a.i./ha	At 14 day intervals	May cause fruit damage on dessert varieties when applied during petal fall to early fruitlet

(particularly apple after apple or pear, and cherry after cherry or plum) often occurs. The reason for this is not known but retardation of growth may be avoided in most 'problem' soils if the sites of the grubbed trees are first treated with chloropicrin.

Treatment of large areas is done with a machine (e.g. modified Egedal soil injector) which dribbles the fluid behind special tines attached to a 1·4 m wide drawbar. The machine is drawn along each row to be replanted. A 22·7 litre drum of chloropicrin will treat 630 m of row when applied at 280 litres/ha.

For gapping up a single grubbed tree and for other small areas, a hand-operated injector may be used. With this a square area with a side of 1·8 m—tree site in centre—should be treated at 230 mm staggered intervals and a depth of 150 mm. Each area requires 100 ml.

Chloropicrin is the only chemical which has proved effective in solving the replant problem but because it is unpleasant to use and it is included in Part I of Schedule 2 of the Health and Safety (Agriculture) (Poisonous Substances) Regulations 1975, special precautions must be observed. These include the use of protective clothing and a respirator fitted with a type C or CC canister when the container is opened and when attaching to machine or filling hand injectors; also when washing out the injector and container.

Before treatment the soil must be worked to a fine tilth and have a slight moisture content. Excessive moisture or dryness reduces the effectiveness of the treatment.

The Ministry of Agriculture, Fisheries and Food has published a 'Code

Table 9.11   Showing hours of wetness needed for ascospores to infect leaves on the tree*

Average temperature during wet period °C	Time wetness must persist for scab infection (hours)
0·6–5·0	48 or more
5·6	30
7·2	20
10·0	14
12·8	11
14·4	10
16·7	9

* From figures from Western New York State, USA (Mills, L.D. and Laplante, A.A., 1951, *Cornell Ext. Bull.* No. 711, pp. 21–7).

Table 9.12   Apple and pear—sprays for scab control

(Unless certain that the fungicide is safe to use on a particular variety in a particular orchard, an Advisory Officer should be consulted. Always read the label on the container.)

Compound(s)	Bud burst	Green cluster	Petal fall and after	Remarks
Benomyl w.p.	560 g a.i./ha (in not less than 220 litres/ha	—	—	Apply every 14 days. When trees are in full leaf, material should be applied in not less than 560 litres/ha.
Captan w.p.	100 g a.i./100 litres (HV) 2·8 kg a.i./ha (LV)	As at bud burst As at bud burst	As at bud burst As at bud burst	The following varieties are sensitive under some conditions: Bramley's Seedling, Monarch, King Edward, Kidd's Late Orange, Winston and also D'Anjou pears.
Carbendazim w.p.	27–33 g a.i./100 litres (HV) 0·55 kg a.i./ha (LV)	—	—	Apply at 20 day intervals or where disease is severe apply every 10 days.
Copper oxide w.p. Bordeaux mixture w.p.	150–200 g metallic copper/ 100 litres (HV) 5·6 kg metallic copper/ ha (LV)	No	No	(1) At bud burst or early bud stage only. (2) Russeting may be caused unless safeners are added. (3) Some pear varieties, including Doyenné du Comice, susceptible to copper damage.

Table 9.12 cont.

Compound(s)	Bud burst	Green cluster	Petal fall and after	Remarks
Dithianon liquid suspension	56 g a.i./100 litres (HV) 1·26 kg a.i./ha (LV)	As at bud burst As at bud burst	As at bud burst As at bud burst	Last treatment at least 8 weeks before harvest.
Dodine w.p.	*Protective* *7-day interval* 24 g a.i./100 litres (HV) 540 g a.i./ha (LV) *or* *10-day interval* 32 g a.i./100 litres (HV) 730 g a.i./ha (LV) *or* *14-day interval* 49 g a.i./100 litres (HV) 1·09 kg a.i./ha (LV) *Curative* 49 g a.i./100 litres (HV) 1·09 kg a.i./ha (LV)	As at bud burst As at bud burst As at bud burst As at bud burst As at bud burst As at bud burst As at bud burst As at bud burst	*7-day interval* 16 g a.i./100 litres 360 g a.i./ha *or* *10-day interval* 24 g a.i./100 litres 540 g a.i./ha *or* *14-day interval* 32 g a.i./100 litres 730 g a.i./ha As at bud burst	For pears, concentration as at bud burst until 2 weeks after petal fall.
Dodine liquid	*Protective* 31 g a.i./100 litres (HV) 730 g a.i./ha (LV) *Curative* 50 g a.i./100 litres (HV) 1·12 kg a.i./ha (LV)	As at bud burst As at bud burst As at bud burst As at bud burst	As at bud burst As at bud burst As at bud burst As at bud burst	
Mancozeb with zineb w.p.	According to label	According to label	According to label	Last treatment at least 1 week before harvest.
Manganese/zinc dithiocarbamate w.p.	250 g/100 litres (HV) 5·6 kg/ha (LV)		200 g/100 litres (HV) 4·5 kg/ha (LV)	Apply at 10–14 day intervals Last treatment at least 1 week before harvest.
Propineb w.p.	105 g a.i./100 litres (HV) 2·4 kg a.i./ha (LV)	— —	— —	Apply at 7-day intervals. Should be used only on

Compound(s)	Bud burst	Green cluster	Petal fall and after	Remarks
Lime sulphur liquid	3 litres/100 litres (HV) 33·6 litres/ha (LV)	2 litres/100 litres (HV) 22·4 litres/ha (LV)	1 litre/100 litres (HV) 11·2 litres/ha (LV)	(?) Lime sulphur may cause damage to several varieties, including Cox's Orange Pippin, Lane's Prince Albert, Beauty of Bath and Newton Wonder. (2) Apply lime sulphur to pears at pre-blossom only. (3) All types of sulphur cause damage to some apple varieties and to pear Doyenné du Comice.
Sulphur w.p. or equivalent as Colloidal sulphur	450 g sulphur/100 litres (HV) 8·5 kg sulphur/ha (LV)	300 g/100 litres (HV) 8·5 kg/ha (LV)	300 g/100 litres (HV) 8·5 kg/ha (LV)	(1) Apply at 14-day intervals. (2) Do not apply to dessert apples from petal fall to fruitlet unless disease is severe.
Thiophanate-methyl	1·1 kg a.i. in 300 to 2000 litres/ha			
Thiram w.p. or liquid	160 g a.i./100 litres (HV) 3·6–4·5 kg a.i./ha (LV) (pear only)	160 g a.i./100 litres 3·6–4·5 kg a.i./ha (pear only)	160 g a.i./100 litres 3·6–4·5 kg a.i./ha (pear only)	(1) Apply to pears only. (2) Should not be used on fruit for canning or quick freeze. (3) Last treatment at least 7 days before harvest.
Compound(s)	*Bud burst*	*Green cluster*	*Petal fall and after*	*Remarks*
Organomercury w.p. or liquid	Protective and Curative (apple) 1·5–3 g mercury/100 litres (HV) 42–84 g mercury/ha (LV)	As at bud burst As at bud burst	As at bud burst As at bud burst	(1) Last treatment at least 42 days before harvest. (2) Some varieties of apples, particularly Cox's Orange Pippin, and some pear varieties, are sensitive and a lower rate must be used. Do not apply to Doyenné du Comice.

For post-harvest/pre-leaf fall sprays on apple see p. 200.

of practice for the fumigation of soil with chloropicrin'. Recommendations include the covering and sealing of treated areas with sheets of low gas permeability, e.g. polythene of at least 150 gauge, as the application proceeds. The treated area should remain covered for at least 4 days; the fumigator must inform the police in advance of the intention to apply chloropicrin. Notices giving warning of the danger from the fumigant must be posted around the area to be treated.

***Scab (*Venturia inaequalis*) (A.L. 245). The scab fungus overwinters within the tissues of fallen apple leaves. Spore cases (perithecia) develop in early spring and, during wet periods, spores (ascospores) are ejected. When the weather is warm and humid the spores germinate on young leaves and fruitlets causing scab infections. Diseased parts soon produce summer spores (conidia) which themselves continue the spread. Infected fruits become spotted, distorted and unsaleable. In some varieties, including Cox's Orange Pippin, young extension shoots become infected and the following spring produce cushions of conidia beneath blister-like swellings.

Routine scab control by the application of fungicides is an essential part of apple growing. Two methods of determining the intervals between applications are recognized according to the type of programme (preventive or curative). With preventive spraying, the aim is to ensure that there is always a sufficient deposit of fungicide on the tree during the vulnerable period. This object is achieved by spraying at 10-day intervals from bud burst until late June, though the time intervals between sprays may be modified according to the maker's instructions. With a curative programme the aim is to apply a suitable fungicide immediately after weather conditions have been favourable for infection, that is, as soon as possible after an infection period assessed from a Mills table (Table 9.11, p. 196) giving periods of leaf wetness needed at various temperatures. Fungicides containing dodine acetate or an organomercury compound are recommended for curative spraying.

Combinations of a fungicide with an insecticide or with another fungicide are commonly used. Before adopting one of these, it is necessary to ensure that the products to be mixed are compatible and guidance from the manufacturers should be sought. Suitable spray schedules are given in Table 9.12 (p. 197).

In apple orchards where traditional control measures have not been sufficiently effective, applications of phenylmercury chloride in autumn and again in spring have much reduced infection by ascospores and so made the normal routine sprays more effective. It should be applied (a) after picking but before leaf fall at 6·2 g mercury/100 litres at not less than 2200 litres/ha; this spray should be directed at the leaves on the tree and any fallen leaves: (b) before bud burst at 12 g mercury/100 litres at not less than 1100 litres/ha; this spray is directed mainly at the leaves on the ground but the trees must also be sprayed.

9.2   Pear

9.2A   Pests

Annual sprays are usually applied for aphids, and pear sucker in areas where this is a pest. Control measures for codling moth and red spider mites may be required locally.

***Aphids.** The pear-bedstraw aphid (*Dysaphis pyri*) is the most important. It overwinters as eggs on the tree and they have finished hatching by the white bud stage. The pinkish aphids cause severe leaf curling and may spread over the tree, persisting into July and then departing for bedstraw (*Galium* spp.), returning to pear in the autumn. The pear-coltsfoot aphid (*Anuraphis farfarae*) causes the leaves to fold upwards and turn red; adults are brown, the young yellow-green. Other aphids occurring include the apple-grass aphid—light infestations are often present—and green apple aphid (see under apple aphids, p. 173), and the pear-grass aphid (*Longiunguis pyarius*).

Control may be achieved either by (*a*) winter washes against the eggs, or (*b*) spring sprays against the newly hatched aphids.

(*a*) Winter washes. The following are effective if thoroughly applied: tar oil, 5 per cent (e.c) or 5–6 per cent (stock emulsion), when the buds are dormant in December-mid-February, or DNOC-petroleum oil, 4·5 per cent, in February-March at bud break. Do not apply washes in windy or frosty weather or when the trees are wet. Do not use DNOC-petroleum oil after the buds have reached the breaking stage.

(*b*) Spring sprays. Spray at green cluster or preferably petal fall with one of the insecticides listed in Table 9.13, (p. 202). The petal fall timing has the advantage that, by choosing an appropriate chemical, bryobia mite (p. 179), pear sucker (p. 205) and caterpillars (p. 189) may be controlled with the same spray.

*Apple blossom weevil (*Anthonomus pomorum*) (A.L. 28)—see under Apple, p. 175.

*Apple twig cutter (*Rhynchites coeruleus*)—see under Apple, p. 179.

*Bryobia mite (*Bryobia rubrioculus*) (A.L. 305)—see under Apple (p. 179) for a brief outline of life history.

Where required spray at petal fall with one of the acaricides listed in Table 9.14, p. 203; with heavy infestations a second spray may be necessary about 3 weeks later. The best results will be obtained by spraying under warm conditions (15°C or over) when the mites are active on the foliage. By choosing appropriate chemicals at petal fall, control of aphids (p. 201 and Table 9.13) caterpillars (p. 189) and pear sucker (p. 205 and Table 9.16) may be obtained at the same time. Annual sprays should be unnecessary.

Tetradifon affects eggs and young stages and could also be used where later sprays are required and both these stages are present.

Table 9.13   Pear—green cluster or petal fall* sprays for aphid control

Compound(s)	Rate* HV g a.i./100 litres	LV g a.i./ha	Applied at petal fall may also control†		
Azinphos-methyl + demeton-S-methyl sulphone	25 + 7·5	560 + 170	B	C	S
Chlorpyrifos	21	475	B	C	S
Demephion	7·5	170	—	—	—
Demeton-S-methyl	7·5	170	B	—	S
Dichlorvos	—	1050‡	B	C	S
Dimethoate	30	680	B	—	S
Dioxathion	60	1320	B	C	S
Fenitrothion§	31	700	B	C	S
Formothion	37	830	—	—	S
Gamma-HCH*	12·5–25	280–560	—	—	—
Malathion	56	1260	B	C	S
Mevinphos‖	6–9	140–210	—	C	—
Nicotine [a]	50 ml (95–98%)	560–700 ml (95–98%)	—	—	—
Oxydemeton-methyl	7·4	160	B	—	S
Phosalone	31	700	B	C	—
Phosphamidon	30	675	B	C	S
Vamidothion**	40	900–1120	B	—	S

* Follow the precautions against poisoning bees, and for avoiding taint of nearby crops with gamma-HCH given under apple aphids (p. 175).

† Sprays at petal fall may also control: B = bryobia mite (p. 201 and Table 9.14); C = caterpillars (p. 189); S = pear sucker (p. 205 and Table 9.16). Use a higher rate if so stated for these pests.

‡ Apply in at least 225 litres water. HV sprays give the best results if applied in calm weather and at air temperatures of 15°C or over.

§ Mixtures of fenitrothion with binapacryl, dodine, or with more than 1 other insecticide or fungicide, may be incompatible or cause phytotoxicity—consult product label.

[a] Air temperature of 15°C or over required.

‖ If caterpillar control is also required use 72 g a.i. (HV) or 280 g a.i. (LV).

** Maximum 2 applications per season.

*Clay-coloured weevil (*Otiorrhynchus singularis*) (A.L. 154)—see under Apple, p. 180.

*Codling moth (*Cydia pomonella*) (A.L. 42). This is less common as a pest of pear than apple; one spray of lead arsenate (300 g a.i./100 litres HV) at petal fall, can be effective, or if required follow the spray programme for apple (see Table 9.4, p. 182).

**Common green capsid (*Lygocorus pabulinus*) (A.L. 154)—see under Apple (p. 182) for brief outline of life history.

To prevent damage where this pest occurs spray at petal fall with a treatment selected from Table 9.5 (p. 183).

*Fruit tree red spider mite (*Panonychus ulmi*) (A.L. 10). This is not a common pest on pears but, where required, follow the summer spray

Table 9.14   Pear—petal fall* sprays for control of bryobia mite active stages

Compound(s)	Rate† HV g a.i./100 litres	LV g a.i./ha	Applied at petal fall may also control*		
Azinphos-methyl	33	740	—	C	S
Azinphos-methyl + demeton-S-methyl sulphone	25 + 7·5	560 + 170	A	C	S
Chlorpyrifos	42	945	A	C	S
Cyhexatin	18·7	420	—	—	—
Demeton-S-methyl	22	490	A	—	S
Dichlorvos	—	1120‡	A	C	S
Dimethoate	22·5	500	A	—	S
Dioxathion	60	1320	A	C	S
Malathion	112	2530	A	C	S
Oxydemeton-methyl	22	480	A	—	S
Phosalone	31	700	A	C	—
Phosphamidon	30	675	A	C	S
Tetradifon§·	12·5	280	—	—	—
Vamidothion‖	40	900–1120	A	—	S

* Sprays at petal fall may also control: A = aphids (p. 201 and Table 9.13); C = caterpillar; S = pear sucker (p. 205 and Table 9.16). Use a higher rate if so stated for these pests.

† Follow the precautions against poisoning bees given under apple aphids (p. 175).

‡ Apply in at least 22 litres water. HV sprays give the best results if applied in calm weather and at air temperatures of 15°C or over.

§ Effective against eggs and young stages, see p. 201.

‖ Maximum 2 applications per season.

programme for Apple, see p. 184, choosing an acaricide from those listed for control of bryobia mite on p. 201 and Table 9.14.

Infestations of *Tetranychus viennensis* sometimes occur on pear. The adult females are very like those of *T. urticae* but in summer are a distinctive plum-red colour. Adult females overwinter on the tree in crevices in the bark, etc., emerging in the spring when they spin a light canopy of silk strands on the undersides of leaves. Eggs are laid in the webbing and along leaf veins from mid-May onwards. The life cycle occupies about 25 days in mid-summer and there may be 5 generations in the year. The mite occurs on several rosaceous hosts, including hawthorn, sloe and mountain ash.

Should control become necessary, one of the organophosphorus acaricides listed for the control of bryobia mite (p. 201 and Table 9.14), applied at petal fall before eggs have been laid, should be satisfactory. For infestations discovered later in the season apply any of the treatments listed in Table 9.14, with a second spray 3 weeks later.

***Leaf weevils** (*Phyllobius* spp.)—see under Apple, p. 186.

***Mussel scale** (*Mytilococcus ulmi*) (A.L. 36)—see under Apple, p. 187.

*Oystershell scale (*Quadraspidiotus ostreaeformis*)—see under Apple, p. 187.

*Pear leaf blister mite (*Eriophyes pyri*) (A.L. 35). Adult mites hibernate under bud scales, invading leaves at bud burst. Leaves become dotted with yellowish or reddish blisters, and reddish or brown pustules occur on fruitlets. The mite is mostly a pest of wall trees, and is controlled by 5 litres lime sulphur/100 litres (HV) in March just before the buds open. In Canada good control has been given by carbaryl, 50 g a.i./100 litres (HV).

*Pear leaf midge (*Dasyneura pyri*). Midges lay eggs in folds of young leaves in May; larvae feed on leaves, the edges of which become rolled upwards. There are several overlapping generations a year. The midge is sometimes a pest of nursery trees. Vamidothion, 40 g a.i./100 litres (HV), and carbaryl, 85 g a.i./100 litres (HV), have given promising results with one spray in June. A maximum of 2 applications of vamidothion are allowed on pear per season. Use of carbaryl in early June could lead to fruit thinning.

*Pear midge (*Contarinia pyrivora*) (A.L. 26). Midges lay their eggs in the flower buds, and the yellowish white larvae feed inside the fruitlets which become swollen and deformed. The standard control has been DDT, 100 g a.i./100 litres (HV) or 2240 g a.i./ha (LV), at white bud. Carbaryl, 87·5 g a.i./100 litres (HV), at late green cluster is probably a superior alternative, and phosphamidon at 20 g a.i./100 litres (HV) is also tentatively recommended. The spray should be repeated in the following season, but routine annual sprays are unnecessary.

*Pear sawfly (*Hoplocampa brevis*). This is a local pest, similar to the apple sawfly in habits and damage caused. Where required it can be controlled with a treatment selected from Table 9.15 (p. 204). Nicotine, 50 ml (95–98 per cent)/100 litres (HV), at petal fall can also be effective, but air temperatures of 15°C or over are required. Dimethoate, and other organophosphorus materials, mentioned for control of sawfly on apple (p. 177), are probably also effective.

Follow the precautions against poisoning bees, and for avoiding taint of nearby crops with gamma-HCH, given under apple aphids on p. 175.

Table 9.15   Pear—sprays* for pear sawfly control

Compound	Rate HV g a.i./100 litres	LV g a.i./ha
Demeton-S-methyl	22	490
Gamma-HCH	12·5	280
Oxydemeton-methyl	22	480
Phosphamidon	30	675

* Apply within 7 days following petal fall.

***Pear slug sawfly** (*Caliroa cerasi*) (A.L. 84). The adults appear in May and June: eggs are laid in slits cut in the leaf. The caterpillars are yellowish white at first but turn dark greenish or black and are slug-like in appearance. They feed on the upper leaf surface and leaves may become skeletonized. There are 2–3 generations a year.

If necessary spray in June with one of the insecticides suggested for control of pear sawfly (p. 204 and Table 9.15), *or* derris (3·5 g rotenone/100 litres HV), *or* lead arsenate at usual dosage rates. Follow the precautions against poisoning bees, and for avoiding taint of nearby crops with gamma-HCH, given under apple aphids on p. 175.

****Pear sucker** (*Psylla simulans*) (A.L. 96). Adults hibernate at rest on the bark, or among dead leaves or other shelter. Eggs are laid on the shoots and spurs from March to petal fall, and the young suckers feed on the buds and blossom trusses. Two further generations follow and if the insects become numerous they render the leaves sticky with honeydew on which sooty moulds grow.

In some orchards 1 control spray is sufficient and the best time is probably about 3 weeks after petal fall when eggs are fewest. Where heavy infestations occur, 2 sprays are probably necessary for full control; one at petal fall (which can be combined with aphid control) and a second 3 weeks later. Use a treatment selected from Table 9.16.

Table 9.16   Pear—sprays for pear sucker control

Compound(s)	Rate HV g a.i./100 litres	LV g a.i./ha	Applied at petal fall may also control*		
Azinphos-methyl	33	750	—	B	C
Azinphos-methyl + demeton-S-methyl sulphone	37·5 + 11	840 + 250	A	B	C
Chlorpyrifos	42	945	A	B	C
Demeton-S-methyl	22	490	A	B	—
Dimethoate	30	680	A	B	—
Dioxathion	60	1320	A	B	C
Fenitrothion‡	31	700	A	—	C
Formothion	37	830	A	—	—
Malathion	112	2530	A	B	C
Oxydemeton-methyl	22	480	A	B	—
Phosphamidon	30	675	A	B	C
Triazophos	40	900	A	B	C
Vamidothion§	40	900–1120	A	B	—

* Sprays at petal fall may also control: A = aphid (p. 201 and Table 9.13; B = bryobia mite (p. 203 and Table 9.14); C = caterpillar (p. 206). Use a higher rate if so stated for these pests.

‡ Mixtures of fenitrothion with binapacryl, dodine, or with more than 1 other insecticide or fungicide, may be incompatible or cause phytotoxicity—consult product label.

§ Maximum 2 applications per season.

The timings of sprays for pear sucker, aphids (p. 201 and Table 9.13) and bryobia mite (p. 201 and Table 9.14) are sufficiently similar for the 3 pests to be controlled at the same time if required and using the same pesticide.

A 5 per cent tar oil wash in December-mid-February will give some control.

*Pear thrips (*Taeniothrips inconsequens*). The insects feed on the flowers and young foliage, and may cause russeting of the fruit of the apple, pear and plum. Thrips are controlled by DDT, 50 g a.i./100 litres (HV), at white bud. There is limited information on which to recommend as alternatives to DDT, but malathion or phosphamidon, at dosage rates as for aphid control Table 9.13 (p. 202), are tentatively suggested.

**Red bud borer (*Thomasiniana oculiperda*)—see under Apple, p. 187.

*Tortrix caterpillars. Several species occur on pear, including *Archips podana* (see under Apple, p. 188), but are rarely important.

**Wasps (*Vespula* spp.) (A.L. 451)—see under Apple, p. 189.

**Winter moths (A.L. 11)—see under Apple, p. 189.

9.2B   Diseases

**Canker (*Nectria galligena*) (A.L. 100). Common canker of the pear is caused by the same fungus as that causing apple canker. Biology and control measures are the same as for Apple (p. 192).

**Fireblight (*Erwinia amylovora*) (A.L. 571). First recorded in Great Britain in 1957 on pears, this disease has spread rapidly and has since infected a number of apple varieties and also other members of the sub-family Pomoideae. These include hawthorn, cotoneaster, pyracantha, stranvaesia, white beam and mountain ash. In pears, infection by the causative bacterium generally occurs through late summer (secondary) blossom, killing this and progressing via the stalk into the twig, branch and finally the trunk. Because the pear variety Laxton's Superb is particularly prone to produce secondary blossoms over a long period, it is very liable to attack.

In summer and autumn, parts of the bark containing active bacteria usually show a red discoloration on cutting but during winter months the discoloration may be dark brown. In a mild winter the bacteria may continue to advance along the affected part but in certain circumstances a limited canker is formed and this may crack at the margin thus becoming isolated from healthy tissue external to it. The bacteria in some of these 'holdover cankers' may survive until the following spring when they can initiate new outbreaks.

Although infection is usually through the blossom, shoots and leaves may also be attacked.

Statutory regulations under the Fireblight Disease Order require that where fireblight occurs, or is believed to occur, it should be immediately reported to the Plant Health Branch of the MAFF. This Order prohibits the propagation of pear trees of the susceptible variety Laxton's Superb, and their planting on agricultural land in England and Wales. It also requires that all trees of that variety in Kent, Essex and East Sussex on agricultural land should be top worked to other varieties or grubbed.

The grubbing of badly infected trees may be required, but where cutting out of branches is permissible, this should be done at not less than 60 cm below visible signs of the disease within the bark. To reduce the chance of reinfection, the cut surface should be painted immediately with white lead finishing paint conforming to B.S. 2526/7.

The cutting parts of all pruning tools should be immersed in 3 per cent lysol between making cuts on both affected and healthy trees.

***Scab (*Venturia pirina*) (A.L. 245). Pear scab is caused by a fungus different from that causing apple scab. Nevertheless, the life histories of the two fungi are almost identical and the reader is referred to the description of apple scab (p. 200).

Spray protectively at 10–14 day intervals from bud burst until at least the end of May with captan, dithianon, dodine, mancozeb/zineb, manganese/zinc dithiocarbamate *or* wettable sulphur, as for apples; later sprays at 14 day intervals may be necessary. Thiram is also suitable. Mercury- or copper-containing sprays are not recommended after green cluster because of possible damage. In general the same strengths of fungicide as recommended for apples (Table 9.12, p. 197) are suitable for most pear varieties. Dodine has been recorded as damaging in cold spring weather. Doyenné du Comice must not be sprayed with copper, sulphur or mercury.

9.3   Quince

9.3A   Disease

Leaf blight and **fruit spot** (*Fabraea maculata = Entomosporium maculatum*). The fungus infects foliage, fruit and young twigs. On the leaves, the spots are at first reddish becoming nearly black; they are irregular in shape and usually up to 2·5 mm across. Spores are produced from these spots. If spots run together the leaves turn yellow and fall early. On fruits, the spots are very dark brown, slightly sunken and produce spores; in severe cases deformity may occur. The disease is common on fruiting trees and may occur in nursery rows where quinces are propagated as rootstocks for pears. Malling Quince A is fairly resistant.

On fruiting trees infected twigs should be removed and burnt while the foliage should be sprayed with Bordeaux mixture or other copper-containing fungicide recommended for this purpose by the makers.

9.4   Apricot, nectarine and peach

9.4A   Pests

***Aphids.** Several species overwinter as eggs on these trees. The peach-potato aphid (*Myzus persicae*) is green, causing severe leaf curl in the spring, and dispersal to summer host plants occurs in May and June. The peach aphid (*Appelia schwartzi*) is dark brown and infests the shoots, causing severe leaf curl; it disperses to other peach trees in the summer. The black peach aphid (*Brachycaudus persicaecola*) overwinters as aphids on the roots, and infests the shoots during the summer. The mealy peach aphid (*Hyalopterus amygdali*) also occurs, dispersing in the summer to reeds.

Either (*a*) winter washes against the eggs or (*b*) spring sprays against the aphids may be used to control these pests. (*a*) In winter, 5 per cent tar oil (HV), in the first half of December for outdoor trees, is effective if thoroughly applied. Under glass, use at 4 per cent in December.

(*b*) In spring, use one of the treatments listed in Table 9.17, p. 208, at the end of the blossom period.

***Common green capsid** (*Lygocoris pabulinus*) (A.L. 164). See under Apple for a brief outline of the life history. If necessary, spray with dimethoate or malathion, at rates as for aphids (Table 9.17, p. 208). The optimum time to spray is when most of the winter eggs have hatched, which is usually at the time of full bloom of Bramley's Seedling apple.

****Fruit tree red spider mite** (*Panonychus ulmi*) (A.L. 10). This mite may occur on outdoor trees and its life history is as described under Apple (p. 184).

The timing of control sprays is as recommended for apple. With heavy infestations, spray about the time of petal fall of Cox's Orange Pippin, and again 3 weeks later if necessary, using an acaricide effective against the active stages.

Table 9.17   Apricot, nectarine and peach—sprays* for aphid control

Compound	Rate HV g a.i./100 litres	LV g a.i./ha
Demeton-S-methyl	22	490
Diazinon†	20	—
Dimethoate	30	680
Malathion	75	1685
Oxydemeton-methyl	22	480

* Follow the precautions against poisoning bees, given under Apple on p. 175.

† For use under glass only.

Table 9.18   Apricot, nectarine and peach—sprays* for control of fruit tree red spider mite (*Panonychus ulmi*) and red spider mite (*Tetranychus urticae*)

Compound	Susceptible stages	Rate HV g a.i./100 litres	LV g a.i./ha
Demeton-S-methyl	active stages	22	490
Oxydemeton-methyl	active stages	22	480
Tetradifon	eggs and young stages	12·5	280

* Follow the precautions against poisoning bees given under Apple on p. 175.

With light infestations spraying can be delayed until mid-June, repeated 3 weeks later if necessary, when it is an advantage to use an acaricide effective against eggs and young stages.

Where only a single spray is given, the best time is about 10 days after petal fall of the apple Cox's Orange Pippin and an acaricide effective against active stages should be used.

Table 9.18 lists the recommended acaricides for the control of red spider on peach, etc., and the mite stages which they affect.

Red spider mite (*Tetranychus urticae*) (A.L. 226). This is the red spider mite common on many crops outdoors and under glass, sometimes infesting top and soft fruit. It overwinters as adult females in the soil and other shelter, and these emerge in spring to feed and breed on the leaves.

Where required apply one of the acaricides listed under fruit tree red spider mite Table 9.18, about mid-May and again 2–3 weeks later if necessary.

*Scale insects.** The brown or peach scale (*Parthenolecanium corni*) (A.L. 88), mussel scale (*Mytilococcus ulmi*) (A.L. 36) and oyster-shell scale (*Quadraspidiotus ostreaeformis*) may occur. Use of 5 per cent tar oil (HV) in the first half of December is effective. Where control during the growing season is required, malathion can be effective: see under Apple—mussel scale, p. 187, and Blackcurrant—brown scale, p. 221. With mussel scale under glass, spray 3–4 weeks earlier than for outdoor trees.

Wasps (*Vespula* spp.) (A.L. 457)—see under Apple, p. 189.

*Winter moths** (A.L. 11)—see under Apple, p. 189. Caterpillars of this group may occur on outdoor trees and are controlled by applying DDT, 50 g a.i./100 litres (HV) or 1120–1400 g a.i./ha (LV), in April when caterpillars are first seen. There is no information on alternative insecticides for use on peaches.

9.4B   Diseases

Peach leaf curl (*Taphrina deformans*) (A.L. 81). Young diseased leaves are thick and yellow, tinged with red. Older leaves are crumpled and

IFH—8

redder. Premature defoliation weakens growth and may render nursery stock useless. The causative fungus overwinters in the bark and between the bud scales.

Apply a copper compound or lime sulphur before the buds begin to swell, usually in late February or early March and repeat immediately before leaf fall in the autumn. Suitable copper sprays are: Bordeaux mixture (w.p.) or copper oxychloride (w.p.) or copper oxide (w.p.) or cuprammonium compound (soluble), each applied at 100–150 g metallic copper/100 litres (HV). The lime sulphur wash should contain 3 litres lime sulphur/100 litres (HV).

The removal of affected leaves before a bloom (spores) appears on them will help control.

Peach powdery mildew (*Sphaerotheca pannosa* var. *persicae*). Powdery (sporing) patches appear on infected leaves and young shoots. The fungus overwinters in the buds and these produce stunted shoots which bear narrow leaves. To control the disease apply, when mildew first seen and at 14 day intervals, sulphur sprays at 300 g sulphur/100 litres (HV).

9.5   Cherry

9.5A   Pests

Routine sprays are required for cherry blackfly and, where necessary, insecticides are also included in spring sprays for control of winter moth caterpillars.

***Aphids.** The cherry blackfly (*Myzus cerasi*) is the only species on cherry. It overwinters as eggs on the bark and these hatch in March and April, hatching being complete by the white bud stage. Successive generations of black aphids are produced, then winged forms which disperse in June and July to bedstraws. Leaves are severely curled and new growth checked.

Either (*a*) winter washes against the eggs or (*b*) spring sprays against the aphids, may be used.

(*a*) Winter washes. Tar oil, 5 per cent (e.c.) or 5–6 per cent (stock emulsion) (HV), when the buds are dormant (December–January), or DNOC-petroleum oil, 4·5 per cent, up to bud break are effective.

(*b*) Spring sprays. Any of the treatments in Table 9.19 (p. 211) are effective at the white bud stage but the organophosphorus insecticides, especially in years of heavy infestations, are more efficient.

Cherry-bark tortrix (*Enarmonia formosana*). The pinkish white caterpillar tunnels under the bark of the trunk, often just below the crotch; successive generations use the same galleries which may become extensive.

Table 9.19   Cherry—sprays* for aphid control

Compound(s)	Rate HV g a.i./100 litres	LV g a.i./ha
Azinphos-methyl + demeton-S- methyl sulphone	25 + 7·5	—
Demeton-S-methyl	22	490
Dimethoate	30	680
Formothion	37	830
Gamma-HCH*	25	560
Malathion	75	1685
Mevinphos	6–9‡	140–210‡
Nicotine†	50 ml (95–98%)	560–700 ml (95–98%)
Oxydemeton-methyl	22	480
Vamidothion§	40	—

* Follow the precautions against poisoning bees, and for avoiding taint of nearby crops with gamma-HCH, given under apple aphids on p. 175.
† Air temperature must be at least 15°C.
‡ Use the higher rate in cold weather.
§ Maximum 2 applications per season.

Heavy infestations sometimes occur in old orchards, and there is some evidence that extensive bark injury can kill large branches or even whole trees. Apple, pear, plum and cherry may be attacked. The moths appear from mid-May until early September.

Control measures have been studied on apple. Drenching sprays of the trunk and main branches in late May, when the caterpillars are near the surface for pupation, of trichlorphon at 100 g a.i./100 litres (HV) have given good results, though such sprays would probably have to be repeated for several seasons. Rather more effective is liberal brushing of the trunk and main branches with creosote, or drenching with undiluted tar oil, after scraping off loose bark in March while the trees are still dormant.

**Cherry fruit moth (*Argyresthia curvella*). Moths appear in late June and in July, and lay eggs under the bud scales, in crevices in the bark, etc., especially towards the tips of branches. Most of the eggs hatch in the autumn, the caterpillars hibernating in a silk cocoon in bark crevices, though some do not hatch until the spring. The caterpillars bore into flower buds in the spring, feeding on the flowers and later the young fruitlets, dropping to the ground when fully fed.

A winter wash of 7–8 per cent tar oil in December or January, if thoroughly applied, will destroy the winter eggs but will not reach hibernating caterpillars. The best control is to spray in late March when the buds are breaking, and again at white bud with heavy infestations,

Table 9.20   Cherry—sprays for cherry fruit moth and winter moth control

Compound(s)	Rate HV g a.i./100 litres	HV g a.i./ha
Azinphos-methyl	27·5	—
Azinphos-methyl + demeton-S- methyl sulphone	25 + 7·5	—
DDT	50–100	1120–2240
Lead arsenate	200	—
Mevinphos	12·5	280

using one of the insecticides listed in Table 9.20 (p. 212). It is important to wet the tops of the trees.

Fruit tree red spider mite (*Panonychus ulmi*) (A.L. 10). This is not usually a pest on cherry. See under Plum, p. 215 and Table 9.22, for control.

*Leaf weevils (*Phyllobius* spp.)—see under Apple, p. 186.
*Pear slug sawfly (*Caliroa cerasi*) (A.L. 84)—see under Pear, p. 205.
*Wasps (*Vespula* spp.) (A.L. 457)—see under Apple, p. 189.
*Winter moths (A.L. 11). If required, use one of the treatments listed in Table 9.20 (p. 212) at the white bud stage.

9.5B   Diseases

***Cherry bacterial canker** (*Pseudomonas mors-prunorum*). During the autumn, the bacteria enter wounds and fresh leaf scars. Until spring the disease develops from the points of entry causing death of bark and young wood. In summer the bacteria become almost inactive but stems, branches or shoots which have been girdled by the 'cankers' die, often after producing pale green or yellow leaves. Where girdling has not occurred, a limited flat canker remains and the margins of this may produce callus. In some years, buds may be killed and sometimes small holes (shot-holes) appear in leaves near to cankers.

In order to reduce the numbers of bacteria during the most vulnerable period, a drenching spray of Bordeaux mixture should be applied 3 times at intervals of about 3 weeks from the end of August. An extra application at petal fall may be advisable on young trees of susceptible varieties in order to build a canker-free framework. The petal fall application may cause damage.

Bordeaux mixture: (1) August: 400 g copper sulphate (pentahydrate) + 600 g hydrated lime + 750 ml cotton seed oil/100 litres. (2) September: 600 g copper sulphate (pentahydrate) + 900 g hydrated lime + 750 ml cotton seed oil/100 litres. (3) October: 1·0 kg copper sulphate, 1·5 kg hydrated lime/100 litres water. (4) Petal fall: as August application (see above). Ready prepared forms of copper may be used but at equivalent copper content they are likely to cause damage and should be used according to the maker's recommendations. Pruning or cutting back of trees should be done between May and the end of August.

Varieties differ markedly in their susceptibility to bacterial canker.

The risk of crotch infection is reduced by using seedling mazzards or F.12/1 rootstocks for frameworking.

Specific cherry replant disease. Where cherry is replanted after cherry, retardation of growth is likely to occur. This has been shown to be due to colonization of the roots of the replanted trees with *Thielaviopsis basicola*. Recently trials have shown that the disease can be controlled in cherry layer beds by applying benomyl, 100 g a.i. in 240–480 litres water/100 m of row, in early May. This application may severely diminish earthworm populations in the treated soil.

Chloropicrin applied as for specific apple replant disease (see p. 194) will also give control of specific cherry replant disease in nursery beds and in established orchards.

9.6   Plum and damson

9.6A   Pests

Annual treatments, as winter washes or spring sprays, for control of aphids are usually required, with occasional sprays against caterpillars. Regular control measures for plum sawfly and fruit tree red spider mite may be required locally.

***Aphids** (A.L. 34). The 3 common species are leaf-curling plum aphid (*Brachycaudus helichrysi*), mealy plum aphid (*Hyalopterus pruni*) and damson-hop aphid (*Phorodon humuli*) which is mainly harmful to damsons. All overwinter as eggs laid in autumn on the twigs.

Leaf-curling plum aphid hatches early, usually by January, the young aphids feeding on the dormant buds; in the spring successive generations feed on the foliage and produce severe leaf curl. From May to July winged forms disperse to various summer host plants. Mealy plum aphids hatch later, but have done so by the white bud stage; from the end of June winged forms disperse to grasses or reeds and other plum trees. Damson-hop aphid hatches in early spring, dispersing to hops from mid-May until late July.

Either (*a*) winter washes or (*b*) spring sprays may be used for control.

(*a*) Winter washes. Thorough HV spraying gives a good kill of winter eggs or young leaf-curling aphids; tar oil, 5 per cent (e.c.) or 5–6 per cent (stock emulsion), when buds are dormant in December to early January for early cultivars, to end of January for main crop, or DNOC-petroleum oil, 4·5 per cent, as delayed dormant wash when the buds are breaking, in

Table 9.21   Plum and damson—spring sprays* for aphid control

Compound(s)	Rate HV g a.i./100 litres	LV g a.i./ha
Azinphos-methyl + demeton-S-methyl sulphone	25 + 7·5	560 + 170
Chlorpyrifos	21	470
Demeton-S-methyl	22	490
Dimethoate	30	680
Fenitrothion‡	31	700
Formothion	37	830
Gamma-HCH*	25	560
Malathion	75	1680
Mevinphos	6–9§	140–210§
Nicotine†	50 ml (95–98%)	560–700 ml (95–98%)
Oxydemeton-methyl	22	480
Vamidothion‖	40	900–1120

* Follow the precautions against poisoning bees, and for avoiding taint of nearby crops with gamma-HCH, given under apple aphids on p. 175.

† Air temperature of 15°C or over required.

‡ Mixtures of fenitrothion with more than 1 other insecticide or fungicide may be incompatible or cause phytotoxicity—consult product label.

§ Use the higher rate in cold weather.

‖ One application at petal fall is recommended; maximum 2 per season.

February or March. This wash will also give some control of fruit tree red spider mite and winter moth eggs.

The buds of some plum cultivars are very subject to tar oil damage. Tar oil should not be used at all on Myrobalan, and not after about mid-January on Belle de Louvain, Victoria, Yellow Egg and the Gages. Do not spray in frosty weather or when the trees are wet.

(*b*) Spring sprays. Any of the treatments listed in Table 9.21 is effective at the white bud stage. The organophosphorus compounds, especially if some leaf growth has occurred, are more efficient than gamma-HCH.

Mealy plum aphid infestations are often overlooked and, as they can persist until August, may require additional sprays of organophosphorus insecticides in May or June is not properly controlled in the spring, or if reinfestation occurs in the summer.

Damson-hop aphids resistant to organophosphorus insecticides have occurred locally on plum trees in hop growing areas. Apart from tar oil winter wash, or nicotine in spring, only propoxur, 50 g a.i./100 litres (HV), can be suggested as an alternative.

Table 9.22   Plum and damson—sprays* for fruit tree red spider mite control†

Compound	Susceptible stages	Rate HV g a.i./100 litres	LV g a.i./ha
Chlorpyrifos	active stages	42	945
Demeton-S-methyl	active stages	22	490
Dimethoate	active stages	22·5	500
Malathion	active stages	112	2530
Oxydemeton-methyl	active stages	22	480
Phosalone	active stages	31	700
Tetradifon	eggs and young stages	12·5	280
Vamidothion‡	active stages	40	900–1120

* Do not spray open blossom; cut down flowering weeds under the trees before spraying.

† For timing see text, p. 215.

‡ Maximum 2 applications per season.

*Common green capsid (*Lygocoris pabulinus*) (A.L. 154). See under Apple (p. 182) for a brief outline of life history. Although there is no experimental evidence for plums, a cot-split spray of dimethoate, fenitrothion, formothion or malathion at rates suggested for apple and pear (Table 9.5, p. 183) should be effective.

***Fruit tree red spider mite (*Panonychus ulmi*) (A.L. 10). For life history see under Apple, p. 184.

Winter washes based on petroleum oil (see under Apple, p. 186) give some control of winter eggs, but it is usually better to restrict winter washes to tar oil for aphids, and to use summer sprays for red spider as these are more efficient.

Summer sprays can be timed to correspond with those recommended for apple. Where infestations are heavy, use an acaricide effective against the active stages (see Table 9.22) and apply at about the time of petal fall of Cox's Orange Pippin apple, with a second spray 3 weeks later. If only one spray is to be given the best time is about 10 days after petal fall of Cox, when most of the winter eggs will have hatched. With light infestations spraying can be delayed until mid-June, as on apple, using an acaricide affecting eggs and young stages. The acaricides given in (Table 9.22) are recommended for use on plum.

*Pear thrips (*Taeniothrips inconsequens*)—see under Pear, p. 206.

**Plum fruit moth (*Cydia funebrana*). Moths are on the wing over a similar period to that of codling moth. Eggs are laid from mid-June to early August, and the whitish caterpillar bores into the fruit towards the stone; in its final growth stage it is reddish in colour ('red plum maggot'). When fully fed in late August and September it leaves the fruit and builds a cocoon for the winter in crevices in the bark, etc.

For control apply azinphos-methyl, 33 g a.i./100 litres (HV) or 750 g a.i./ha (LV); *or* azinphos-methyl + demeton-S-methyl sulphone, 37·5 + 11 g a.i./100 litres (HV) *or* 840 + 250 g a.i./ha (LV), in late June and again 2–3 weeks later. Annual sprays should not be necessary once good control is achieved. Observe minimum interval of 3 weeks between last application and harvest.

Plum sawfly (*Hoplocampa flava*). The adults appear at blossom time, the female laying eggs in the flowers. The creamy white caterpillars bore into the fruitlets; one caterpillar attacking as many as 4 before it is fully fed and drops to the ground, where it builds a cocoon in the soil for the winter. Marked preference for cultivars is shown, Czar being particularly susceptible.

To control, spray at the cot-split stage (7–10 days after petal fall) with one of the insecticides listed in Table 9.23. These compounds will also control early hatching fruit tree red spider mites, but for full control of red spider the timing should be as given on p. 215 (Table 9.22).

Red-legged weevil (*Otiorrhynchus clavipes*) (A.L. 57). This wingless weevil is about 13 mm long, shining black with reddish legs. It appears in late April or May and shelters by day under clods and in rough vegetation. It feeds at night on foliage, flowers and fruitlets, and gnaws at bark, often causing young shoots to break off. Peach, apricot, nectarine, raspberry, gooseberry, etc., are also attacked.

The control measures recommended for clay-coloured weevil on Apple (p. 180) can be used.

Scale insects. The brown or peach scale (A.L. 88), mussel scale (A.L. 36) and oystershell scale may occur on plum. Control is effected by tar oil, 5 per cent (HV), applied when the buds are dormant in December to early January for early cultivars, to end January for main crop, or by DNOC-petroleum oil, 4·5 per cent (HV), up to bud break in late January or early February.

Table 9.23   Plum and damson—sprays* for plum sawfly control

Compound	Rate HV g a.i./100 litres	LV g a.i./ha
Chlorpyrifos	42	945
Demeton-S-methyl	22	490
Dimethoate	30	680
Fenitrothion†	31	700
Oxydemeton-methyl	22	480

* Apply at the cot-split stage (7–10 days after petal fall).
† Mixtures of fenitrothion with more than 1 other insecticide or fungicide may be incompatible or cause phytotoxicity—consult product label.

Table 9.24 Plum and damson—spring sprays* for tortrix moth and winter moth control

Compound	Rate HV g a.i./100 litres	LV g a.i./ha
Azinphos-methyl	33	750
Chlorpyrifos	21	470
DDT	50–100	1120–2240
Fenitrothion†	31	700
Lead arsenate	200	—
Mevinphos	12·5	280
Tetrachlorvinphos	75	—

* To reduce danger to bees, do not spray open blossom; cut down flowering weeds beneath the trees before spraying.

† Mixtures of fenitrothion with more than 1 other insecticide or fungicide may be incompatible or cause phytotoxicity—consult product label.

The buds of some plum cultivars are very subject to tar oil damage. Tar oil should not be used at all on Myrobalan, and not after about mid-January on Belle de Louvain, Victoria, Yellow Egg and the Gages. Do not spray in frosty weather or when the trees are wet.

See also under mussel scale on Apple (p. 187) and brown scale on Currant (p. 221) for control by spring sprays.

**Tortrix moths. Caterpillars of the plum tortrix (*Hedya pruniana*) feed on foliage and tunnel in the shoots from April to June. Several of the tortrix caterpillars found on Apple (p. 188) also occur on plum. The caterpillars are kept in check by spring sprays against winter moth caterpillars (see below and Table 9.24).

**Wasps (*Vespula* spp.)—see under Apple, p. 189.

**Winter moths (A.L. 11)—see under Apple, p. 189, for life histories. For control, use one of the treatments listed in Table 9.24, at the white bud stage.

9.6B   Diseases

***Blossom wilt, brown rot and allied diseases of plum (*Sclerotinia fructigena* and *S. laxa*) (A.L. 248). *Sclerotinia fructigena* and *S. laxa* cause brown rot of plum fruits, while the latter species also causes blossom wilt sometimes followed by spur blight and canker and wither tip of shoots. Both species overwinter in mummified fruits and cankers which produce spores in spring so continuing the cycle. *S. laxa* can enter shoots and spurs via damaged leaves.

Diseased trusses, shoots and cankered spurs and branches should be cut out and burnt, preferably in the current spring or summer when their presence can more easily be recognized. Mummified fruits should also be collected and burnt. Tar oil spraying as for aphids (see p. 173) late in the following dormant period gives partial control by destroying spore cushions that arise on any diseased parts overlooked in the earlier cutting out.

9.7   Hazelnut, cobnut and filbert

9.7A   Pests

Nut weevil (*Balaninus nucum*). The adult weevils emerge in May. The female bores a hole in the side of the young nut and deposits an egg, and the white grub feeds on the kernel. When full grown in late July or August the grub drops to the ground, and builds a cocoon in the soil for the winter, pupating in the spring. Holes drilled by the female weevil also permit the entry of the brown rot fungus (nut drop). Control is effected by DDT, 62·5–100 g a.i./100 litres (HV), or 5 per cent DDT dust, 3 kg a.i./ha, in late May or early June and again 2–3 weeks later. Carbaryl, 75 g a.i./100 litres (HV), should also be effective.

Winter moths (A.L. 11). Caterpillars of the winter moth group (see under Apple, p. 189) may also infest nut trees, feeding on the foliage. Apply DDT, 50 g a.i./100 litres (HV), about the middle of April, when caterpillars are seen. As an alternative, carbaryl at 75 g a.i./100 litres (HV) should be effective, although there is limited experience on the use of alternatives to DDT on this crop.

9.7B   Diseases

Gloeosporium bud rot and **twig canker** (*Gloeosporium* spp.) of cobnut and filbert. In early spring, infected buds become brown and die. The causative fungus penetrates the twig causing a canker in which the fungus may overwinter. The disease is present in most plantations and is a common cause of crop loss.

Spray after harvest but before leaf fall with an organomercury compound using w.p. formulations at rates equivalent to 3·1 g mercury/100 litres (HV) or 70–87 g mercury/ha (LV).

9.8   Currant and gooseberry

9.8A   Pests

Annual control measures are required for aphids and blackcurrant gall mite, and locally for sawflies and capsid.

***Aphids** (A.L. 176). Several species overwinter as eggs on currant and gooseberry. The eggs hatch in spring, the aphids breed and in the summer disperse to various other host plants, returning in the autumn to lay eggs.

The currant-sowthistle aphid (*Hyperomyzus latucae*) causes the leaves of red and blackcurrants to curl downwards and stunts young growth; it disperses to sowthistle. The lettuce aphid (*Nasonovia ribis-nigri*), a darker green species, normally infests gooseberry and disperses to lettuce. *Cryptomyzus* are pale yellow green and delicate-looking species; the red-currant blister aphid (*C. ribis*) causes leaf blisters which are red on red and white currant and yellowish green on blackcurrant; the currant aphid (*C. galeopsidis*), which causes no obvious injury, has 3 forms, 2 of which remain respectively on black and redcurrants. The third, and redcurrant blister aphid, disperse to hedge woundwort and hemp nettle. The gooseberry aphid (*Aphis grossulariae*) is deep green in colour and causes severe curling and distortion of young leaves of currant and gooseberry; some of the aphids remain on the bushes all summer. The permanent currant aphid (*A. schneideri*) is blue-green in colour, and is responsible for similar damage on red and blackcurrant. Other species occur less commonly.

The currant root aphid (*Schizoneura ulmi*), covered with white waxy material like the woolly aphid, infests the roots of currant and gooseberry. It winters as eggs on elm, from which aphids disperse to currant and gooseberry in the summer.

Except with the currant root aphid, either (*a*) winter washes against the eggs or (*b*) spring sprays against the aphids can be used.

(*a*) Winter washes. These are effective if thoroughly applied. Use tar oil, 5 per cent (e.c.) or 5–6 per cent (stock emulsion) (HV), when buds are dormant, in December to January, not later, or DNOC-petroleum oil, 4·5 per cent (HV), up to bud break, January or early February.

Do not spray in windy or frosty weather or when the bushes are wet. The cultivar Raby Castle is liable to oil injury and should not be treated with winter washes.

(*b*) Spring sprays. Apply one of the compounds in Table 9.25 (p. 221) just before the first flowers open.

If endosulfan is used just before flowering for blackcurrant gall mite (see below) this will also give some control of leaf midge, aphids and partial control of common green capsid. Little is known of control measures for currant root aphid.

***Blackcurrant gall mite (Big bud mite)** (*Cecidophyopsis ribis*) (A.L. 277). The microscopic colourless mites feed and multiply inside the buds of blackcurrant, which become swollen and in the following spring many produce much reduced flower trusses or fail to open. Mites disperse from the swollen buds from March until the buds have dried up, which may be late in July, the main emergence period being from early April to the end of June, with a peak usually in May. Emergence is accelerated by rising temperatures, and mites may be dispersed to fresh sites by air currents or transportation by insects or rain.

The difficulty in chemical control is to protect the young buds against mite infestation for a sufficiently long period without leaving residues on the fruit. Endosulfan is now used almost exclusively to control gall mite, and this material will also give some control of leaf midge, aphids and capsid. Endosulfan may be used alone or in combination with benomyl, where this fungicide is used for disease control. Lime sulphur is less effective and its use has greatly declined. Lime sulphur sprays need to be used with caution, especially on sulphur-shy cultivars, as (particularly in south and east England) its application may lead to leaf scorch and some reduction in growth and yield.

(*a*) On fruiting bushes: apply endosulfan, 50 g a.i./100 litres (HV), at first open flower and again 3 weeks later. Thorough spraying is essential and for large bushes 2000–2500 litres spray/ha should be applied. Medium or LV applications will not be effective.

Benomyl is toxic to gall mite, although less so than endosulfan, and repeated applications of this fungicide greatly reduce mite populations. In lightly infested blackcurrant plantations where benomyl is used for a full disease control programme, only one spray of endosulfan need be used for mite control. This should be applied 3 or 4 weeks after first open flower. With the latter timing there is less risk to pollinating insects, but endosulfan should not be applied later than this and there must be an interval of 6 weeks between the last application and harvest.

If lime sulphur is used, this is applied at 0·5–1·0 per cent at first open flower and again 3 weeks later. A further spray 2 weeks later is permitted, providing there is little or no risk of leaf scorch. Sulphur sprays should be avoided on susceptible cultivars, e.g. Amos Black, Davidson's Eight, Edina, Goliath, Monarch, Victoria, Wellington XXX, Westwick Triumph.

(*b*) On nursery stock and non-fruiting bushes: apply endosulfan, 50 g a.i./100 litres (HV), when growth starts and follow with up to 4 further applications, each at an interval of 10–14 days.

Blackcurrant leaf midge (*Dasyneura tetensi*). There are three, sometimes four, generations a year; midges of the first appear in April to June, the second late June and July, and the third late July and August. Eggs are laid between folds of young leaves, and feeding by the white or orange maggots causes the leaves to become tightly twisted and folded. Shoot growth may be checked and lateral branches develop, but more

Table 9.25   Currant and gooseberry—spring sprays for aphid control

Compound	Rate HV g a.i./100 litres	LV g a.i./ha
Demeton-S-methyl	22	490
Dimethoate	30	680
Formothion	37	—
Malathion	75	1680
Oxydemeton-methyl	22	480
Phosphamidon*	20	—

* Only 1 application may be made during the season (see footnote ‡ Table 9.28).

important is the masking of symptoms of reversion in nursery stock. Pupation occurs in the soil. Some cultivars are more susceptible, e.g. Goliath, Seabrook's Black, Baldwin and Wellington XXX.

Endosulfan sprays for gall mite, *or* carbaryl (75 g a.i./100 litres HV) for winter moth caterpillars, *or* demeton-S-methyl *or* dimethoate sprays for aphids (Table 9.25), will also control midge, and the last 2 can be used later in the season if attacks are discovered.

Good control is possible by DDT sprays against emerging midges (50 g a.i./100 litres (HV), or 100 g a.i./100 litres (HV) for heavy infestations), when the first flowers are about to open and again 3 weeks later, but timing is difficult owing to the variability of times of emergence. DDT may also encourage development of red spider mite infestations by reducing predator populations.

Blackcurrant sawfly (*Nematus olfaciens*) (A.L. 30). There are 2 or more overlapping generations a year. Adults emerge from overwintering cocoons in the soil in May and June, and those of the later broods from mid-June to mid-September. Eggs are laid on the undersides of the leaves, especially near the middle of the bush. The green black-spotted caterpillars at first feed gregariously but later spread through the bush and may cause defoliation.

Bushes should be examined in May and early June for caterpillars; if present, spray in early June while the caterpillars are still small, with one of the treatments indicated in Table 9.26. It is important to see that the middles of the bushes are well sprayed. A second spray can be applied 2 weeks later if necessary.

Brown or **Peach scale** (*Parthenolecanium corni*) (A.L. 88). The full grown female scales are about 3–6 mm long, tortoise-shaped, and chestnut brown in colour. Eggs are laid beneath the scales in the summer, and in the autumn the young escape. They overwinter on the branches, often under loose bark, and after a short period of activity in the spring settle down, becoming adult in June. After laying eggs the female dies.

Table 9.26 Blackcurrant and gooseberry—sprays for control of blackcurrant sawfly and gooseberry sawfly

Compound(s)	Rate HV g a.i./100 litres	LV g a.i./ha	Suggested for use on*	
Azinphos-methyl	33	750	B	G
Azinphos-methyl†	22	500	B	G
Azinphos-methyl + demeton-S-methyl sulphone	37·5 + 11	840 + 250	B	G
Azinphos-methyl + demeton-S-methyl sulphone†	25 + 7·5	560 + 165	B	G
Carbaryl	75	—	B	G
Derris	4 (rotenone)	—	—	G
Fenitrothion	45	1050	B	G
Malathion	112	2530	—	G
Nicotine	50 ml (95–98%)	—	—	G
Tetrachlorvinphos	75	—	B	G

* For timing of sprays see text: B = Blackcurrant p. 221; G = Gooseberry, p. 224.
† Reduced rates stated may be used for later 'knock-down' sprays when most of the eggs have hatched.

Thorough application of winter washes can give adequate control: DNOC-petroleum oil, 4·5 per cent (HV), when buds are breaking, or tar oil, 5 per cent (e.c.) or 5–6 per cent (stock emulsion) (HV), when the buds are dormant, in December to January, not later. Do not treat in windy or frosty weather or when the bushes are wet. The redcurrant cultivar Raby Castle is liable to oil injury and should not be included.

If control is required out of the dormant season, spray with malathion, 187 g a.i./100 litres (HV), either in late summer or early spring. This spray is directed against the young 'crawlers' before they have settled down and formed protective scales; good cover of the wood and undersides of the leaves is essential. With heavy infestations a second spray 14 days later may be necessary.

*Clay-coloured weevil (*Otiorrhynchus singularis*) (A.L. 75)—see under Apple, p. 180.

***Common green capsid (*Lygocoris pabulinus*) (A.L. 154). This pest overwinters as eggs on currant, gooseberry and other shrubs, and on apple and pear. Hatching of the eggs appear to be a little later than with apple capsid and begins at the time of early pink bud of Bramley's Seedling apple, ending at about petal fall. The capsid nymphs puncture leaves and shoots; when adult, they fly away to various herbaceous plants and give rise to a second generation in July and August, the adults of which return to the woody host plants in the autumn to lay winter eggs.

Full bloom of Bramley's Seedling can be taken as a guide to optimum timing of sprays which are suggested in Table 9.27, p. 223.

Table 9.27 Currant, gooseberry, strawberry, raspberry and blackberry—sprays for common green capsid control

Compound	Rate HV g a.i./100 litres	Suggested for use on*				
DDT	100	S				
Dimethoate	30	C	G	S	R	B
Fenitrothion	45	C†	G†	S	R	B
Formothion	37	C	G	S	R	B
Malathion	112	C	G	S	R	B

* For timing of sprays see text: C = currant, p. 223; G = gooseberry, p. 223; S = strawberry, p. 236; R = raspberry, p. 230; B = blackberry, p. 230.
† Or 1050 g a.i./ha (LV).

Currant and gooseberry—apply when the first flowers are about to open on currant and again 3 weeks later if necessary, and at the end of flowering on gooseberry. Endosulfan used against gall mite on blackcurrant will give moderate control of capsid.

Strawberry—dimethoate, *or* formothion, *or* malathion used against aphids just before flowering should give some control of common green capsid, but sprays may have to be repeated either during or just after flowering to prevent damage to fruitlets. If it is necessary to spray during flowering use malathion under dull weather conditions and late in the day to minimize the risk of killing pollinating insects. Alternatively, DDT, *or* fenitrothion, applied just before flowering, should have sufficient persistence to give effective control.

Raspberry, blackberry, etc.—apply in late April or early May. On blackberry, endrin for 'redberry' control (see p. 231) would also control capsid.

*Currant clearwing moth (*Synanthedon salmachus*). A local pest mainly of blackcurrant, though redcurrant and gooseberry may also be infested. The fly-like moths, which are active in sunny weather, lay eggs in June and the caterpillars tunnel in the pith of the branches until full grown in the following April. Azinphos-methyl, 33 g a.i./100 litres (HV), after picking will give some control.

*Eelworm. The leaf and bud eelworm (*Aphelenchoides ritzemabosi*) may also attack blackcurrant, especially the cultivars Daniel's September Black and Westwick Choice. The eelworms live in the buds, which fail to open in the spring, and lengths of bare branch develop. Damage may be worse following a wet season. There is no effective recommendation for control by nematicides.

*Gooseberry bryobia (*Bryobia ribis*) (A.L. 305). At one time this mite was a serious pest in many gooseberry plantations, but it is now uncommon. Although it will feed on other plants, such as apple, it is only

Table 9.28   Sprays for control of gooseberry bryobia on gooseberry and red spider mite on currants, strawberry and gooseberry

Compound	Rate HV g a.i./100 litres
Demeton-S-methyl	22
Dichlorvos*	1120/ha
Dimethoate	22·5
Formothion	37
Oxydemeton-methyl	22
Phosphamidon	20 (blackcurrant)‡
Quinomethionate	25 (blackcurrant)†
	12·5 (gooseberry)†
Tetradifon	12·5

* Requires calm weather conditions with air temperature at least 15°C; useful if spray required near harvest (pre-harvest interval 24 hours).

† See text (p. 225); consult product label for susceptible cultivars.

‡ Only 1 application may be made during the season; a different compound will have to be chosen if phosphamidon has already been used to combat aphids.

known as a pest on gooseberry. It overwinters as eggs under loose bark; hatching begins near the beginning of March and continues well into April. There is only 1 generation, the laying of winter eggs beginning in May. The mites feed on the leaves in warm sunny conditions, retiring to the wood and under bud scales in cold weather, and also for moulting. On severely infested bushes the foliage becomes yellowed, and leaves may wither and drop.

Organophosphorus pesticides applied in warm weather in April when the mites are active on the leaves are effective. This treatment could be combined with aphid control by spraying just before first open flower (see Table 9.25 and p. 221). The use of these materials against aphids may account for the decline of bryobia mite as a pest. Should heavy infestations be allowed to develop, an additional spray may be needed earlier, as soon as mites are seen on the opening buds.

Suitable acaricides, which also control red spider mite, are listed in Table 9.28 p. 224).

A winter wash of 4·5 per cent DNOC-petroleum oil, thoroughly applied when the buds are breaking, is effective against the winter eggs of gooseberry bryobia.

***Gooseberry sawfly** (*Pteronidea ribesii*) (A.L. 30). The adult sawflies first appear in April and May. Eggs are laid on the undersides of the leaves, especially near the centre of the bush. The black-spotted green caterpillars feed together for a few days, later spreading through the bush which they may defoliate. When fully grown they are green with an orange patch behind the head and another near the tail and lack the black spots. There are 3 overlapping generations in the year. *P. leucotrocha* also occurs, and both species may attack red and whitecurrant but not black.

Examine the bushes in May and early June so that spraying can be started as soon as appreciable numbers of caterpillars are seen. The usual time for treatment is shortly after fruit set.

Use one of the treatments listed in Table 9.26 (p. 222), spraying the centres of the bushes thoroughly. As the eggs hatch over a period of 2–3 weeks, a second spray may sometimes be necessary.

With light infestations, when spraying can be delayed until most of the eggs have hatched, lower rates (Table 9.26) can be used giving a good knockdown of caterpillars present and 1 spray may often give adequate control.

Magpie moth (*Abraxas grossulariata*) (A.L. 30). The moth flies in July and August and lays its eggs on the leaves of currant and gooseberry. The caterpillar feeds for a while and then finds suitable shelter for hibernation, reappearing at bud burst to feed on the leaves, pupating in June. A small second brood may occur. It is now an uncommon pest.

As alternatives to DDT, azinphos-methyl at 33 g a.i./100 litres (HV); *or* carbaryl, 75 g a.i./100 litres (HV); *or* fenitrothion, 45 g a.i./100 litres (HV), should be effective, applied when the first flowers are about to open. By using the azinphos-methyl + demeton-S-methyl sulphone formulation, 25 + 7.5 g/100 litres (HV), control of both caterpillars and aphids should be obtained.

Red spider mite (*Tetranychus urticae*) (A.L. 226). This is the red spider mite common on many glasshouse and outdoor plants, often infesting soft fruit crops, usually blackcurrant and strawberry and especially in warm summers. Straw mulches in blackcurrant and strawberry plantations favour hibernating mites and may encourage infestations. The use of DDT may also encourage a build-up of populations by reducing predator numbers.

Where necessary spray just after flowering, or just after picking with one of the treatments in Table 9.28.

Quinomethionate will control red spider mite (and Bryobia mite) when applied against American gooseberry mildew. It can be used on most blackcurrant cultivars, but some (French Black, Seabrook's Black, Laxton's Giant, Noire de Bourgogne, Royal de Naples) are intolerant. It can be used on the gooseberry cultivars Careless, Leveller and Keepsake, but information about other cultivars is limited—consult product label for both crops.

Snails. Snails of several species (*Helix aspersa, Cepaea nemoralis, C. hortensis, Hygromia striolata*) sometimes cause trouble by finding their way into trays of picked fruit. They may do so by crawling from ground vegetation, and also by individuals 'roosting' in the bushes and falling into picking containers. There are indications that populations have increased in plantations where post-picking copper sprays for control of blackcurrant leaf spot have been discontinued.

An important source of trouble is the indiscriminate placing of fruit trays on the ground, allowing snails to enter them for shelter, and this can be avoided by placing the trays on a hard standing, or on polythene sheeting to separate them from the soil or vegetation.

Control of weeds by the efficient use of herbicides, particularly on headlands, is a sure way of depleting snail populations in blackcurrant plantations.

*Winter moths (A.L. 11). Caterpillars of the winter moth group (see under Apple, p. 189) may attack currant. Control as for magpie moth (p. 225).

9.8B   Diseases

***American gooseberry mildew** of gooseberry (*Sphaerotheca morsuvae*) (A.L. 273). The white powdery (sporing) fungal growth is seen on young leaves, fruits and shoots. In some seasons the disease becomes epidemic. In late summer and autumn, the fungal growth becomes a brown felt-like layer containing black spore cases (perithecia). Some fall to the ground and spores are ejected in the following spring and so start a new cycle.

Control with a treatment selected from Table 9.29 (p. 227). Benomyl *or* carbendazim *or* thiophanate-methyl should be applied at first open flower and on 2 subsequent occasions at 14-day intervals. Sulphur *or* dinocap, *or* quinomethionate should be sprayed on the foliage (1) just before the flowers open, (2) at fruit set and (3) fruit swelling, 14–21 days later but do not use sulphur on the varieties Careless, Leveller, Lord Derby, Early Sulphur, Golden Drop, Roaring Lion and Yellow Rough which are sulphur-sensitive. Further, sulphur must not be used when fruit is for canning.

Infected shoot tips should be removed as soon as the wood is ripe in late August or September, and excessive nitrogenous manuring should be avoided.

***Blackcurrant leaf spot** (*Pseudopeziza ribis*). From about May onwards, brown spots or patches appear on the leaves; on these patches spores (conidia) are formed which spread infection through the plantation. When the disease is unchecked, premature defoliation occurs, weakening subsequent growth and reducing the yield. The fungus overwinters in dead leaves and spores (ascospores) from these restart the cycle the following spring.

Spray with one of the treatments indicated in Table 9.29 (p. 227). Benomyl, carbendazim, copper, dodine, mancozeb mixtures, manganese/zinc dithiocarbamate mixtures, thiabendazole, thiophanate-methyl and zineb should be applied from grape stage at 14-day intervals, drazoxolon should be applied in early May and repeated at 14-day intervals until 4 weeks before harvest; quinomethionate should be applied

Table 9.29 Blackcurrant and gooseberry—sprays for control of blackcurrant leaf spot, and *sphaerotheca mors-uvae* (American gooseberry mildew and powdery mildew of blackcurrant)

Compound(s)	Rate HV a.i.	Suggested for control of*		
Benomyl	560 g/ha†	AGMG	BLS	PMB
Carbendazim	550–660 g/ha†	AGMG	BLS	PMB
Dinocap	25 g/100 litres	AGMG‡	—	PMB
Dodine	32·5 g/100 litres	—	BLS	—
Drazoxolon	100 g/100 litres	—	BLS	PMB
Mancozeb mixtures	see	—	BLS	—
Manganese/zinc	product			
dithiocarbamate	labels	—	BLS	—
Quinomethionate	25 g/100 litres	—	BLS	PMB
	12·5 g/100 litres	AEMG	—	—
Sulphur (w.p. or colloidal)	150–200 g/100 litres	AGMG§	—	—
Thiabendazole	240 g/100 litres	—	BLS	PMB
Thiophanate-methyl	1·1 kg/ha†	AGMG	BLS	PMB
Zineb	140 g/100 litres	—	BLS	—

* See text for timing of sprays. AGMG = American gooseberry mildew of Gooseberry, p. 226; BLS = Blackcurrant leaf spot, p. 226; PMB = Powdery Mildew of Blackcurrant, p. 228.
† In not less than 1000 litres/ha.
‡ Or 560 g a.i./ha (LV).
§ See text (p. 226) and product label for list of susceptible varieties.

at intervals commencing mid to late May or as soon as mildew appears on the crop.

One or more sprays of these chemicals applied immediately after harvest will improve control.

If the fruit is to be processed, it is important to consult the processors before applying any spray.

Blackcurrant rust (*Cronartium ribicola*). This rust spends part of its life cycle on blackcurrants and part on five-needled pines, particularly the Weymouth pine (*Pinus strobus*). Spores (aecidiospores) from pine trees infect nearby currant bushes, the fungus appearing as yellow outgrowths on the underside of the leaf in early summer. From these outgrowths, the spores (uredospores) are produced to spread the disease within the currant plantation. Later, yet other kinds of spores (teleutospores and basidiospores) are produced on the leaves and these cause reinfection of pines.

Where rust is troublesome, the bushes should be sprayed post harvest with copper prepared from w.p. formulations containing copper oxychloride, copper oxide or Bordeaux mixture to contain 150–250 g copper/100 litres (HV) or with zineb w.p. 140 g a.i./100 litres (HV) or with mancozeb with zineb or with manganese/zinc dithiocarbamate according to the maker's instructions.

*European gooseberry mildew (*Microsphaera grossulariae*) is much less serious than American gooseberry mildew (see p. 227). It is seen as a delicate sporing mould, mainly on the upperside of the leaf; rarely on berries. Overwintering spore cases fall to ground with leaves and restart cycle by ejecting spores (ascospores) the following spring. Spray with one of the sulphur preparations or with dinocap or quinomethionate as listed for American gooseberry mildew (Table 9.29, p. 227).

**Gooseberry cluster cup rust (*Puccinia pringsheimiana*) (A.L. 198). This is seen in early summer as dark red or orange coloured blister-like swellings on the leaves and fruits. Later the blisters become covered with minute pits with raised, reflexed, yellow margins (cluster cups). The spores which develop within the pits can infect sedges (*Carex* spp.) but not gooseberries. On sedges, the disease appears in a different form but in the following summer spores are developed which infect gooseberries and so start the cycle again.

A single application of a copper-containing spray should be applied about 14 days before flowering, using Bordeaux mixture w.p., 150–200 g copper/100 litres (HV).

Copper containing fungicides should not be used on the varieties Careless, Early Sulphur, Freedom, Golden Drop, Leveller, Lord Derby, Roaring Lion and Yellow Rough. With canning and other forms of processing the manufacturers should be consulted before any sprays are used.

Because the disease cannot occur without the presence of infected sedges, those in the neighbourhood should be destroyed.

**Powdery mildew of blackcurrant (American gooseberry mildew) (*Sphaerotheca mors-uvae*) (A.L. 273) is usually first seen in May or June as a white powdery (sporing) fungal growth on both surfaces of the leaves and on the fruit. In late summer and autumn, the fungal growth becomes a felt-like layer containing black spore cases (perithecia) which overwinters on shoots and fallen leaves; it is possible that these can start infection the following year.

Spray with one of the treatments indicated in Table 9.29 (p. 227). Dinocap or quinomethionate should be applied at intervals commencing mid to late May or as soon as mildew appears in the crop. Benomyl, carbendazim, thiabendazole and thiophanate-methyl should be applied at grape stage and on 3 subsequent occasions at 14-day intervals. Drazoxolon should be applied in early May and repeated at 14-day intervals until 4 weeks before harvest.

Sprays of the above fungicides should also be applied immediately after picking and repeated if necessary at 10–14 day intervals.

Several of the treatments applied for American gooseberry mildew will also give control of blackcurrant leaf spot (see Table 9.29, p. 227).

Shoots with dead or heavily infected tips should be pruned out. If the fruit is to be processed it is important to consult the processors before any spray is applied.

9.9   Blackberry, loganberry and raspberry

9.9A   Pests

Annual sprays are required for control of raspberry beetle and aphids, and on susceptible cultivars for cane midge.

***Aphids.** There are several species, overwintering as eggs on the canes. The rubus aphid (*Amphorophora rubi*), which begins hatching in March, is a large pale green aphid and causes slight leaf curling, dispersing between *Rubus* spp. in the summer. The raspberry aphid (*Aphis idaei*) is smaller and grey green. It causes pronounced curling of young leaves, and infests fruiting laterals, dispersing to raspberry and raspberry hybrids in June and July. Both are important as virus vectors.

The blackberry aphid (*Macrosiphum fragariae*) hatches in February and March, feeding on the tips of the buds, and heavy infestations cause severe leaf curling. Dispersal to grasses occurs in May and June. Only blackberry is severely attacked.

For control either (*a*) winter washes or (*b*) summer sprays may be used.

(*a*) The winter washes are: tar oil, 5 per cent (e.c.) *or* 5–6 per cent (stock emulsion) (HV), when the buds are dormant *or* DNOC-petroleum oil, 4·5 per cent (HV), up to bud break.

(*b*) For spring sprays, use one of the treatments given in Table 9.30 in late April.

***Bramble shoot moth** (*Notocelia uddmanniana*). Moths appear from late June into August, laying eggs on the leaves and shoots of blackberry, loganberry and phenomenal berry. The caterpillars which hatch hibernate in the second instar, reappearing in April to feed on the blossom and web together young leaves on the shoots. Pupation occurs in the webbed leaves in June.

The caterpillars can be controlled by spraying at bud burst with azinphos-methyl, 16·5 g a.i./100 litres (HV); *or* azinphos-methyl +

Table 9.30   **Blackberry, loganberry and raspberry—sprays for aphid control**

Compound	Rate HV g a.i./100 litres	LV g a.i./ha
Demeton-S-methyl	22	490
Dimethoate*	30	—
Formothion*	37	—
Malathion*	75	—
Oxydemeton-methyl	22	480

* Probably also effective LV.

demeton-S-methyl sulphone, 25 + 7·5 g a.i./100 litres (HV), which will control aphids as well as caterpillars; *or* carbaryl, 75 g a.i./100 litres (HV).

*Clay-coloured weevil (*Otiorrhynchus singularis*) (A.L. 57)—see under Apple, p. 180.

*Common green capsid (*Lygocoris pabulinus*) (A.L. 154)—see under Currant, p. 222.

***Raspberry beetle (*Byturus tomentosus*) (A.L. 164). This is the most important pest of cane fruits. The beetles hibernate in the soil, emerging in April and May, and are active in sunny weather frequenting flowers of apple, hawthorn, raspberry, etc. Eggs are laid in the flowers of raspberry and other *Rubus* spp., hatching in 10–12 days. The larvae feed on the surface of the fruit, and as it begins to ripen tunnel into the plug. They may leave one fruit and attack a neighbouring one. When full grown they pupate in the soil. The adult beetles will also feed on the flower buds and tips of young canes, especially on raspberry.

To control this pest spray at the times given below with azinphos-methyl, 310 g a.i. in 100 litres/ha (LV); *or* azinphos-methyl + demeton-S-methyl sulphone, 280 + 84 g a.i. in at least 225 litres/ha (LV); *or* derris, 2 g rotenone/100 litres (HV); *or* fenitrothion, 45 g a.i./100 litres (HV); *or* malathion, 112 g a.i./100 litres (HV); *or* tetrachlorvinphos, 75 g a.i./100 litres (HV).

Raspberry—at first pink fruit. One spray at this time is often adequate, though to avoid slight damage to the basal druplets of the earliest berries, growers of high quality dessert fruit sometimes find an additional earlier spray worthwhile when about 80 per cent of the blossom is over.

Loganberry—when about 80 per cent of the blossom is over and again when the first fruits are colouring, usually about 2 weeks later.

Blackberry—just before first open flower.

Where infestations of adult beetles are sufficiently heavy to cause serious damage to flower buds, especially on raspberry, spray also at the white bud stage.

Avoid spraying open flowers so far as possible because of the danger to bees; where it is necessary to apply insecticides to open flowers, spray under dull conditions and late in the evening when fewest bees are active, and use either derris, malathion (e.c. formulations) or tetrachlorvinphos.

Observe minimum intervals between last application and harvest; consult product labels.

**Raspberry cane midge (*Thomasiniana theobaldi*). Midges normally emerge in late April and early May, though not until 2–3 weeks later in cold springs, and lay eggs on the young spawn of raspberry, the females seeking breaks in the rind, such as growth splits, for the purpose. The pink larvae feed under the rind, and the damaged tissues are susceptible to fungal attack which may lead eventually to the death of the cane ('cane blight'). Two further generations of midges appear in July and August, but these overlap considerably and a partial fourth brood may occur. The later generations are larger and considerable damage may result. The winter is passed as larvae in cocoons in the soil, pupating in the spring. Cultivars such as Bath's Perfection, Malling Enterprise, Malling Jewel and Malling Promise, the rind of which splits freely, are most susceptible. Norfolk Giant, Malling Landmark, and Exploit are lightly attacked, while Lloyd George is intermediate.

The standard control measure is to spray the young spawn growth with gamma-HCH, 12·5 g a.i./100 litres (HV), with added wetter, to kill emerging and egg-laying midges. The spray is applied in the first week of May, when most spawn growth on Malling Promise is 25–30 cm high, and repeated 2 weeks later. In cold springs, sprays should be delayed a week or two.

In recent years the insect has chiefly been a pest of Malling Jewel occurring in a few plantations. With this cultivar, spawn growth and development of growth splits occur relatively later, and 3 sprays at 10-day intervals are suggested. In all cases the spawn and soil must be thoroughly wetted; it is not necessary to spray the fruiting canes.

Annual sprays should not be necessary on the lightly attacked cultivars.

In experiments fenitrothion, 45 g a.i./100 litres (HV), has given very good control of young midge larvae in growth splits and is a promising alternative to gamma-HCH.

Do not spray with gamma-HCH during or after flowering because of the risk of tainting fruit. See also, above, regarding danger to bees.

*Raspberry moth (*Lampronia rubiella*) (A.L. 66). The moths fly in May and June and eggs are laid in the flowers. The young caterpillars feed in the plug, causing little damage, and leave as the fruit ripens to spin cocoons for the winter in the soil, in crevices in canes, stakes, etc. In April they emerge and tunnel in the shoots which shrivel and die.

To control, spray with 8 per cent tar oil in late February or in March, with at least 1700 litres/ha, directed to wet thoroughly the soil, and bases of canes and stakes.

As a spring spray to kill emerging caterpillars, azinphos-methyl in early April, at 16·5 g a.i./100 litres (HV) *or* carbaryl, 75 g a.i./100 litres (HV), should be effective.

**'Redberry'. In this condition of blackberries, which appears to be associated with the gall mite *Aceria essigi,* some of the berries fail to ripen normally, only some of the druplets turning black while the remainder stay red. The mites overwinter as eggs, mature and young stages under the shoot scales near the buds; in spring and summer they move on to the expanding shoots, and later into the flowers and fruit.

A useful reduction of the incidence of 'redberry' can be obtained by spraying with endrin, but this acaricide may only be used on blackberry before flowering. Apply at 37·5 g a.i./100 litres (HV), with added wetter, before flowering in late April or early May and again 3–4 weeks later. Annual sprays should not be necessary.

Red spider mite (*Tetranychus urticae*) (A.L. 226)—see under Currant, p. 225.

Strawberry rhynchites (*Caenorhinus germanicus*). This pest may also attack raspberry and blackberry; the weevils girdle the tips of new growth and the stems of blossom trusses. Azinphos-methyl, 16·5 g a.i./100 litres (HV); *or* carbaryl, 75 g a.i./100 litres (HV); *or* malathion, 112 g a.i./100 litres (HV), should be effective, applied when damage is first seen.

9.9B   Diseases

Blackberry purple blotch (*Septocyta ramealis*). New infections are seen in early spring as small, light green blotches usually near the base of the canes. These enlarge, coalesce and quickly turn purple. In severe attacks, infections occur along the length of the canes and the canes may be killed. Spores of the fungus are released from the purple blotches spreading the disease during the growing season.

Effective control of the disease has been obtained by applying sprays equivalent to 15 g copper/100 litres (HV) in mid-May and on 2 subsequent occasions at 10-day intervals followed by sprays of benomyl, 12·5 g a.i./100 litres (HV) in mid-June and mid-July, followed by another copper spray immediately after harvest.

Alternatively, apply copper at the above rate immediately before blossom, at fruit set, immediately after harvest and again 14 days later. It is important to direct the spray at the young growing canes and ensure good cover.

If fruit is to be used for processing, the processor should be consulted before applying any spray.

Loganberry cane spot (*Elsinoe veneta*). Spotting of stems, leaves, flower stalks and fruits occurs showing symptoms like those caused on raspberry by the same fungus (see below).

Control measures are listed in Table 9.31, p. 223. The times of application are (1) pre-blossom, (2) as soon as the fruit has set.

***Raspberry cane spot** (*Elsinoe veneta*). New infections are seen as small purple spots on the young canes from early June onwards. The spots enlarge later becoming elliptical up to 6 mm long having a light grey centre with a purple border; the centres of the spots split leaving cavities which give the fruiting canes a rough and cracked appearance. Where spots have coalesced, the tips of canes may be killed. Leaves and fruits are sometimes attacked; the latter become distorted.

Spores (conidia) of the cane spot fungus are released from the spots, so spreading the disease during the growing season. The fungus overwinters in the canes and produces a second type of spore (ascospore) in spring to restart the cycle.

In raspberry plantations which are not severely affected by cane spot, an application of benomyl *or* copper *or* dichlofluanid *or* lime sulphur *or*

Table 9.31 Blackberry, loganberry and raspberry—sprays for control of cane spot on loganberry and raspberry; raspberry spur blight; and grey mould on blackberry, loganberry and raspberry

Compound	Rate at Bud burst	White bud	Suggested for use against*
Benomyl	560 g a.i. in at least 1120 litres/ha	15 g copper/100 litres/ha	LCS RCS RSB GM
Bordeaux (w.p.)	20 g copper/100 litres (HV)	15 g copper/100 litres (HV)	LCS RCS RSB
Carbendazim	300 g a.i. in at least 2000 litres/ha		RCS RSB GM
Copper oxychloride	20 g copper/100 litres (HV)	15 g copper/100 litres (HV)†	LCS RCS RSB
Dichlofluanid	2·5 kg a.i./ha in at least 2000 litres/ha		LCS RCS‡ RSB GM‡
Lime sulphur (with wetter)	5 litres/100 litres (HV)	2·5 litres/100 litres (HV)	RCS
Thiophanate-methyl	1·1 kg/ha in at least 1000 litres/ha		LCS RCS RSB GM
Thiram§	16 g/100 litres (HV)	16 g/100 litres (HV)	LCS RCS RSB GM

* For timing see individual crop/disease combination and product label; LCS = Loganberry cane spot, p. 232; RCS = Raspberry cane spot, p. 232; RSB = Raspberry spur blight, p. 234; GM = Grey mould, p. 234.
† Add cotton seed oil as for Cherry-Bacterial canker, p. 212.
‡ Not more than 7 sprays per season; see product label.
§ Do not use when fruit is to be canned or quick frozen.

thiram (Table 9.31, p. 233) should be made when the buds are not more than 12·5 mm long and subsequently at manufacturer's recommendations.

In plantations where cane spot is established, spray with benomyl *or* dichlofluanid *or* thiram at fortnightly intervals from bud burst until pink bud.

Badly cankered and spotted canes should be cut out and burnt.

***Raspberry spur blight** (*Didymella applanata*). Dark purple blotches arise at the nodes around the leaf petiole bases. These extend longitudinally and can coalesce to form long discoloured lengths of cane. Buds arising at infected nodes are weakened or killed. The leaves are frequently attacked.

Spores (conidia) are released from spore sacs on the discoloured areas so spreading the disease during the season. The fungus overwinters within the canes and produces another type of spore (ascospore) in the spring to restart the cycle of infection.

Sprays suggested in Table 9.31, p. 233, should be applied at the times of application given for cane spot control.

Seriously affected canes should be cut out and burnt.

***Grey mould** (*Botrytis cinerea*). The fungus invades the floral parts and not infrequently attacks the canes. Under moist conditions the fruits become infected, the fungus producing a grey furry mould. Large black bodies, the sclerotia, develop in the bark of the canes and these fall on to the soil. Under suitable conditions the sclerotia produce spores which spread the disease.

Spray with dichlofluanid 4 times at 10-day intervals starting at early flower or with benomyl, thiophanate-methyl *or* thiram at 14-day intervals from early flower. Rates are given in Table 9.31, p. 233. Alternatively dichlofluanid can be applied in a 2 or 4 spray programme at rates and times according to the makers' instructions. Not more than 7 sprays of dichlofluanid should be applied to raspberries.

Strains of *Botrytis cinerea* tolerant to benomyl, carbendazim and thiophanate-methyl have been shown to be present in some commercial crops. Where these strains occur, satisfactory control may not be achieved and alternative fungicides should be used. If fruit is to be used for processing, the processor should be consulted before applying any spray.

9.10   Strawberry

9.10A   Pests

***Aphids.** Several species are found on strawberry, but the strawberry aphid (*Chaetosiphon fragaefolii*) is the most important as it is the main vector of virus diseases. It is a creamy white aphid with knobbed hairs on its back. It occurs on the plants all the year, with peak numbers in early

Table 9.32   Strawberry—treatments* for aphid control

Compound	Rate for HV g a.i.	LV g a.i./ha	Suggested spray volume litres/ha
Demephion†	22·5/100 litres	330	
Demeton-S-methyl†	22/100 litres	325	at least 200 (LV)
Derris†	4 (rotenone)	—	
Dichlorvos†‡	1050/ha	—	at least 1125 (HV)
Dimethoate†	335/ha	—	at least 225
Dioxathion	60/100 litres	—	
Disulfoton (granules)†§	1120/ha	—	
Formothion†	37/100 litres	—	
Malathion†	112/100 litres	—	
Mevinphos†	140/ha	—	2250
Nicotine	50 ml (95–98%)/100 litres	—	
Oxydemeton-methyl†	22/100 litres	320	at least 200 (LV)
Phorate (granules)†§	1700/ha	—	
Phosphamidon†	20/100 litres	—	
Pirimicarb	2·5/100 litres	—	
Thiometon†	25/100 litres	—	

* Do not spray open blooms.
† May also control red spider mite (see p. 236).
‡ Calm weather conditions and an air temperature of at least 15°C give best results.
§ Do not use in glass- or plastic-houses.

summer on established fruiting plants, and in September on first year plants. Winged forms appear in May and June, dispersing to other strawberry plants, and small numbers also occur in October to December.

The shallot aphid (*Myzus ascalonicus*), a light greenish brown species, sometimes colonizes the plants in autumn and may cause severe damage in the following spring, especially after a mild winter, distorting leaves and blossom and destroying the crop.

The following aphicides (Table 9.32) are included in the Agricultural Chemicals Approval Scheme for use on strawberry; the organo-phosphorus compounds are more effective than derris, nicotine or pirimicarb. Observe the statutory minimum intervals between last application and harvest; note these may differ for crops grown in glass- or plastic-houses.

Treat fruiting plants shortly before flowering, when there is a good growth of foliage, in mid or late April, or early April for plants under cloches. If red spider mites are also present, HV sprays of suitable organophosphorus compounds (Table 9.32) should be used if control of both pests is required from a single pesticide.

Treat runner beds in late May and again in early July.

*Chafer grubs (A.L. 235). The large white grubs attack the roots, causing wilting and death of plants. Damage usually only occurs where crops are planted after pasture (see also p. 120). If chafer grubs are seen during cultivation, work gamma-HCH into the soil before planting, as for wireworm control (see p. 241).

*Common green capsid (*Lygocoris pabulinus*) (A.L. 154)—see under Currant, Table 9.27, p. 223.

**Cutworms (*Agrotis* spp., etc.) (A.L. 225). In some seasons these plump greenish brown caterpillars feed on the roots and crowns, and may eat away the growing point. They feed at night and are are most likely to be troublesome from late July to September (see also p. 138).

To control, drench with DDT, 50 g a.i./100 litres (as emulsion), at 150 ml/plant. Information is lacking on the value of alternative insecticides.

*Eelworms (A.L. 414). The leaf and bud eelworms (*Aphelenchoides fragariae* and *A. ritzemabosi*) feed in the crowns and in the folds of young unopened leaflets. Damaged leaves may be puckered, some showing on expansion a pale grey or silver patch near the base of the midrib. The main crown may become blind, secondary crowns developing. The stem eelworm (*Ditylenchus dipsaci*) causes a marked corrugation of leaves, with shortening and thickening of the stalks of leaves and blossom trusses. Strawberries are attacked by the stem eelworm races affecting onions, oats, red clover, narcissus, parsnips and other vegetables.

Runners should only be taken from healthy crops. If need be, control may be obtained by hot water treatment: 46°C for 10 minutes for leaf and bud eelworms, 46°C for 7 minutes for stem eelworm. Hot water treatment needs to be undertaken with great care to avoid killing the plants. In experiments, drenching established plants (already infested with stem eelworm) with nematicides (e.g. parathion or thionazin) has given disappointing results, although good results have been achieved using aldicarb granules.

For control of *Xiphinema diversicaudatum*, vector of arabis mosaic, see p. 242.

*Leafhoppers (*Aphrodes* and *Euscelis* spp.). These insects are vectors of green petal virus disease, which may be carried from clover and certain weeds.

Their control has not been studied and there is no standard recommendation but spraying of valuable runner beds in areas where the disease occurs with dimethoate, 30 g a.i./100 litres (HV), *or* malathion, 112 g a.i./100 litres (HV), in July and twice more at 14-day intervals is suggested.

***Red spider mite (*Tetranychus urticae*) (A.L. 226). This is the glasshouse red spider mite which also occurs on a variety of outdoor plants, including strawberry. It overwinters as adult females on the undersides of the strawberry leaves, in the soil and other shelter. They become active in April, feeding on the foliage. Eggs are laid on the lower leaf surfaces and up to 7 generations may follow during the summer.

Damaging infestations chiefly occur in warm summers but tend to be more frequent in some intensive strawberry growing areas. They are more likely to occur in the second and third years of fruiting beds, and in outdoor crops do not usually become severe until after the crop has been picked. In crops protected by cloches or plastic tunnels severe infestations are more common and may occur during flowering and fruiting. Some cultivars, e.g. Cambridge Favourite, are particularly susceptible.

To reduce the risks of infestations arising it is important to plant runners as free as possible of red spider mites.

Chemical control has become complicated in recent years by the widespread development of resistance to organophosphorus compounds and in some areas also to other materials such as dicofol or tetradifon. Choice of acaricide on any particular farm must be guided by previous experience; where organophosphorus resistance has not developed, such pesticides used against aphids will also control red spider mite, but where resistance to organophosphorus acaricides is known to occur alternative compounds must be chosen.

Table 9.33   Strawberry—treatments* for red spider mite control

Compound	Rate (HV) g a.i.	Suggested spray volume litres/ha
1. Organophosphorus compounds		
Demeton-S-methyl	22/100 litres	2250
Dichlorvos†	600/ha	1000
Dimethoate	335/ha	220–1100
Dioxathion	60/100 litres	1320
Disulfoton (granules)‡	1100/ha	—
Oxydemeton-methyl	22/100 litres	2250
Phosphamidon	20/100 litres	2250
Triazophos	40/100 litres	2250
2. Other acaricides		
Dicofol§	40/100 litres	2250
Quinomethionate‖	12·5/100 litres	2250
Summer petroleum oil	0·8/100 litres	2250
Tetradifon	12·5/100 litres	2250

* Avoid spraying open blooms because of danger to pollinating insects.

† Non-persistent so useful if a spray is required near harvest (pre-harvest interval, 24 hours); best results are obtained with HV sprays under calm weather conditions and with air temperatures of 15°C or over.

‡ When used for aphid control, see Table 9.32 (p. 235). Do not use in glass- or plastic-houses.

§ Gives best results with air temperature of 15°C or over but should not be used during blossom as it may damage flowers.

‖ May cause slight leaf scorch but this will have little or no effect on the growth of the crop.

Acaricides suitable for use on strawberry are listed in Table 9.33, p. 237.

Crops in the open. Where organophosphorus compounds are effective, rely on those applied for aphid control (Table 9.32, p. 235) just before flowering; where resistance to these compounds occurs include an acaricide from one of the other groups if infestations warrant it. If infestations develop, a second spray after picking and mowing or burning off, and when new foliage growth has appeared, will help to decrease mite populations occurring in the following spring. Endosulfan used at this time against strawberry mite will also give some control of red spider mites.

Crops under glass or plastic. Routine sprays are especially important with these crops; one should be applied in early April with a second 10–14 days later, and where necessary a third after (not during) flowering. All sprays should be allowed to dry before replacing cloches or plastic tunnels to reduce risk of chemical damage. Where resistance to organophosphorus compounds occurs, dicofol, which is active against all mite stages, is one of the most effective alternatives.

A further spray after picking and mowing or burning off, when new growth has appeared, is particularly important with protected crops to minimize subsequent infestations, especially where a second crop is taken in the same year.

Runner beds. Routine sprays of organophosphorus pesticides against aphids will also control red spider mites, but a close watch should be kept for signs of failure. If alternative acaricides appear necessary give at least 2 sprays, in early July and 2–3 weeks later, or at other times if infestations are discovered.

**Strawberry blossom weevil (*Anthonomus rubi*). Adult weevils emerge from hibernation in April and May. The female lays eggs in the unopened flower buds, then partially severs the flower stalk below. The larvae develop inside the flower bud, adult weevils emerging in July. The damage is often less serious than it looks; slight thinning of the blossom may in fact result in larger fruit.

For control apply azinphos-methyl, 375 g a.i./ha (HV); *or* DDT (as w.p. or slurry), 100 g a.i./100 litres (HV), or DDT as 5 per cent dust at 2·8 kg a.i./ha; *or* malathion, 112 g a.i./100 litres (HV), as soon as the first damage is seen and again about 10 days later if necessary. Damage is usually noticed after flowering has begun, and to minimize the danger to pollinating insects treatments should be made in dull weather or late in the evening.

***Strawberry mite (*Tarsonemus fragariae*) (A.L. 584). The minute mites feed amongst the young folded leaflets, which may remain undersized, wrinkled and turn brown, and with heavy infestations plants are stunted. The mites breed and overwinter on the plants. The pest occurs sporadically, and is usually more in evidence in hot dry summers. Except sometimes with crops under cloches, attacks do not usually become severe until after cropping, and are more likely to occur in the second and subsequent years in fruiting beds. It is important to plant runners as free of these mites as possible.

Where the pest occurs, spray thoroughly after picking, when new growth appears after mowing or burning off, with dicofol, 40 g a.i./100 litres (HV), *or* endosulfan, 50 g a.i./100 litres (HV). Endosulfan gives quicker results than dicofol and is to be preferred with heavy infestations. Good cover is important and drenching sprays should be used with either of these acaricides. If a second crop is to be taken in the autumn, endosulfan may be applied within 1 week of removal, by mowing, of the old crop. Otherwise, only dicofol can be used before harvest. Dicofol is the only suitable acaricide which can be used if an infestation is discovered shortly before picking, e.g. on crops under cloches; it can be applied up to 7 days before picking but should not be used during blossom as the flowers may be damaged.

Strawberry rhynchites (*Caenorhinus germanicus*). The weevils bite through the stems of leaves and fruit trusses in May and June, causing them to wither. For control, use DDT, 50–62·5 g a.i./100 litres (HV) or 1400 g a.i./ha (LV), *or* 5 per cent DDT dust at 2·25–3·35 kg a.i./ha, in mid to late April at first sign of damage. Repeat after 10 days if required. Information is lacking on alternatives to DDT but it is probable that malathion, 112 g a.i./100 litres (HV), would be effective.

If an insecticide is applied during flowering choose a dull day or late evening to minimize the danger to pollinating insects.

Observe the minimum interval between last application and harvest: DDT, 2 weeks; malathion, 24 hours (4 days to avoid taint, 7 days if crop for processing).

Strawberry seed beetle (*Harpalus rufipes*). The adult beetles overwinter under rough vegetation, and may enter strawberry fields when the fruit is forming. They bite the seeds from the fruit, spoiling its appearance and market value. Eggs are laid in summer in the soil in rough weedy places, and the larval stages feed upon seeds, with a partiality for fat-hen, pupating the following spring and giving rise to further adults.

Other ground beetles, e.g. *Pterostichus* (*Feronia*) spp., eat holes in the fruit which resemble slug damage, but are less important.

Surveys have shown that strawberry seed beetle is a local and sporadic pest and that much of the damage attributed to it is in fact caused by linnets. Linnets pick the seeds out cleanly, usually only from the upper surfaces of exposed fruits, and cause little damage to the flesh. The beetles also remove the seed or bite it open on the spot and in so doing cause severe injury to the adjoining flesh of the fruit; they may damage any part of the fruit but more often the lower surface next to the ground.

It is not possible to predict in which crops damage may occur though it is more likely in crops that were allowed to become weedy in the previous season or are surrounded by rough vegetation. Maiden crops also seem more likely to be attacked.

Damage can be considerably reduced by using poison baits, either methiocarb 4 per cent slug pellets broadcast over the rows at 220 g a.i./ha, or a bait made with 840 g a.i. malathion/11·2 litres water/45 kg crushed oats/ha. The oat bait is prepared by applying the diluted malathion through a sprayer or can with a fine rose to the heap of oats, which is turned until the mass is evenly moistened. The bait is placed in small handfuls between the plants or between the rows at about 1 m intervals, scuffing aside the straw if necessary. It remains effective for nearly a week in damp weather and 10–12 days in dry weather. Wear rubber gloves while handling the bait and avoid dropping any on the fruit.

Either bait may be applied early in the fruiting period as soon as damage is noticed (methiocarb pellets at least 7 days before picking). An earlier application, immediately before strawing, may also be useful, especially in fields where the beetles are known to occur soon after strawing, but a second application may be needed as it is unlikely that the baits would remain effective long enough to give protection against beetles entering during fruiting.

Strawberry tortrix moth (*Acleris comariana*). The moth has 2 broods, in June to July and August to September. It overwinters as eggs on the plants, these hatch in April to early May and the caterpillars feed on leaves and flowers. Damage is usually visible before the flowers open. Caterpillars of other species may sometimes occur, e.g. *Cnephasia interjectana*, *Olethreutes lacunana* and *Clepsis spectrana*. Caterpillars of these species overwinter on the plants and can cause damage quite early in the season, especially on protected crops.

Where control is needed spray with a treatment selected from Table 9.34 (p. 240) before flowering, as soon as caterpillar damage is seen. A second spray may be needed in late summer after picking, against second generation caterpillars. If a spray during flowering is found necessary, choose a dull day or late evening to minimize danger to pollinating insects.

Table 9.34   Strawberry—sprays for strawberry tortrix moth control

Compound(s)	Rate g a.i.	Suggested spray volume litres/ha
Azinphos-methyl	375/ha	at least 500
Azinphos-methyl +	25 + 7·5/100 litres	2250 HV
demeton-S-methyl sulphone	280 + 84/ha	at least 500 LV
DDT	100/100 litres	2250 HV
	2240/ha	at least 500 LV
Fenitrothion	550/ha	at least 500
Malathion	112/100 litres	
Mevinphos	280/ha	2250 (HV)
Trichlorphon	800/ha	at least 500

*Swift moths (*Hepialus* spp.) (A.L. 160). The whitish caterpillars live in the soil and are occasionally injurious, biting through roots and excavating cavities in the crowns. They are more likely to occur in grass or very weedy land, and may also occur in hop gardens. If the caterpillars are seen during preparation of land for strawberries, work gamma-HCH into the soil before planting as for wireworm control (see p. 241). With attacks on established crops drenching with DDT, 50 g a.i./100 litres (as emulsion) at 150 ml/plant, is effective. There is no information on alternative insecticides.

*Vine weevil (*Otiorrhynchus sulcatus*)—see Wingless weevils, below.

**Wingless weevils (*Otiorrhynchus* spp.) (A.L. 57). Several species occur as pests, the larvae killing or weakening plants by feeding on the roots. Two common species are the vine weevil (*O. sulcatus*) and the strawberry root weevil (*O. rugosostriatus*). Eggs are found mostly in late July, August and September, in the surface soil beneath the canopy of leaves. The larvae hatch in the late summer and autumn and feed on the roots during the winter and spring, then pupate in the soil. Vine weevil adults emerge in May to June and strawberry root weevils in June to July. Plants may collapse during the fruiting period. The adults also feed on the foliage at night but damage is not serious. A few adults overwinter in the soil under the plants but most disappear in the late summer.

In parts of the south-west the red-legged weevil (*O. clavipes*) is a serious pest. Adults appear in two waves, emerging *en masse* in the spring from pupae formed in the autumn, and in a succession from mid-June to end of August from pupae formed in late spring and summer. Eggs are laid from late May to end of August, and the larvae feed on the roots.

It is not worth trying to treat plants when an attack is discovered during flowering or fruiting. The object is to eradicate as infestation and prevent it spreading to adjoining strawberry crops. After fruiting, apply DDT dust 5·6 kg a.i./ha to the crowns and surrounding collar of soil to kill grubs hatching from the eggs and before they can penetrate deeply in the soil. This should be done about the second week of August for vine weevil and strawberry root weevil, and the first week of September for red-legged weevil. Drenching the crowns and surrounding collar of soil with DDT at 1 g a.i./litre (as emulsion), 250–500 ml/plant, is also effective.

Before replanting in infested land work gamma-HCH into the soil as for wireworm control (see below).

Information is limited on the value of insecticides other than these organochlorine compounds.

**Wireworms (*Agriotes* spp.) (A.L. 199). Strawberries are susceptible to damage by these pests which may occur where crops are grown in broken-up grassland. Roots are bitten through and holes drilled into the crowns (see also Chapter 5). When breaking up grassland for strawberries work gamma-HCH at 0·84–1·4 kg a.i./ha into the soil at least 2–3 weeks before planting. Do not plant potatoes, carrots or onions for at least 18 months because of risk of taint.

9.10B   Diseases

*Arabis Mosaic (virus) (A.L. 530). Yellow spots or blotches usually appear on the foliage and in some varieties the spots become bright red. Leaf crinkling and stunting may occur. The leaf symptoms are most pronounced in late spring and autumn, tending to fade during summer. The causative virus is transmitted by the eelworm *Xiphinema diversicaudatum* which lives in the soil and attacks the roots.

Unless plants were infected at the time of planting, the disease is usually seen in well-defined patches corresponding to pockets of soil infested with *X. diversicaudatum*. In the latter instance, control of the disease has been obtained by fumigating the affected soil with dichloropropane-dichloropropene mixture. This should be applied at the rate of 448 kg/ha by injecting 3·5 ml at a depth of 15 cm at 30 cm spacings. Best results are obtained if application of the nematicide is made in the autumn.

***Grey mould (*Botrytis cinerea*) (MAFF Bull. 95). The fungus invades the floral parts and is also common on decaying plant remains and, under moist conditions, infects strawberry fruits on which it is seen as a grey furry mould. Loss of crop can be very high. Spray with a treatment from Table 9.35 (p. 242) and repeat every 10 days until harvest. Alternatively spray with dichlofluanid in a 2 or 4 spray programme at rates and times according to the manufacturers' instructions. The flower parts must be well covered, applying at least 2250 litres/ha; corresponding rates for spraying 3 times are given on the labels of some products. Lance spraying should be considered if good cover from boom spraying is doubtful.

Strains of *Botrytis cinerea* tolerant to benomyl, carbendazim and thiophanate-methyl have been shown to be present in some commercial

Table 9.35   Strawberry—sprays for control of grey mould and mildew

Compound	Rate (a.i.)	For use against*	
Benomyl	560 g/ha in at least 1000 litres/ha	GM	M
Captan†	200 g/100 litres	GM	—
Carbendazim	560–600 g/ha	GM	M
Dichlofluanid‡	1·7 kg/ha	GM	—
Dinocap	25 g/100 litres	—	M
Sulphur†	160–480 g/100 litres	—	M
Thiophanate-methyl	1·1 kg/ha	GM	M
Thiram†	16 g/100 litres	GM	—

* For timing see product label and text; GM = Grey mould, p. 242; M = Mildew, p. 243.
† Liable to render fruit unsuitable for processing, consult processors before deciding spray programme.
‡ See product label for various spray programmes; do not use on strawberries under glass or plastic.

crops. Where these strains occur, satisfactory control may not be achieved with these fungicides.

Mildew (*Sphaerotheca macularis*) (MAFF Bull. 95). The disease is seen during spring as dark patches on the upper side of the leaf. These patches correspond to a whitish-grey sporing growth on the underside. Affected leaves may curl upwards as if with drought, and the mildew spreads to other leaves, to the blossoms and the berries. The latter may become shrivelled or otherwise unmarketable. The fungus overwinters on old green leaves.

Royal Sovereign, Cambridge Vigour and Gorella are very susceptible to mildew; Cambridge Favourite is resistant.

Where mildew is likely to occur, applications of benomyl *or* carbendazim *or* thiophanate-methyl made for *Botrytis* control will provide control of mildew also. Alternatively spray with dinocap *or* sulphur just before flowering and at 10−14 day intervals.

Foliage should be burnt off after harvesting. If this is not done, post-harvest applications of fungicide may be needed.

9.11   Grape vine

9.11A   Diseases

*Downy mildew** (*Plasmopara viticola*), mainly on outdoor grapes. Light green patches occur on the upper surfaces of leaves and these correspond to growth of the sporulating fungus on the undersides; diseased areas later become dry and brittle. Berries can also be affected and, when severe, may shrivel. Overwintering spores are produced in affected leaves and these can renew the disease the following spring. Where infection is expected to occur a protective spray of zineb at 140 g a.i./100 litres HV should be applied at 10−14 day intervals. Diseased leaves and tendrils should be burnt.

***Powdery mildew** (*Uncinula necator*) (A.L. 207). Powdery mildew is common on vines growing both outdoors and in the glasshouse. The mildew forms white sporulating patches on young leaves and shoots but its development may be so sparse that the grey or purplish discoloration of the diseased parts is the most obvious symptom. Flowers and berries may also become infected causing them to drop. At a later stage berries become distorted and cracked.

With vines, both outdoors and indoors, sulphur dusts or sprays at 200−300 g a.i./100 litres HV *or* dinocap (w.p. or e.c.) at 12·5 g a.i./100 litres HV should be applied to the foliage as soon as the mildew appears. Where it has been troublesome in previous years the first application should be made 10−14 days before it is expected. Outdoors a total of up to 4 applications may be needed to prevent a build-up.

***Grey mould** (*Botrytis cinerea*). The fungus attacks the developing or ripe grapes particularly after they have been damaged by wind, hail or other agencies. In wet weather the typical furry grey growth of the fungus is produced on infected fruit and the spores are readily disseminated by wind and rainsplash. Infection of the stalks leads to premature fruit fall.

The vines should be sprayed HV with dichlofluanid 1·7 kg a.i./ha immediately after blossom and at 10–14 day intervals until 3 weeks before harvest.

9.12   Hop

9.12A   Pests

Annual routine control measures are necessary against damson-hop aphid.

Clay-coloured weevil (*Otiorrhynchus singularis*)—see under Apple (p. 180).

***Damson-hop aphid** (*Phorodon humuli*) (MAFF Bull. 164). This aphid is one of the main limiting factors to hop production, and routine protection of the crop is essential. It overwinters as eggs on the twigs of *Prunus* spp., especially blackthorn, bullace, damson and plums, and the eggs begin hatching in February or March. After 1 or 2 generations of wingless aphids, winged forms begin appearing in the latter half of May and disperse to hops. Individual aphids may visit several hop plants and most eventually settle at the tips of the bines or laterals. The migration usually begins in earnest in early June, reaching a maximum in the second half of the month. It then declines to end in late July or early August (or sometimes later). A return flight to the winter hosts occurs in September and October. There is no evidence that winged aphids bred on hops spread infestations to other hops.

Crops should be continuously protected by insecticides from when the first adult wingless aphids ('cows') mature (usually in the second week of June) until the infestation is completely controlled after migration ends in late July or early August (or sometimes later). The time at which migration into hops is completed is critical for gaining control of the aphid. From mid July the canopy of foliage near the top wires becomes very dense and on some cultivars the mature leaves on the lower part tend to curl downwards so that thorough spraying at this stage is extremely difficult. At the same time, growth of the bine slows down and movement of systemic insecticides appears to be restricted. Thus if migration continues into August it is considerably more difficult to finally control the infestation than if migration is completed in July. If hot dry weather follows during August, surviving aphids are able to multiply at a phenomenal rate and there may be severe infestation of the cones when they are harvested in September. Control may be achieved with either foliar sprays or soil drenches, or a combination of the two. It is cheap and easy to include an insecticide in some of the frequent fungicidal sprays which are applied to most cultivars. (There is no evidence that mixing compatible materials affects the protection given by either material). However, good control may not be achieved with sprays after mid-July because of the difficulty of obtaining good spray cover when the growth has become dense. Most of the spray materials now used against this aphid are effective for 10–12 days from application. The application of soil drenches is an additional operation and the materials used are relatively expensive. However, they may give a longer period of protection. The time of application of soil drenches must be related to rainfall, as they may not work if applied to dry soil. There is limited evidence that if sprays are avoided in July and August parasites and predators may be able to control any infestation developing after the use of a well-timed soil drench, thus avoiding the difficult problem of chemical control at this stage.

Where soil drenches are to be used they should be applied to protect the crop up to the end of migration, using sprays when necessary early in the season. In a spray programme, applications should be at 10- to 12-day intervals up until migration has finished and the crop is clean.

Sprays. Since about 1962 damson-hop aphid has become progressively more resistant to organophosphorus insecticides so that most materials in this group, including demeton-S-methyl and menazon, are no longer effective. The resistance pattern in the aphids is not fully understood and certain organophosphorus compounds continue to give good control. Some of the more recently introduced carbamates are also effective.

Early sprays Table 9.36 (p. 245) may, with care, be applied along alternate alleys, but once the bines have reached the top wire every alley should be sprayed. The volume applied should be increased progressively from 340–560 litres/ha for the first spray to 1125–1350 (or 1750, according to cultivar) litres/ha for later sprays. Above

Table 9.36   Hop—sprays for damson-hop aphid control

Compound (and class)	Early sprays (before bines reach top wire)		Subsequent sprays
	HV g a.i./100 litres	MV g a.i./50 litres	HV g a.i./100 litres
Acephate (organophosphorus)	56	56	56
Endosulfan* (organochlorine)	65	33	70
Methidathion (organophosphorus)	45	45	60–90
Methomyl (carbamate)†	110	55	110
Omethoate† (organophosphorus)	57	57	57–115
Propoxur (carbamate)	75	—	75

* Maximum 770 g a.i./ha per application and 3500 g a.i./ha per season.

† If adverse growing conditions occur after burr stage replace by propoxur.

1350–1750 litres/ha (according to cultivar) a comparatively large increase in effort and cost is needed for small improvements in cover, although it may sometimes be worthwhile to increase the rate up to 2250 litres/ha. The first spray can be applied at a ground speed of up to 6·4 km/h (4 mile/h), but ground speeds should be reduced to 4·0–4·8 km/h (2·5–3·0 mile/h) once the spray target becomes more difficult to cover. Lower speeds give further improvement in spray cover. To achieve maximum penetration with air blast sprayers, the recommended PTO speed on the tractor must be maintained even if this restricts the ground speed. A high standard of maintenance and adjustment of spraying machines is essential if good cover of this difficult spray target is to be achieved. Nozzle and boom settings should be set to give the best possible cover of the undersides of the leaves and adjusted as necessary during the season.

Drenches. For soil drenches use dimefox at 120 or 240 g a.i./100 litres water (rate according to control programme adopted—consult manufacturer's label), applying 110 ml of the diluted insecticide to the crown of each hill. Alternatively, apply mephosfolan at 1 kg a.i./110 litres/1000 hills, using 114 ml dilute solution/hill. Dimefox should not be applied to newly planted hops, except bedded sets, and the concentration should be reduced to 60 g a.i./100 litres water on first and second year hops. Where only 4 bines are taken from a hill, 750 g a.i. mephosfolan/1000 hills are adequate; this concentration should also be used on newly planted hops. As before, 114 ml of the diluted material should be applied to the crown of each hill, using a hand lance or mechanical applicator designed for this purpose.

Much of the active ingredient from soil drenches is absorbed by the base of the bines, rather than through the roots, so it is important to ensure that the base of the bines is treated, including that of any bine emerging away from the crown of the plant. Neither dimefox nor mephosfolan should be applied to hops within 4 weeks of harvest. Do not handle hops treated with either of these materials for at least 4 days after application, unless rubber gloves are worn. When soil drenches are applied to moist soil, dimefox protects the crop for 2–3 weeks and mephosfolan for 5 weeks or more. If the soil is moist when a soil drench is applied the aphids will be killed over the whole plant within a few days. However, if the soil is dry the drench may have little effect unless there is fairly heavy rainfall within a week or two. Increasing the volume of drench, or applying water before drenching, gives little or no improvement in absorption of insecticide.

*Hop capsid (*Calocoris fulvomaculatus*). This pest overwinters as eggs laid in softer parts of hop poles; these hatch in late May and the buds puncture the leaves, which become perforated and split open; the bine may also be scarred and become twisted in growth. This is no longer a serious pest, and has probably been eliminated by sprays used against damson-hop aphid.

*Hop flea beetle (*Psylloides attenuata*). Adults hibernate under rough vegetation, emerging in the latter part of April and in May; they eat holes in the leaves and damage young shoots. The second brood in late summer feeds on the leaves and cones. This has become a relatively uncommon pest. If required, spray when the beetles are first seen with DDT, 1·1 kg a.i./ha in not less than 340 litres. Malathion, 112 g a.i./100 litres, is probably also effective.

*Hop root weevil (*Epipolaeus caliginosus*). Adult weevils are found throughout the year, being most numerous in late summer and autumn. Eggs are laid in punctures made in the rootstock from mid-August to May, and the grubs tunnel in the rootstock and base of the bine for 9–18 months. The bine is weakened or killed, and feeding by adults in spring may also weaken growth. It is now an uncommon pest.

Control measures under no-cultivation conditions have not been studied. The previous recommendation was that if required hills should be dusted with gamma-HCH dust (0·2–0·65 per cent a.i.) at about 22·5 g a.i./10 hills in late May or early June immediately before earthing up.

*Hop strig midge (*Contarinia humuli*). Midges emerge from pupae in the soil in the latter part of July and in August, and lay eggs in the cones from early August. The maggots feed at the bases of the bracts and along the strig, and the cones become brown and undersized. The maggots are fully fed by the end of August and drop to the soil. This pest is now rarely seen, but if control is necessary spray with DDT, 100 g a.i./100 litres (HV) in late July.

**Red spider mite (*Tetranychus urticae*). The mite winters as adult females in the soil, crevices in poles and wirework. It emerges in late April and feeds on the leaves, where eggs are laid and up to 7 generations may follow during the summer.

This pest appeared to have been virtually eliminated from hop gardens by the organophosphorus pesticides used against damson-hop aphid, but in the last few years infestations of organophosphorus-resistant mites have become increasingly numerous. HV sprays of dicofol, 40 g a.i./100 litres, *or* tetradifon, 12·5 g a.i./100 litres, will control such infestations until strains resistant also to these materials appear on hops. Tetradifon is slow acting since, although it kills eggs and young mites, its effect on adult females is to render them sterile. Dicofol kills all stages and is therefore quicker acting. Best results will be obtained by spraying when egg hatching is completed in May, when the bines are 1·5–1·8 m high, with repeated sprays later where necessary.

Young hops, especially those produced by mist propagation, should be watched closely for red spider mite infestations in their first season.

**Rosy rustic moth (*Hydraecia micacea*). This common insect may be found in many hop gardens, especially on and near the headlands. Serious damage to mature hops is probably uncommon but damage is found more frequently on young hops. The moths are on the wing from late July to late October, being most numerous in September. Eggs are laid in the autumn and these hatch in the following spring. The young caterpillars begin to ascend the bine in late April or early May; they bore into the stem and tunnel in the pith. During June they leave the aerial part of the bine to enter the soil and they complete their growth by feeding on the crown or the base of the bine. All appear to have descended by the end of June. From July onwards they enter the pupal stage. The aerial part of the bine may split open or break at the exit hole left by a descending caterpillar. There is evidence of increased susceptibility to Fusarium canker and a reduction in yield following caterpillar damage. Growth of young hills can also be severely checked.

Where control is required, the attack on the bines in the spring can be reduced by applying a drenching spray to each hill of DDT, 1050 g a.i. in 560 litres/ha, wetting the soil and young shoots. A first spray should be applied about the end of the first week in May when young caterpillars are most numerous; a second spray should be applied 1 week later. If severe damage to the bines is first noticed in May, a spray directed to the base of the bines in early June, with a second 10 days later, will reduce the number of caterpillars descending to damage the rootstock.

Where rosy rustic moth is a recurring problem weed growth, which attracts egg-laying moths, should be prevented in late summer and autumn. It is also worth delaying training of the hops as long as possible in the spring so that many of the caterpillars will attack bines which are to be discarded.

Slugs (A.L. 115). Several species, such as the field slug (*Agriolimax reticulatus*) and garden slug (*Arion hortensis*) are often common in hop gardens. Suppression of weeds by herbicides may tend to concentrate their attacks on young hop shoots, which may be severely damaged. Where the use of copper fungicides has been reduced this may also favour an increase in slug populations.

Metaldehyde or methiocarb pelleted slug baits should be used where necessary.

*Wireworms (A.L. 199). Hop sets planted in land ploughed from old grass may be attacked by wireworms, and these are sometimes troublesome when replanting in old gardens.

When breaking up grassland for hops, work gamma-HCH into the soil at 0·8–1·7 kg a.i./ha at least 2–3 weeks before planting. An alternative measure is the application of gamma-HCH wireworm dust at about 0·35 g a.i. to each planting hole before planting. Where established hops are attacked sprinkle this amount of dust over the soil surrounding each hill.

Xiphinema diversicaudatum. This dagger nematode is the vector of arabis mosaic virus, the hop strain of which appears to be an essential component of nettlehead and split leaf blotch diseases. Where possible, growers planning new hop plantings should choose sites which appear to be free from *X. diversicaudatum*. (Soil samples can be examined for *X. diversicaudatum* by ADAS). If the use of a *Xiphinema*-infested site is unavoidable, and particularly if virus-infested hops have been grubbed, the site should be fallowed for 2 years. As an alternative, a dichloropropane-dichloropropene mixture injected at 560–840 litres/ha will kill about 95 per cent of the *X. diversicaudatum* population to a depth of about 60 cm. This treatment should considerably delay the build-up of the infection, but the benefit from a 2 year fallow will last much longer.

9.12B   Diseases

***Downy mildew** (*Pseudoperonospora humuli*). The disease begins each spring from systemically infected shoots (basal spikes) arising from the crown of the rootstock. Spores from these result in more basal spikes. In suitable weather, the disease progresses upwards throughout the growing season. Terminal and lateral shoots become 'spikes', young leaves are killed, old leaves are spotted, and cones may become infected at any stage of development. If routine preventive methods are not adopted, the whole crop may become worthless. Downy mildew too is probably the most common cause of death of rootstocks, being introduced through short shoots (secondary basal spikes) and through the base of the bine; some varieties are very prone to this form of the disease.

After hand removal of spikes, Bordeaux powder, copper oxychloride or other specially formulated dust, based on 6 per cent metallic copper, should be applied at about 1·0 kg copper/ha as the first shoots appear. Alternatively use cufraneb 0·54 kg a.i. in 45–54 litres as a spray. The new shoots should be well covered. Two or 3 further dustings or sprays

Table 9.37   Hop—dusts and sprays for control of downy mildew

Compound*	Rate† HV	LV
Bordeaux mixture‡	100–150 g copper/100 litres	2·24–3·36 kg copper/ha
Copper oxychloride‡	100–150 g copper/100 litres	2·24–3·36 kg copper/ha
Cufraneb	540–1400 kg a.i./ha in 450–900 litres/ha depending on stage of growth	
Cuprous oxide	75–150 g copper/100 litres	1·68–2·24 kg copper/ha
'Polyram'	160 g a.i./100 litre	1·8 kg a.i./ha up to burr stage 3·6 kg a.i./ha from burr stage
Propineb	140 g a.i./100 litres	1·6–3·2 kg a.i./ha
Zineb	140 g a.i./100 litres	2·9–3·7 kg a.i./ha

* Some of these fungicides are specially formulated for use on hops.
† For timing of applications see text, p. 250.
‡ Dust applications should be at 2·0–3·3 kg copper/ha.

should be applied at intervals of 10–14 days. Alternatively streptomycin w.p. as a spray may be applied to the young shoots so that their removal from the hill may be safely delayed for 2 weeks or more. It is especially important to use streptomycin in accordance with the maker's instructions.

Fungicides listed in Table 9.37 (p. 249) formulated as wettable powders or liquids for use on hops, should be applied to the foliage within 10 days of the last hill treatment and then at 10–14 day intervals until immediately after burr. More frequent and later applications may be given on susceptible varieties if the weather is warm or humid.

Copper-containing dusts, as for hill treatment, may be used for foliar application instead of a spray where the latter is inappropriate. Intervals between dust applications should be shortened to 7 days in rainy weather.

To control downy mildew, a protective coating of fungicide is needed throughout growth so that a full programme is necessary in most years. Coverage of all growth, particularly the highest, is of utmost importance.

Cones with fungicidal deposit are not always acceptable and before late applications are made, the grower is recommended to consult his hop factor.

Additional measures include the removal of all spikes (basal, terminal and lateral) as soon as identifiable. Applications of copper to control downy mildew will also give some reduction in the amount of powdery mildew developing.

***Hop mould** or **Powdery mildew** (*Sphaerotheca macularis*). Mould, a single fungus, appears in 2 forms. From May onwards the white powdery stage develops on the leaves and sometimes on the young shoots. The burr and cones may likewise be attacked and become distorted and useless. If effective treatment has not been given, and if the weather is humid and warm, the late summer stage develops. This is seen as foxy-red spots or patches on leaves and cones. On these red patches, minute black spore cases (perithecia) form. When cones shatter, spore cases fall to the ground, where they overwinter. In spring, the spores within the cases mature and are ejected to the shoots or lower leaves of the plant where they germinate and so begin the cycle again. The fungus may also overwinter in the form of mycelium within the bud scales and in April a small number of shoots smothered with white powdery fungal growth appear.

Sulphur, as a fine dust (28–56 kg sulphur/ha) or w.p. (300–400 g sulphur/100 litres HV; 8·4 kg sulphur/ha LV) or colloidal formulations (rates as for w.p.), *or* dinocap w.p. (25 g a.i./100 litres HV; 560–700 g a.i./ha LV), *or* pyrazophos (according to manufacturer's instructions) should be applied in mid-May and then at intervals of 10–14 days according to the proneness of the garden to mould. A wetting agent may be needed when using some proprietary preparations at high volume. If the full programme has been carried out and if no mould can be detected, applications may cease at burr. Otherwise they may be continued for another few weeks.

When a garden has not been picked because of excessive mould, the bines should be cut and burnt before the cones shatter. The variety Northern Brewer is particularly susceptible.

Chapter 10
Pest and disease control in glasshouse crops

10.1 General

10.1.1 Chemical sterilants for border soils

Soil-borne pests and diseases frequently become limiting factors in the economic production of glasshouse crops, for these crops are usually grown without rotation. Moreover, pathogenic fungi such as *Phytophthora* spp., *Rhizoctonia solani* and *Pythium* spp., and pests such as nematodes, leatherjackets and wireworms may occur in maiden loams used in propagating and potting composts causing serious losses of seedlings and young plants.

Soil-borne pathogens and pests can be eliminated from composts by heat treatment, most commonly by 'steaming' (MAFF Bull. 22), but their complete eradication in glasshouse border soils by steaming is seldom accomplished as the heat does not penetrate to a sufficient depth.

Chemical soil sterilants can be used as substitutes for steaming but are more specific in their fungicidal, nematicidal and insecticidal action than heat, and as they, or their decomposition products may be highly phytotoxic, a lengthy time period between treatment and planting is usually necessary to ensure their release from treated soils. Methyl bromide, however, disappears rapidly. Vapours from chemical sterilants may diffuse through partitions, or be carried via heating ducts or ventilators into adjoining glasshouses in sufficient amounts to injure plants.

For the successful use of chemical soil sterilants the soil must be thoroughly worked before treatment and should be evenly moist but not wet. Soils with a high organic content absorb chemicals more strongly than light sandy soils and retain the chemical longer so that longer intervals are required between treatment and planting. Adequate distribution of chemical sterilants in the soil is essential and exceptionally heavy soils which cannot be broken down to a reasonable tilth are unsuitable for treatment by chemicals as clods are not penetrated. Where the chemicals are mixed into the soil by rotary cultivation, L-shaped tines should be used with a rotor speed set between 120–150 rev/minute and a forward speed not exceeding 4·6 m/minute. Most chemical sterilants are volatile or decompose to produce volatile compounds, and with these it is necessary to 'seal' the soil surface for a few days by compacting and wetting it to a depth of 13–25 mm or covering the soil with 'polythene' sheeting. Methyl bromide, which is gaseous above 5°C (40°F), diffuses into the soil from the surface when released from canisters or cylinders under suitable plastic sheets. These have their edges buried in the soil to a depth of about 130 mm to prevent the escape of the gas into the atmosphere, and are kept in place for 4 days. The rate at which steriliza-tion is accomplished and the speed of elimination of the chemical from treated soil are greater at higher soil temperatures, consequently it is inadvisable to use chemicals for soil sterilization during the late autumn and winter. Release will be slower for unheated houses and is hastened by forking or rotary cultivation of the soil to a depth of 250–300 mm. In autumn and winter when soil temperatures may be lower cultivation should be repeated 2–3 times at fortnightly intervals when the 'slower release' compounds are used.

The following chemicals are used for general soil disinfection, and the rates and methods of application are those which have been found satisfactory under most conditions.

Where applications are made to normal loam soil the *minimum* intervals to be allowed between treatment and planting are generally shown on the manufacturer's labels and should be adhered to. When chloropicrin or methyl bromide are used the absence of hazardous con-tamination should be decided by the trained operators.

Chloropicrin. Adhere to the *Code of Practice for the Fumigation of Soil with Chloropicrin* (MAFF). Inject chloropicrin 3·5 ml a.i. at a depth of 150 mm at 300 mm spacings (630 kg/ha). Immediately seal soil surface with sheets of low gas permeability, e.g. polythene of at least 150 gauge, for not less than 4 days. The soil temperature should be at least 13–16°C (55–60°F).

Chloropicrin with dichloropropane-dichloropropene and methyl isothiocyanate. Adhere to the *Code of Practice for the Fumigation of Soil with Chloropicrin* (MAFF). Apply as described for chloropicrin. Establish freedom from phytotoxic residues by the cress germination method (see dazomet). The next crop should only be carnation, tomato or lettuce.

Cresylic acid. Wear neoprene gloves and dilute 1 litre concentrate to 100 litres and apply at 38 litres/m². Ventilate house during application and for at least 24 hours. Delay planting for at least 3 weeks in light soil or 4–5 weeks in heavy soil.

Dazomet. On light to medium soils apply by sprinkling 380 kg prill/ha evenly on the surface and immediately admixing with the soil by rotary cultivation or by forking twice to a depth of 250–300 mm followed by surface sealing. The temperature at a depth of 150 mm should be not less than 7°C (45°F) at treatment and should not fall below 4°C (40°F) for 4 weeks.

Treated soil should not be planted until freedom from phytotoxic residues has been established by the cress germination test. Half fill a jar with a representative 1 kg sample of moistened soil, then scatter the cress seed on the soil surface and seal the jar. If the seed germinates normally it is advisable to flood the treated soil before proceeding with cultivations and planting.

Dichloropropane-dichloropropene mixture may be used before planting to control potato-root and root-knot eelworms, but the soil temperature should be at least 5°C (41°F). Polyethylene protective boots and gloves must be worn during the application, and ventilate the house during and for at least 24 hours after treatment. Inject 3·5 ml of product at a depth of 200 mm at 300 mm spacings (340–450 kg/ha) followed by surface sealing. Its regular use at the full rate is inadvisable on heavy soils.

Dichloropropene may be used to control potato-root and root-knot eelworms before planting. For details see dichloropropane-dichloropropene mixture.

Formaldehyde is applied usually as a 1 in 50 dilution of commercial (*c.* 40 per cent) formalin at 27 litres diluted liquid/m². Better results are obtained on light open soils by applying an additional 22–27 litres of the liquid/m² to the second spit at double digging.

Metham-sodium. The soil temperature should be at least 7°C (45°F) and ventilators left open during application and then closed for 1 week. Metham-sodium can be applied as a soil drench at 12 litres of 33 per cent solution in 1200 litres water/100 m² but do *not* seal the soil surface. Or apply 12 litres of 33 per cent solution in 200 litres water/100 m² followed by soil rotavation to a depth of 200 mm and seal the soil surface with water 200 litres/100 m². Alternatively, metham-sodium can be injected undiluted at 11 ml per injection at a depth of 200 mm and 300 mm spacings; rotavate and seal the soil with water as above. Before planting in treated soil, establish the absence of phytotoxic residues by the cress germination test (see dazomet, p. 253).

Methyl bromide is widely used in Britain for the control of pathogenic fungi and nematodes in glasshouse soils. It has the advantage of rapid diffusion into and from the soil, consequently the interval between treatment and planting is short (6–8 days). Trials have confirmed the efficacy of methyl bromide at 470–700 kg/ha. Methyl bromide has a high mammalian toxicity and special equipment, including impermeable sheeting to retain it in the soil, is required. Soil temperature must be at least 10°C (50°F). Chloropicrin (2 per cent) is added as a warning agent and at present methyl bromide must be applied only by contractors or trained operators.

10.1.2   For the treatment of small quantities of soil for propagating and potting (MAFF Bull. 22)

Dazomet. Thoroughly mix dazomet 220 g prill/m³ of warm, moist soil, cover for 3 weeks, then turn 3 times at fortnightly intervals and apply the cress germination test (p. 253) before using the soil. Beware of phytotoxic fumes affecting nearby plants.

Formaldehyde. Apply 150–300 litres of a 1 in 50 dilution of commercial (*c.* 40 per cent) formalin per m³ of soil to saturate the soil in 150 mm layers as the heap is built. Cover with plastic sheeting for at least

2 days, uncover and turn. Allow 4–6 weeks before use depending on temperature, lightness of soil, and number of turnings.

Metham-sodium. Apply 1 litre of a 33 per cent solution of metham-sodium in 100 litres water at 30 litres/m³ to moist soil in 150 mm layers as heap is built. After 14 days turn at least 3 times at 7–14 day intervals. Beware of fumes affecting nearby plants and apply the cress germination test (p. 253) before using the soil.

10.1.3   Sprays (high volume and low volume)

Foliar sprays may be applied to glasshouse crops as described in Chapter 3, p. 79.

10.1.4   Atomized fluids (aerosols), fogs and smokes

Atomized solutions, fogs and smokes are convenient to use but are generally less effective than high volume sprays of the same pesticide. Fogs and smokes are usually not as efficient in windy weather.

Aerosols and products suitable for ultra low volume application using a compressed-air paint gun are listed under 'atomization' in the text. In addition, there are several machines available which will apply a wide range of pesticides as fogs or atomized sprays. Information on products which are suitable for use with these machines can be obtained from the equipment manufacturer.

Smoke canisters are available in sizes designed for various glasshouse volumes and the manufacturer's instructions should be followed for the quantities to be used.

10.1.5   Soil drenches

Drenches are conveniently applied by hosepipe from a tank containing the diluted product, but dispensers, giving measured doses, are available for some materials. Equipment for automatic watering and liquid feeding may be unsuitable for the application of insecticides and fungicides because of the risks of uneven distribution and the settling of wettable powders in the pipes or tubes.

Insufficient information is available at present to make firm recommendations for methods of soil drenching for all types of irrigation equipment.

A systemic compound applied as a soil drench to the rooting region of the crop can generally protect the plants for longer than the same compound applied as a foliar spray. However, soil drenches use more chemical than foliar sprays so the relative merits of the two methods will depend *inter alia* on the crop and the pesticide to be applied. An application rate of 11 litres/m² of diluted material corresponds to 85–115 ml/150 mm pot and to 485 ml/250 mm pot.

10.1.6   Granules and prills

At present, few pesticides for use on glasshouse crops are formulated as granules or prills. The manufacturers' instructions must be observed as some materials should be distributed on the soil surface and watered-in and others require thorough incorporation into the border soil or compost. Hand-operated and motorized applicators are available for distributing the granules or prills. Some of the materials are highly toxic and, therefore, it is important to avoid contamination of paths etc.

On mushrooms, granules are usually either incorporated with the compost by the spawning machines or mixed with the casing, ideally, with a rotary mixer.

10.1.7   Dusts

Dusts suitable for glasshouse crops are usually formulated with a light, inert carrier and applied with equipment which disperses the material as a fine cloud. Dusts may be preferred to sprays when the glasshouse humidity is very high.

10.1.8   Susceptibility of plants to chemicals

In addition to the major glasshouse crops which will be considered, many other species of plants are grown or propagated commercially in glasshouses but, because of their relatively low economic importance or, in some cases, their novelty, little reliable information is available regarding the tolerance of such plants to pesticides. Factors such as stage of growth, climatic conditions, nutrition and the particular cultivar can influence crop susceptibility to injury.

Plants grown in glasshouses during the winter under adverse light conditions have a 'soft' growth which is abnormally susceptible to damage by chemicals applied at concentrations which may be tolerated by plants grown in the summer. High volume sprays, dusts and smokes are usually less phytotoxic than atomized solutions (aerosols); but avoid high air humidity and water droplets on foliage when using ultra low volume techniques. Considerable caution should be exercised in applying pesticides, in any form, to seedlings and recently potted or planted young plants or cuttings. Damage by pesticides is most likely to occur if they are applied in hot, bright sunshine or to plants suffering from water stress. On the other hand conditions should be such that HV sprays dry within a few hours.

Varieties (cultivars) of certain species, particularly chrysanthemums, differ considerably in their tolerance to chemicals and preliminary tests should be made on new varieties or with new compounds and formulations.

10.1.9   Glasshouse hygiene

Strict attention to hygiene is important in reducing the incidence of pests and diseases in glasshouses. When diseases have occurred, the internal structure of glasshouses and mushroom houses (including any staging) and the soil surface should be disinfected, after the removal of any debris, by spraying rapidly with a 1 in 50 dilution of 40 per cent formaldehyde and closing the house for at least 24 hours. Protective clothing and a respirator should be worn for comfort. Alternatively, fumigate by adding potassium permanganate to the formalin but as this reaction is violent, observe the following procedure.

Close the ventilators and space containers (each not less than 10 litres), one for each 100 m³ of the glasshouse. Add to each container 500 ml of c. 40 per cent formalin and place alongside on a sheet of paper 100 g potassium permanganate. Commencing furthest from the door quickly add the permanganate to each container and leave the glasshouse. Keep the temperature at 10°C (50°F) for at least 24 hours and then ventilate thoroughly.

Boxes, pots, etc., may be disinfected by steam, boiling water or formaldehyde. Glass partitions cannot be relied on to prevent ingress of the highly phytotoxic formaldehyde vapour into adjoining houses.

10.1.10   Resistance of pests and diseases to chemicals

In the continuous monoculture systems often used in modern glasshouse practice, pests and fungi sometimes develop resistance to chemicals. Always apply the recommended dose of any pesticide by the method stated on the label and, when possible, from time to time change to another product that has a different mode of action.

10.1.11   Biological and integrated control

Biological control techniques have been developed for: red spider mite on cucumbers, tomatoes and chrysanthemums using the predatory mite *Phytoseiulus persimilis*; whitefly on cucumbers and tomatoes using the parasite *Encarsia formosa*; the aphid *Myzus persicae* on chrysanthemums using the parasite *Aphidius matricariae*; and caterpillars on chrysanthemums using the bacterium *Bacillus thuringiensis*. These techniques are shown in Table 10.1, p. 258, together with some treatments for controlling other pests and diseases using materials which are non-injurious to the parasite/predator. Further information on biological control is given in Grower's Bulletins Nos. 1

Table 10.1 Integrated control of some chrysanthemum, cucumber and tomato pests and diseases

Pest/disease	Chrysanthemums	Cucumber	Tomato
1. Pests			
Aphid—biological control for *Myzus persicae* only.	Before planting add to every 10 cuttings 1 *M. persicae* which has previously been exposed to the parasite for 4 days.	Apply soil drench (5 litres/m²) of pirimicarb 19 g a.i./100 litres.	Apply soil drench (5 litres/m²) of pirimicarb 19 g a.i./100 litres. Localized infested patches can be controlled by HV sprays of pirimicarb 25 g a.i./100 litres applied as a 'spot treatment'.
Red spider mite	Before planting, add 50 red spider mites and 1 predator per 50 cuttings.	Two weeks after planting place 20 red spider mites on each plant and 10—14 days later place 2 predators on alternate plants.	Three weeks before planting infest 1/5th of pot plants with 30—40 red spider mites. 10 days later place 4 predators on the infested plants. At planting, distribute infested plants evenly throughout crop.†
Whitefly	No 'integrated' control method yet available. See p. 000 for chemical control.	Infest every fifth plant with 10 whitefly scales and 14 days later add 100 parasitized black scales to the same plants.	Introduce parasites at the rate equivalent to 1 per plant 2 weeks after the *first* whiteflies appear on the plants and repeat every 2 weeks until 2 weeks after black scales appear on the crop plants.
Caterpillars	Spray HV 5000 litres/ha with *Bacillus thuringiensis* at manufacturers' dosage.	—	No 'integrated' control method yet available. See p. 307 for chemical control.

Leaf miner	Apply dioxathion 100 g a.i./100 litres as a mist above the tops of the plants or spray pirimicarb 25 g a.i./100 litres HV.	—	At pricking out, incorporate dimethoate granules 0·02 g a.i./kg compost.
Thrips	Apply soil drench (5 litres/m² bed) of gamma-HCH 20 g a.i./100 litres. diazinon 40 g a.i./100 litres.	As for Chrysanthemums—leaving minimum of 2 days between treatment and harvest (diazinon).	Apply soil drench (5 litres/m² bed) of gamma-HCH 20 g a.i./100 litres or
2. Diseases Grey mould	Spray dichlofluanid* 50 g a.i./100 litres HV.	Apply soil drench of benomyl, or carbendazim or thiophanate-methyl at rates given in Table 10.5, p. 282.	Spray dichlofluanid 50 g a.i./100 litres HV allowing 3 days between treatment and harvest. or apply soil drench of benomyl or carbendazim, or thiophanate methyl, see Tomato—Grey mould, p. 308.
Leaf mould	—	—	
Powdery mildew	Spray dinocap* 6·3 g a.i./100 litres HV or triforine* 20 g a.i./100 litres HV.	As for Grey mould above, or soil drench of dimethirimol at rates given in Table 10.5, p. 282.	—

* Varieties of chrysanthemums may differ in their susceptibility to spray damage and as these materials have not been tested or approved for use on all varieties. first assess the treatment on a few plants.

† If preplanting treatment has been omitted and localized infestation occurs introduce 10–20 predators on each plant within the infested area.

and 2 obtainable from the Glasshouse Crops Research Institute, Littlehampton.

10.2   Specific crops

ARUM (CALLA) LILY

Pests

**Aphids (*Myzus persicae, Macrosiphum euphorbiae, Aphis gossypii*). These, and occasionally other spp., cause stunting and disfiguration of the young growth and severe spotting of the flower spathes. Control before the flowers open by adopting methods recommended for Carnation—Aphid, Table 10.2, p. 266.

**Red spider mite (*Tetranychus urticae*). For control measures see Carnation—Red spider mite, Table 10.2, p. 266.

*Thrips (*Heliothrips femoralis, H. haemorrhoidalis, Thrips tabaci*) breed on the foliage and flowers of arum causing pale, whitish spots on the leaves and brown spots on the flowers. For control see Carnation—Thrips, p. 264.

Diseases

*Corm rot (*Pectobacterium carotovorum*). Infected plants are attacked at the collar and the corms become invaded causing yellowing and death of the foliage. Badly diseased corms should be destroyed. Slightly affected corms may be saved by washing away all the soil when they are dormant, scraping away all diseased tissue and steeping for 2 hours in formaldehyde, 20 ml 40 per cent formaldehyde/litre. Ensure that the corms are not recontaminated and pot without delay into soil sterilized by heat or formaldehyde. Disinfect the glasshouse when vacant, see p. 257.

*Root rot (*Phytophthora richardiae*). Leaves of attacked plants first become yellowish and streaky, and gradually turn brown and die. Flowers become brown at their tips and may be deformed. Treat as for corm rot, p. 260.

ASPARAGUS FERN (*Asparagus plumosus, A. sprengeri*)

Pests

*Caterpillars. Foliage-eating caterpillars may be controlled with DDT sprays, dusts, atomized solutions and smokes, as for Carnation—Tortrix, p. 264.

***Red spider mite (*Tetranychus urticae*) causes browning and abscission of leaflets. Control with demeton-S-methyl, *or* diazinon, *or* dicofol, *or* dimethoate, *or* naled, *or* oxydemeton-methyl, *or* tetradifon as in Table 10.2, p. 266.

*Thrips (*Thrips tabaci*). Severe attacks cause a silvering of the leaflets. Control, as for Carnation—Thrips, p. 264, but not with malathion.

*Whitefly (*Trialeurodes vaporariorum*) (A.L. 86). For description see Cucumber—Whitefly, p. 281. Control as for Chrysanthemum—Whitefly, Table 10.2, p. 266, but not with malathion.

Diseases

*Basal rot (*Pythium* spp.). The fungus attacks the stems near soil level. Control as for root rot (p. 261). It is possible that watering with captan 100 g a.i./100 litres will check the spread of the disease.

*Root rot (*Fusarium* spp.). Branches become yellow and finally desiccated, basal shoots wither and roots decay. To prevent re-occurrence, sterilize the soil, preferably by steaming.

AZALEA

Pests

*Red spider mite (*Tetranychus cinnabarinus, T. urticae*). Mites may be a pest on azaleas if plants to be forced are placed in houses containing infected plants or hibernating mites of the second species. To control use demeton-S-methyl, *or* diazinon, *or* dicofol, *or* oxydemeton-methyl, see Table 10.2, p. 266.

*Thrips (*Heliothrips haemorrhoidalis*). This thrip causes silvery or whitish patches with reddish-brown spotting on the undersides of the leaves and on the petals of flowers. Attacks almost invariably result from existing infestations within the glasshouse and are rare in commercial houses. Control as for Carnation—Thrips, p. 264.

Diseases

*Grey mould (*Botrytis cinerea*). Rotting of flowers by *Botrytis* occurs under excessively damp conditions; ample ventilation and prompt removal of all diseased flowers will usually prevent its spread. But it is advisable, especially in unheated houses, to apply a suitable treatment listed in Table 10.3, p. 274.

BEANS (FRENCH AND CLIMBING FRENCH)

*Aphids. Control with pirimicarb 25 g a.i./100 litres HV and allow 7 days between spraying and harvest *or* treat with dichlorvos (aerosol) *or* gamma-HCH (aerosol, smoke, or HV spray) *or* nicotine (smoke or HV spray) *or* malathion (HV spray) using the rates and observing the preharvest intervals as for Tomato, Table 10.7, p. 304.

**'French-fly' (*Tyrophagus longior*). For description and control see Cucumber—French-fly, p. 277.

**Red spider mite (*Tetranychus urticae, T. cinnabarinus*). On *young plants* use dicofol (HV spray) and on *established plants* dicofol (aerosol or HV spray) *or* tetradifon (HV spray) using the rates and observing the preharvest intervals as for Tomato, Table 10.7, p. 304.

*Symphylids. For description and control see Tomato—Symphylids, p. 306.

*Whitefly—for description see Cucumber—Whitefly, p. 281. Control with DDT, *or* DDT/gamma-HCH, *or* dichlorvos, *or* malathion, *or* pyrethrin/resmethrin, *or* resmethrin as in Table 10.7, p. 304.

*Wireworms—see Tomato—Wireworms, p. 307.

BEGONIA

Begonia flowers are susceptible to injury by many pesticides, particularly those applied as sprays. Do not spray plants in sunshine or under slow drying conditions.

Pests

**Aphids. A number of species including *Myzus persicae, Aulacorthum solani* and *Aphis gossypii* breed on begonias under glass and may cause serious damage to foliage and flowers. Control before flowering with gamma-HCH, *or* malathion, *or* naled, *or* nicotine, *or* pirimicarb as in Table 10.2, p. 266, or by dipping the plants in gamma-HCH, *or* malathion, *or* nicotine at concentrations as for HV sprays.

*Mealy bug (*Planococcus citri*). This species, and occasionally *Pseudococcus adonidum*, attack plants but are not common in commercial houses. Several treatments, at 10–14 day intervals, with malathion at 110 g a.i./100 litres HV by spraying or dipping will control this pest.

*Mites. The Tarsonemid (cyclamen) mite (*Steneotarsonemus pallidus*) and the broad mite (*Hemitarsonemus latus*) cause stunting, crinkling and curling of the leaves and malformed flowers. The control of both is as for Cyclamen—Tarsonemid mite, p. 285.

*Thrips. Several species, e.g. greenhouse thrips (*Heliothrips haemorrhoidalis*), banded greenhouse thrips (*Hercinothrips femoralis*) and onion thrips (*Thrips tabaci*) infest begonias, the leaves of which may show irregular brownish-red streaks on the upper surface and roughened brownish spots on the underside. Control, before flowering, as for Carnation—Thrips, p. 264.

*Whitefly (*Trialeurodes vaporariorum*). For description see

Cucumber—Whitefly, p. 281. Control with diazinon, *or* dichlorvos, *or* malathion, *or* naled, *or* pirimiphos-methyl as for Chrysanthemum—Whitefly Table 10.2, p. 266 but avoid spraying open blooms.

Diseases

*Grey mould or **Blotch** (*Botrytis cinerea*). Foliage, stems and flowers may be attacked if begonias are grown under excessively damp and cool conditions. Spray at the first sign of attack with a suitable treatment listed in Table 10.3, p. 274.

*Powdery mildew (*Oidium begoniae*). A whitish, powdery mould on the young leaves, stems and flower buds causes severely attacked tissues to be desiccated and is usually associated with large fluctuations in humidity and temperature. Control with chemicals is difficult to obtain without phytotoxic effects. Vaporize sulphur at 0·5 g/100 m³ nightly over a 5–10 day period, *or* spray with benomyl *or* carbendazim *or* dinocap *or* thiophanate-methyl at rates shown in Table 10.3, p. 274, but some cultivars may be slightly damaged by dinocap.

*Root and **Stem rots.** Several fungi including *Rhizoctonia solani, Thielaviopsis basicola* and *Pythium* spp., cause rotting of the roots, tubers or stems, particularly if the plants are frequently overwatered. Prevent by using sterilized composts but if infection occurs water the compost with cupric ammonium carbonate according to manufacturer's instructions. In trials, soil drenches of drazoxolon 19 g a.i./100 litres has given promising control of *Pythium* but check that the material is non-injurious to the particular variety by treating a few plants in the first instance.

CALCEOLARIA

Sprays must not be applied in bright sunshine or to plants requiring watering.

Pests

**Aphids (*Aulacorthum circumflexum, A. solani* and *Myzus persicae*). These, and occasionally other species, breed on calceolaria. Sprays of gamma-HCH, *or* nicotine, *or* pirimicarb, *or* smokes of gamma-HCH, *or* nicotine, *or* pirimicarb, *or* drenches of demeton-S-methyl, *or* oxydemeton-methyl as in Table 10.2, p. 266 will control them, except gamma-HCH will not kill *A. circumflexum.*

*Leaf hopper (*Zygina pallidifrons*). The feeding of these insects on the undersides of the leaves causes mottled or bleached spotting. They are easily controlled with gamma-HCH, *or* malathion, as in Table 10.2, p. 266.

**Whitefly (*Trialeurodes vaporariorum*). Calceolarias are often badly checked in growth and soiled by whitefly. Control as for Chrysanthemum Whitefly, Table 10.2, p. 266, but do not use malathion.

Diseases

*Root and **Stem rots.** Calceolarias are susceptible to several soil-borne fungal diseases. Seed and potting composts should be made with steam-sterilized soil.

CARNATION
(see MAFF Bull. 151)

Pests

***Aphids. The peach-potato aphid (*Myzus persicae*) and the glasshouse potato aphid (*Aulacorthum solani*) are pale to dark yellowish green when wingless, and mainly confined to the young leaves. The young stages feed within the growing tip, thus protected from contact insecticides. Systemic insecticides may give better control. Control measures should be effected before flowers begin to open. See Table 10.2, p. 266, for suitable treatments. If conditions permit the heating of the closed glasshouse without detriment to flower quality, control may be obtained by applying naled to the cold cast-iron pipes at the rate of 5·3 g/100 m^3 and then turning on heat to keep the pipes at a minimum temperature of 60°C (140°F) overnight or 82°C (180°F) for 3 hours. If naled is to be applied to hot pipes, 49°–60°C (120°–140°F), a respirator must be worn.

*Carnation tortrix (*Cacoecimorpha pronubana*). The yellowish, olive-green caterpillars feed on upper leaves, tying them together with silken threads, and later enter flower buds and destroy blooms. For control, use DDT dust at 840 g a.i./ha *or* DDT sprays 625 g a.i./100 litres HV *or* DDT aerosol 11·5 g a.i./100 m^3 *or* DDT smokes. Apply immediately the pest is seen and repeat after 14 days.

Earwigs—see Chrysanthemum—Earwigs, p. 271.

**Red spider mite (*Tetranychus cinnabarinus*) (A.L. 224). A reddish brown mite which causes initially small whitish spots on the leaves; as the infestation increases, the leaves become pale stippled and desiccated. The mite breeds on carnations throughout the year. Occasionally *T. urticae* attacks carnations causing similar damage. Mites may be controlled by treatments selected from Table 10.2, p. 266. The organophosphorus compounds do not kill the eggs so a further treatment will be required 7–10 days later. Complete coverage of the infested foliage is required with sprays of contact acaricides. Control should be effected before the flowers begin to open.

*Thrips (*Thrips tabaci, Taeniothrips atratus*) are slender small insects, brownish-black and very active when adult but yellowish and less active when immature. They feed on petals causing silvery streaks and sometimes distortion, and often invade houses in summer. They are controlled by atomized solutions or HV sprays of diazinon *or* gamma-HCH *or* malathion at the rates given in Table 10.2, p. 266. Naled fumigation (see Carnation—Aphid, p. 264) is also effective.

*Wireworms—see Tomato—Wireworms, p. 307.

*Woodlice—see Tomato—Woodlice, p. 307.

Diseases

***Fusarium wilt (*Fusarium oxysporum* f. *dianthi*) invades the vascular tissues causing whole plants or parts of them to wilt. Use steam-sterilized soil for beds or pots, and propagate from disease-free plants. In border soils, metham-sodium (see p. 254) used prior to the removal of an old, infected crop and followed by steam sterilization gives better control than steaming alone but no firm recommendation can yet be given.

Remove all diseased plants immediately, as the pathogen can sporulate freely on infected shoots and to reduce spread of wilt treat the entire crop with soil drenches of benomyl 75 g a.i./100 litres at 5·5 litres/m² *or* carbendazim 82 g a.i./100 litres at 5 litres/m² *or* thiophanate-methyl at 75 g a.i./100 litres at 11 litres/m². Repeat the soil drench 14 days later avoiding waterlogging. Preventive soil drenches applied before infection occurs are more effective than those applied after symptoms appear.

*Greasy-blotch (*Zygophiala jamaicensis*) appears on leaves as green fibrillate patches which may extend and join to give the leaves an oily appearance; encouraged by excessive humidity. When the disease persists in spite of efforts to reduce humidity, overseas experience suggests that captan 100 g a.i./100 litres HV at 7–10-day intervals is of value but insufficient UK evidence is available for a firm recommendation.

**Grey mould (*Botrytis cinerea*) causes spotting and rotting of buds and flowers and may be serious if lesions become established on old, yellowing leaves and are allowed to sporulate and provide inoculum for the flowers. Glasshouse ventilation and heating should be regulated to prevent high humidity which favours the disease. For fungicide control measures see Table 10.3, p. 274.

Leaf spot or **Alternaria blight (*Alternaria dianthi*) begins as small purple spots which enlarge under moist conditions into typical leaf spots with purple edges and brown-black sporulating centres. The disease may also infect stems. Spray with captan 100 g a.i./100 litres HV, *or* mancozeb 100 g a.i./100 litres HV, *or* zineb 125 g a.i./100 litres HV.

*Mildew (*Oidium* spp.) appears as a whitish powder on the leaves and occasionally on the calyces. It is controlled by dinocap at 8 g a.i./100 litres HV, 2 or more applications at 10–14-day intervals (the addition of a wetter is necessary), *or* sulphur dust at 28 kg a.i./ha at 7-day intervals.

Table 10.2. **Control of aphids, red spider mite and whitefly on carnations, chrysanthemums and roses**

(Note:) Before using materials listed in this table consult product label for information on species/cultivars which should not be treated. Always cut open blooms before treating. If biological control of pests is being used, see also Table 10.1, p. 258.

Compound	Rate	Suitable for use on			Effective against		
		Carnations	Chrysanthemums	Roses	Aphids	Red spider mite	Whitefly
I. HV Sprays	a.i./100 litres						
Cyhexatin	25 g	Yes	Yes	Yes	—	Yes	—
DDT	100 g	Yes	Yes	Not Blooms	—	—	Adults, scales
Demeton-S-methyl*	22 g	Yes	Most varieties	Yes	Yes	Yes	—
Derris	3-7 g	Yes	Yes	Yes	Yes	Yes	—
Diazinon	16 g	Yes	Yes	Yes	Yes	Yes	Adults
Dicofol	10 g	Do not apply to soft growth		—	—	Yes	—
Dicofol/tetradifon	40 g + 12·5 g	Yes	Yes	—	—	Yes	—
Dimethoate*	30 g	Yes	—	Yes	Yes	Yes	—
Formothion*	44 g	Yes	—	Yes	Yes	Yes	—
Gamma-HCH	12·5 g	Yes	Yes	Yes	Yes+	—	Adults
Malathion	113 g	Yes	Avoid open blooms	Yes	Yes	Yes	—
Nicotine	50 ml (95—98 per cent)	Yes	Yes	Yes	Yes	Yes	—
Oxydemeton-methyl*	22 g	Yes	Most varieties	Yes	Yes	Yes	—
Parathion	10 g	Yes	Yes	Yes	Yes	Yes	Adults
Petroleum emulsion (glasshouse grade)	1 litre product	—	Yes	—	—	Yes	—
Pirimicarb*	25 g	Yes	Yes	Yes	Yes	—	—
Pyrethrin/resmethrin	See product label	—	Yes	—	—	—	Adults, scales, eggs
Quinomethionate	12·5 g	Yes	Most varieties	—	—	Yes	—
Tetradifon	10 g	Yes	Yes	Yes	—	Eggs	—

2. Atomizations

Product	a.i./100 m³						Control
Diazinon	5·3 g	Yes	Yes	Yes	Yes	Yes	Adults
Dichlorvos	3·5 g	Yes	Most varieties	—	Yes	Yes	Adults
Dicofol	6 g	Yes	Yes	Yes	Yes+	—	—
Gamma-HCH	2·5 g	Spring or summer only	—	Yes+	Yes	—	Adults
Malathion	8·5 g	Yes	Yes	Yes	Yes	Yes	Adults
Parathion	1·7 g	Yes	Yes	Yes	Yes	Yes	Adults
Resmethrin	See product label	Yes	Yes	Yes	—	—	Adults

3. Smokes (S), Fogs (F), Fumigations (Fu)

Product	Rate						Control
DDT		Yes	Yes	Yes	—	—	Adults
Gamma-HCH (S)		Yes	Yes	—	Yes+	—	Adults
Gamma-HCH + DDT (Fu)†		Yes	Yes	—	Yes+	—	Adults
Naled (S) (Fu)		Yes	Yes	Yes	Yes	—	Adults
Nicotine	See Product label	Yes	Yes	Yes	Yes	Yes	—
Parathion (S)		Yes	Yes	Yes	Yes	—	Adults
Pirimicarb (S)		Yes	Yes	Yes	Yes	—	Adults, scales
Pirimiphos-methyl (F)		Yes	Yes	Yes	Yes	—	Adults
Propoxur (S)		Yes	Yes	—	Yes	—	Adults, young scales
Pyrethrin/resmethrin (F)		Yes	Yes	Yes	—	—	
Quinomethionate (S)		—	Yes	—	—	Yes	—
Sulfotep (S)		Yes	Yes	Yes	Yes	—	Adults

4. Soil drenches‡

Product	a.i./100 litres						Control
Demeton-S-methyl*	14 g	Yes	Yes	Yes	Yes	Yes	—
Dimethoate*	30 g	Yes	—	Yes	Yes	Yes	—
Formothion*	44 g	Yes	—	Yes	—	—	—
Oxydemeton-methyl*	14 g	Yes	Yes	Yes	Yes	Yes	—

5. Granules

Product	Rate	a.i./100 m²	a.i./100 m²	a.i./100 m²			Control
Aldicarb*	For rate see individual crop	56 g	48–56 g	112 g	Yes	Yes	Adults, scales
Oxamyl*		90 g	112 g	—	Yes	Yes	—

* Systemic compounds
+ No control of *Aulacorthum circumflexum*
† See Carnation Aphids p. 264
‡ For volume of diluted pesticide to be applied see product label.

Rust (*Uromyces dianthi* (*caryophyllinus*)) causes small blisters on the lower leaves and stems, which rupture forming pustules filled with brown, powdery spores. The disease develops when the foliage remains wet or dew forms overnight. For control, reduce the humidity within the glasshouse and spray with benodanil 50 g a.i./100 litres HV, *or* mancozeb/zineb 175 g w.p./100 litres HV, *or* oxycarboxin 70 g a.i./100 litres HV, *or* thiram 300 g a.i./100 litres HV, *or* zineb 125 g a.i./100 litres HV and repeat at intervals of 10–14 days. Cut open blooms before spraying, and check the tolerance of the variety to the treatment by spraying a few plants before treating the rest of the crop.

Slow wilt (*Erwinia chrysanthemi*). A systemic bacterial disease which is now rare as a result of modern 'cultured cutting' propagating techniques. Plants wilt over considerable periods before death. To control the disease, plant in steamed soil, for no chemical method is yet available. Stock plants should be tested for freedom from the disease if there is any risk of its introduction.

Stem-rot (*Fusarium roseum*) causes a rotting of the base of the stem and sometimes of branch stems; woody tissues are not discoloured. The pathogen often invades wounds made when removing shoots or picking flowers and may cause basal rot of cuttings taken from diseased plants. As the pathogen is not systemic, it is not detected by cultured-cutting technique. It is common in many unsterilized soils. Control is effected by spraying stock plants with captan 100 g a.i./100 litres HV at 10–14-day intervals before and during cutting-taking, by planting in steam sterilized soil and by propagating in sterilized sand or other sterile media. If established plants are attacked, remove infected shoots and spray with captan several times at 7–10-day intervals.

Verticillium wilt (*Verticillium* (*Phialophora*) *cinerescens*) causes rapid wilting, often on one side initially. Then plants soon turn grey-green and become desiccated with extensive light brown discolouration of the xylem. The fungus persists for many years and penetrates deeply into soil. Remove diseased plants and isolate the area from which they are removed. It may be controlled by steaming the soil in raised beds, but in border soil more effective control may be obtained if metham-sodium is applied prior to steaming, see Carnation—*Fusarium* Wilt, p. 265. Trials in Britain and overseas indicate that metham-sodium 0·8–1·2 litres of a 33 per cent solution/100 litres applied at 1220 litres/100 m², to the subsoil and at the same rate to the upper 300 mm before planting (use cress germination test, p. 253) controls the disease but more evidence is required before a firm recommendation can be made. For use of benomyl *or* carbendazim *or* thiophanate-methyl soil drenches, see *Fusarium* wilt (p. 265).

CELERY (Self-blanching)
(see S.T.L. 169)

Pests

**Aphids may infest the crop at all stages of development. Attacks early in the crop may be controlled by HV sprays of demeton-S-methyl, *or* dimethoate, *or* formothion, *or* oxydemeton-methyl. Later attacks can be controlled by sprays of malathion *or* smoking with nicotine. Rates and preharvest intervals are as given for Lettuce—Aphid, Table 10.6, p. 291.

   *Carrot fly (*Psila rosae*) (A.L. 68) is described under Carrot, p. 323. When carrot fly is a problem drench the soil with diazinon 50 g a.i./100 litre at 130 ml/plant within a week of planting out.

   **Celery fly (*Philophylla heraclei*) (A.L.87) is also known as celery leaf miner. The larvae feed in the leaves causing large blisters. Larvae are controlled by dimethoate 37 g a.i./100 litres HV, *or* malathion 113 g a.i./100 litres HV, *or* trichlorphon 37 g a.i./100 litres HV. Observing the following minimum intervals between treatment and harvest of 24 hours (preferably 4 days to avoid taint) for malathion; 7 days (dimethoate) and 48 hours (trichlorphon).

   **Slugs—see Tomato—Slugs, p. 306.

Diseases

**Celery leaf spot (*Septoria apiicola*) (A.L.241). For description see p. 326 and for control treat seed by the thiram soak method (see p. 314) and spray the growing crop fortnightly with benomyl 50 g a.i./100 litres HV *or* carbendazim 50–60 g a.i./100 litres HV, *or* zineb 140 g a.i./100 litres HV allowing at least 2 days between spraying and harvest.

   ***Damping-off (*Pythium* spp., *Rhizoctonia solani*). Patches of seedlings collapse and when pulled out roots are shrivelled and sometimes reddish-brown. This condition called 'rusty root' is caused by the fungus *Artotrogus hydrosporus* (syn. *Pythium artotrogus*). Propagate seedlings in sterilized compost and avoid overwatering. Soil drenches of etridiazol† applied according to the manufacturer's instructions may decrease the spread of *Pythium* but further work with this material is necessary before firm advice can be given. Quintozene which is usually recommended for controlling *Rhizoctonia* attack in other crops has caused a check to growth of celery with consequent early bolting.

   **Grey Mould (*Botrytis cinerea*). The fungus colonizes damaged petioles and a greyish mould with profuse clusters of spores is produced. Infected tissue rots and eventually the whole leaf stalk collapses. High humidity encourages the disease. For control spray with benomyl *or* carbendazim as for Celery Leaf spot p. 326.

† Proposed BSI common name for 5-ethoxy-3-trichloromethyl-1,2,4-thiadiazole.

***Sclerotinia rot** (*Sclerotinia sclerotiorum*) (A.L. 265). The first symptoms are usually pink lesions at the base of leaf stalks. The lesions become covered with a white, fluffy growth within which the black sclerotia (or resting bodies) are formed. These sclerotia can remain viable in the soil for several years and it is important to destroy carefully all infected plant debris and sterilize the soil by steam or methyl bromide, see p. 254, after an infested crop. In trials benomyl applied as a post-planting drench followed by fortnightly benomyl sprays has given promising control.

CHRYSANTHEMUM
(see MAFF Bull. 92)

There are marked differences in the tolerances of different chrysanthemum varieties to pesticides and their susceptibilities to diseases and pests. In case of doubt it is advisable to make a trial on a few plants of any new variety.

Pests

****Angleshades moth** (*Phlogophora meticulosa*), and other **Noctuid moths**. Caterpillars of these moths, particularly of the first, feed on the foliage and flowers of chrysanthemums. They are controlled by DDT smoke *or* dust at 100 g a.i./ha or spray at 100 g a.i./100 litres HV. Smokes of DDT/HCH can be used where pests susceptible to the second ingredient are present. Repeat treatments after 14 days if necessary.

Specific 'Chrysanthemum Sprays' may be used if red spider mite or mildew are also present.

Spodoptora littoralis (*Prodenia litura*) caterpillars are resistant to DDT. Young larvae are controlled by dichlorvos atomized at 3·5 g a.i./100 m³ *or* carbaryl 75 g a.i./100 litres HV *or* trichlorphon 200 g a.i./100 litres HV.

*****Aphids.** Chrysanthemums in glasshouses are attacked by several species of aphids, the most important being the peach-potato aphid (*Myzus persicae*) whose wingless form is green or greenish yellow; the mottled arum aphid (*Aulacorthum circumflexum*) yellowish green with a well-defined dark horseshoe-shaped mark on the abdomen; the chrysanthemum stem aphid (*Macrosiphoniella sanborni*) blackish red or black, and the dark green leaf-curling plum aphid (*Brachycaudus helichrysi*).

Control methods may be complicated by varying patterns of resistance in *M. persicae* and where this aphid is a problem biological control can be very effective, see Table 10.1, p. 258. Chemical control methods are listed in Table 10.2, p. 266, and treatment should be affected before the buds open. Systemic compounds give the most persistant protection to the foliage but compounds with a good contact or fumigant action are needed to control aphids on flower buds. Demeton-S-methyl drenches have given better control than drenches of oxydemeton-methyl.

Capsid (Tarnished plant bug) (*Lygus rugulipennis*). The bug feeds on stems, leaves and flowers causing puckered, mis-shapen leaves, distorted stems and small, malformed flowers. Very occasionally *Lygocoris pabulinus* and *Calocoris norvegicus* cause similar damage. Control of the pest is achieved by: atomized solutions of DDT 10·5 g a.i./100 m³ *or* gamma-HCH 3·5 g a.i./100 m³; smokes of gamma-HCH *or* DDT; sprays of DDT as for Angleshades moth, see p. 270. Specific 'Chrysanthemum Sprays' containing DDT, petroleum oil and thiram may be used if it is necessary to control red spider mite and mildew.

Earwigs (*Forficula auricularia*) feed on flowers, giving them a ragged appearance and increasing their susceptibility to grey mould. To control, spray with carbaryl 200 g a.i./100 litres HV *or* gamma-HCH 12 g a.i./100 litres HV, *or* trichlorphon 80 g a.i./100 litres HV. Where HV sprays are undesirable, e.g. where grey mould (*Botrytis*) is present, atomize gamma-HCH 3·5 g a.i./100 m³ *or* smoke with gamma-HCH. Repeat at 7-day intervals if necessary.

Leaf and bud eelworm (*Aphelenchoides ritzemabosi*) (A.L. 339, 379). The first sign of attack is a yellowish green or purple blotching on the basal leaves, often delineated by veins; later these blotches darken and become desiccated.

Plants should be propagated from eelworm-free stock. If this is impractical, clean plants may be obtained by applying drenches of thionazin 29 g a.i./100 litres to the cuttings when well rooted and again after 14 days. Alternatively apply aldicarb granules to soil surface at 48—56 g a.i./100 m² (96—112 g a.i./100 m² on stool beds). These chemical treatments have largely replaced hot-water treatment of the washed, cut-back stools of stock plants. Stools were immersed for 5—5·5 minutes in water kept at 46°C (115°F) *or* for 20—30 minutes in water kept at 43°C (110°F), followed by immediate immersion in cold water and growing on until cutting material is obtained. Hot-water treatment may delay cutting production and is severely damaging to some varieties.

Leaf miner (*Phytomyza syngenesiae*) (A.L. 550). The larvae of this small fly disfigure the leaves by eating tunnels (mines) in leaves. Control with sprays of diazinon 16 g a.i./100 litres HV, *or* gamma-HCH 12 g a.i./100 litres HV at 14-day intervals *or* aldicarb granules applied to soil, see Table 10.2, p. 266. Strains resistant to gamma-HCH have been observed. If biological control of other pests is being used see Table 10.2, p. 266.

***Red spider mite** (*Tetranychus urticae*). For a description see A.L. 224 and Cucumber—Red spider mite, p. 280. The mites remain on the undersides of the leaves before the flower buds are well formed and then migrate up to the opening buds where they do considerable economic damage, hence they should be controlled before the buds begin to open. Biological control (see Table 10.1, p. 258) can be used to control this pest and may be preferable where mite resistance to chemicals is a problem. Chemical treatments are listed in Table 10.2, p. 266, but organophosphorus compounds do not kill eggs so retreatment will be required in 7–10 days. Complete spray coverage of the infested regions of the plants must be obtained with contact acaricides.

Stool miner (*Psila nigricornis*). The larvae tunnel into basal tissues and reduce cutting production. They may be controlled by drenching sprays of gamma-HCH 10 g a.i./100 litres in early May and again in early September.

Symphylids (*Scutigerella immaculata*). For description and control see Tomato—Symphylids, p. 308.

Thrips—see Carnation—Thrips, p. 264.

Whitefly (*Trialeurodes vaporariorum*) (MAFF Bull. 86)—see Cucumber—Whitefly, p. 281. Chemical control is complicated by resistance of whitefly strains to many insecticides and a biological control has yet to be perfected on this crop. Aldicarb granules applied for the control of aphids and red spider mites should also control whitefly. Alternative treatments are listed in Table 10.2, p. 266 and frequency of application must be related to the kill of the various stages of this pest. Thus if adults only are killed it may be necessary to repeat the treatment within 5 days, but consult the product label.

Diseases

Blotch (*Septoria chrysanthemella*) causes dark grey-black spots or blotches on older leaves with defoliation if severe. To control, increase ventilation and spray with zineb 125 g a.i./100 litres HV. (For 'tank mixed' zineb see Tomato leaf mould, p. 309.)

Damping-off (*Pythium* spp. and *Rhizoctonia solani*). Damping off is most frequent in propagating beds and young plants and is associated with poor hygiene. After an infected crop sterilize beds and compost by steaming. Copper fungicides *or* drazoxolon *or* etridiazole† watered onto soil according to manufacturer's instructions as soon as *Pythium* infection occurs may prevent disease spread. If damping-off is caused by *Rhizoctonia*, dust stem bases and soil surface with quintozene dust 13 g a.i./m².

Grey mould (Botrytis flower-rot) (*Botrytis cinerea*) causes spotting and rotting of buds and flowers. Infection may also be serious in the leaves and stems, especially in cuttings inserted in the propagating bench after protracted periods of cold storage. Grey mould is most common under warm, humid conditions and the mass of grey spores which are produced at the rotting stage distinguishes this disease from petal blight

† Proposed BSI common name for 5-ethoxy-trichloromethyl-1,2,4-thiadiazole.

and ray blight. With *Botrytis*-susceptible varieties in humid weather, it is advisable to apply a preventive spray as a fine mist to the crop at bud burst, using one of the materials listed in Table 10.3, p. 274. Dichlofluanid at 50 g a.i./100 litres, applied as a fine mist, has given good results but its safety on all varieties has yet to be established. If infection has already occurred, remove diseased blooms promptly and apply one of the spray, dust or smoke treatments listed in Table 10.3, p. 274, and repeat the treatment if humid conditions continue.

Spraying stock plants with dichlofluanid at 100 g a.i./100 litres HV prior to the removal of cuttings greatly decreases the incidence of *Botrytis* on cuttings, and additional control is obtained in mist-propagated cuttings if sprays are applied 5 and again 10 days after insertion.

Alternatively, dusting cuttings and the surface of the rooting medium with captan, *or* dicloran, *or* thiram immediately after inserting cuttings, or watering in with captan 2 g a.i./litre reduces rotting of cuttings by *Botrytis*.

***Mildew** (*Oidium chrysanthemi*). A typical powdery mildew, starting as patches on the upper side of young leaves and progressing in severe cases to cover upper and lower leaf surfaces. It may be controlled by the vaporization of commercially pure sulphur 14 g a.i./100 m^3 at intervals of 7–14 days, or nightly at 1·7–3·5 g a.i./100 m^3, using apparatus designed to vaporize sulphur without risk of ignition, for fumes from burning sulphur are extremely phytotoxic. Alternatively, apply a spray or smoke selected from Table 10.3, p. 274. On year-round pot chrysanthemums drenching the moist compost with dimethirimol (used according to manufacturer's instructions) has given good control when applied at the onset of attack and reapplying when reinfection occurs. If biological control of pests is being used, see also Table 10.1, p. 258.

Petal blight (*Itersonilia perplexans*) is first seen on the petals as minute, reddish-brown spots which under humid conditions coalesce to form water-soaked spots. It usually first appears on the tips of the outer petals and gradually extends inwards to affect the entire bloom. It may be confused with grey mould (p. 272) but the sporulating mycelium is seen as a greyish white sheen on the rotting petals. The disease is more prevalent in unheated houses where water droplets persist on the petals overnight. To control, avoid high humidity by ventilation and heating where possible. Alternatively, preventive sprays of zineb 125 g a.i./100 litres prepared by the tank-mixed method (see Tomato—Leaf mould, p. 309) should be applied as a fine spray at the coloured bud stage. If humid conditions prevail, further applications should be given at 7-day intervals. Although less effective than freshly prepared tank-mixed material, zineb w.p. at 125 g a.i./100 litres HV, *or* zineb dust at 3·3 kg a.i./ha may be used.

Phoma root-rot (*Phoma chrysanthemicola*). The plants are stunted or killed by severe rotting of roots and, ultimately, of base of stem. Lower leaves show interveinal and chlorotic areas which become necrotic.

Table 10.3. Ornamentals—Powdery mildew and grey mould

(Note:) Before using a fungicide listed in this table consult manufacturer's label for information on species and cultivars which may or may not be treated. If biological control of pests is being used, see also Table 10.1, p. 258

Fungicide	Application rate		Usual interval between applications (days)	Remarks
	Powdery mildew	Grey mould		
	a.i./100 litres	a.i./100 litres		
1. HV sprays				
Benomyl*	25 g	25 g	14	
Captan	—	100 g	10	
Carbendazim*	28–30 g	28–30 g	14	
Chlorothalonil	150 g	110 g	7–14	
Dinocap	6·3 g	—	10–14	Some varieties of chrysanthemum, rose and begonia susceptible to damage.
Pyrazophos	15 g	—	10–14	Do not spray rose, aquilegia and Scorzonera.
Quinomethionate	12·5 g	—	14	Some varieties of azalea, begonia, chrysanthemum, hydrangea, pelargonium, rose and some foliage species susceptible to damage. Quinomethionate also controls red spider mites.
Thiophanate-methyl*	50 g	50 g	14	
Thiram	—	300–320 g	10–14	Hydrangea susceptible to damage.
Thiram/petroleum oil/DDT	1 litre	1 litre	14	Asparagus fern, fuschia and pelargonium

	(product)	(product)		
Triforine	25 g	—	10–14	susceptible to damage. This product also controls capsids, red spider mites and caterpillars. A few varieties of roses susceptible to damage.
2. Dusts		*a.i./ha*		
Captan	—	3.4–4.2 g	7–10	See thiram HV sprays above.
Thiram	—	3.4–9.4 g	10–14	
3. Smokes				
Dinocap	‡	—	‡	See dinocap HV sprays above.
Tecnazene	—	‡	‡	
4. Soil drenches	*a.i./100 litres*	*a.i./100 litres*		
Benomyl*	25 g	25 g		Recommended for use on pot plants, applied in place of a watering.
Carbendazim*	28 g	28 g		Recommended for use on pot plants, applied in place of a watering.
Dimethirimol†	10 g	—	When disease reappears	Recommended for pot cineraria and pot chrysanthemum. See manufacturer's label.
Thiophanate-methyl*	50 g	50 g		Recommended for use on pot plants, applied in place of a watering.

*† Where strains of fungi tolerant to systemic fungicides marked * or † occur an alternative fungicide outside the particular group must be used.
‡ Follow manufacturer's instructions.

Varieties differ markedly in susceptibility. Planting in steam-sterilized soil will avoid the disease and pre-planting soil drenches of nabam 140 g a.i./100 litres at 22 litres/m² have given good control.

*Ray-blight (*Mycosphaerella ligulicola*). This disease can be distinguished from petal blight as infection develops first on the inner petals and spreads outwards producing flower distortion and dark brown rotting. The fungus also attacks stems, especially of cuttings and young plants, causing dark lesions and stunting. Remove infected plants and spray HV every 7–10 days with captan 100 g a.i./100 litres *or* mancozeb 80 g a.i./100 litres *or* mancozeb/zineb 175 g w.p./100 litres. In the USA, dipping cuttings in benomyl at 50 g a.i./100 litres prior to insertion in the propagation bench has been successful as a preventive measure. After an infected crop, sterilize the soil by steaming.

*Rust (*Puccinia chrysanthemi*) attacks the leaves under humid conditions causing yellowish spots which develop into rusty, spore-producing pustules on the undersides. Spray with benodanil 50 g a.i./100 litres HV *or* oxycarboxin 71–94 g a.i./100 litres HV *or* thiram 300 g a.i./100 litres HV *or* zineb 125 g a.i./100 litres HV at 10–14-day intervals when the disease first appears. 'Chrysanthemum Spray' formulations containing thiram, petroleum oil and DDT are also effective, and give simultaneous control of caterpillars, capsids, mildew and red spider mite. Preventive spraying may be advisable if the disease is endemic.

*Verticillium wilt (*Verticillium albo-atrum*). The lower leaves turn yellow-brown and discoloration of foliage progresses upwards; the plants are stunted and woody. Care in the selection of cuttings and planting in sterilized soil helps to avoid the disease.

CINERARIA

Pests

**Aphids (*Myzus persicae, Aulacorthum circumflexum, A. solani* and *Aphis gossypii*). Control, with the rates given in Table 10.2, p. 266, using diazinon *or* malathion *or* naled *or* nicotine *or* parathion *or* pirimicarb, *or* gamma-HCH (not effective against *A. circumflexum*). Plants in flower may be treated with smokes of gamma-HCH. Drenches of demeton-S-methyl *or* oxydemeton-methyl have been successfully applied to cinerarias but should be used with care and not in hot, bright weather.

**Leaf miner (*Phytomyza syngenesiae*)—see Chrysanthemum—Leaf miner, p. 271.

*Whitefly (*Trialeurodes vaporariorum*). Control with DDT *or* diazinon *or* malathion *or* naled *or* parathion *or* pirimiphos-methyl as for Chrysanthemum—Whitefly, Table 10.2, p. 266.

Diseases

*Powdery mildew (*Sphaerotheca fuliginea*). For control measures, see Table 10.3 p. 274, avoiding, if possible, applications during hot, bright weather. Alternatively apply soil drenches of dimethirimol according to manufacturer's instructions.

*Grey mould (*Botrytis cinerea*). Attacks flowers, leaves and stems; for control measures see Table 10.3 p. 274.

CUCUMBER AND MELON

Pests

*Aphid (*Aphis gossypii*). This aphid is very variable in colour—yellowish green to black, and feeds on foliage causing yellowing, and on fruitlets which become distorted or fail to develop. Suitable control measures (Table 10.4, p. 278) depend on the stage of growth and season.

*Clover mite (*Bryobia spp.*). Clover mites sometimes invade cucumber houses from clover and grasses in February and March and cause damage to foliage and fruit. HV sprays of dicofol have given good control, see Table 10.4 p. 278.

**French-fly (*Tyrophagus longior*) is a glistening-white globular, sluggish mite with prominent bristles, commonly occurring in straw and undecomposed horse manure. It migrates from beds and feeds on and within the leaves of the shoot tips of young plants causing irregular perforations in the leaves, distortion and sometimes blindness of the shoots. It does not reproduce on cucumbers.

To control, spray the plants, especially growing tips and surface of beds, with parathion 10 g a.i./100 litres HV as soon as damage is seen and not later than 4 weeks before cutting. Incorporate a wetting agent if necessary to ensure that the spray penetrates the growing tips. Repeat after 7 days if the attack continues. Parathion smokes should be used if it is necessary to control the pest within 4 weeks of cutting but 24 hours must elapse before harvest.

**Fungus gnats (root maggots) (*Sciara* spp.). (S.T.L. 110). The larvae are about 6 mm long with black heads and creamy translucent bodies. They are common in beds containing horse manure, and occasionally attack the large roots and stem-bases of cucumbers in beds, especially if the beds become too dry. They also destroy young plants in pots. To control drench the beds with parathion 6 g a.i./100 litres at 5–7·5 litres/m run of bed. One application only per crop is acceptable under official recommendations and is usually sufficient to control an attack, but 4 weeks must elapse before harvest.

***Root-knot eelworm (*Meloidogyne hapla* and *M. incognita acrita*) (A.L. 307) cause swollen, gall-like growths on the roots which reduce the vigour of the plants and greatly decrease cropping. Attacks are most

Table 10.4. **Cucumber—aphid, red spider mite and whitefly**

(Note:) If biological control of pests is being used, see also Table 10.1, p. 258.

Compound	Rate	Minimum interval between treatment and harvest	Suitable for use on		Active against		
			Young plants	Established plants	Aphid	Red spider mite	Whitefly
1. HV Sprays	*a.i./100 litres*						
Demeton-S-methyl*	22 g	21 days	Yes	—	Yes	Yes	—
Derris	3–7 g	24 hours	Yes	Yes	Yes	Yes	—
Diazinon	16 g	14 days	—	Summer only	Yes	Yes	Adults
Dicofol	10 g	48 hours	†	Yes	—	Yes	—
Dicofol/tetradifon	40 g + 12·5 g	48 hours	—	Summer only	—	Yes	—
Dimethoate*	30 g	7 days	—	Yes	Yes	Yes	—
Dinobuton‡	30 g	3 days	—	Summer only	—	Yes	—
Malathion**	113 g	24 hours**	—	Yes	Yes	Yes	Adults
Nicotine	50 ml (95–98 per cent)	48 hours	—	Yes	Yes	—	—
Oxydemeton-methyl*	22 g	21 days	Yes	—	Yes	Yes	—
Parathion§	10 g	28 days	Yes	—	Yes	Yes	Adults
Petroleum emulsion (glasshouse grade)	1 litre (product)	Nil	—	Yes	—	Yes	—
Pirimicarb*	25 g	48 hours	Yes	Yes	Yes	—	—
Pyrethrin/resmethrin	See product label	Nil	Yes	Yes	—	—	Adults, scales and eggs
Quinomethionate	12·5 g	48 hours	—	Summer only	—	Yes	—
Tetradifon	10 g	Nil	†	Yes	—	Yes	—

2. Atomization

		a.i./100 m³						
Diazinon		5.3 g	48 hours	—	Summer only	Yes	Yes	Adults
Diazinon + dicofol		5.3 g + 6.3 g	48 hours	—	Summer only	Yes	Yes	Adults
Dicofol		7 g	48 hours	—	Summer only	—	Yes	—
Dinobuton‡		11.3 g	3 days	—	Summer only	Yes	Yes	—
Malathion		8.5 g	24 hours	—	Summer only	Yes	Yes	Adults
Parathion‡		1.7 g	48 hours	—	Yes	Yes	Yes	Adults
Resmethrin		See product label	Nil	Yes	Yes	—	—	Adults
Tetradifon		3.5 g	Nil	—	Summer only	—	Yes	—

3. Smokes (S), Fogs (F) and Fumigation (Fu)

		a.i./100 m³						
Nicotine	(S) (Fu)	See product label	24 hours	—	Yes	Yes	—	—
Parathion‡	(S)	See product label	24 hours	—	Yes	Yes	Yes	Adults
Pirimicarb	(S)	See product label	Nil	Yes	Yes	Yes	—	—
Pirimiphos-methyl	(F)	See product label	Nil	Yes	Yes	—	—	Adults, scales
Propoxur	(S)	See product label	48 hours	—	Yes	Yes	—	Adults
Pyrethrin/resmethrin	(F)	See product label	Nil	Yes	Yes	—	—	Adults, young scales
Quinomethionate	(S)		48 hours	—	Yes	—	Yes	Adults
Sulfotep	(S)		24 hours	Yes	Yes	—	Yes	Adults

* Systemic compound.
† Risk of 'hardening'. only spray infested leaves.
** Allow 4 day interval to avoid taint.
‡ Max. 5 applications/season.
§ Max 3 applications/season.

serious at high temperatures in light soils and, for control, the borders should be steam-sterilized or treated with dichloropropane-dichloropropene mixture *or* dichloropropene *or* methyl bromide as described under 10.1A (see pp. 253–254). Where endemic attack is encountered soil drench young plants once with parathion 6 g a.i./100 litres at 500 ml/plant.

***Millepede (glasshouse)** (*Oxidus gracilis*) (A.L. 150). The glasshouse millepede is common in cucumber houses and when numerous, destroys plants by feeding on the soft tissues of the stem base and larger roots. To control this pest drench the beds with DDT 95 g a.i./100 litres at 2·5 litres/m run *or* gamma-HCH 12·5 g a.i./100 litres at 2·5 litres/m run *or* parathion 25 g a.i./100 litres at 5 litres/m (only 1 application per crop). Alternatively drench the bed with nicotine 65 ml (95–98 per cent)/100 litres (with a wetting agent as for HV spraying) at 5 litres/m run applied in the evening, the next morning spray the millepedes on the surface of the beds, paths, etc. with the same solution. Do not allow DDT or gamma-HCH to come into contact with the foliage of cucumbers. Minimum intervals between treatment and harvest are 2 weeks for DDT, zero for gamma-HCH and 4 weeks for parathion.

***Red spider mite** (*Tetranychus urticae*) (A.L. 224). Adult mites of this species hibernate within the glasshouse and emerge when heat and light conditions are favourable, to lay eggs on main crop cucumbers. Eggs hatch within a few days producing whitish pink globular larvae (with three pairs of legs) which pass through two nymphal stages to become greenish oval mites. The complete life cycle from egg to adult takes from 24 days at 15°C (60°F) to 6½ days at 30°C (85°F).

Prevent or reduce the hibernation of adult mites by removing an old crop before mid September, because in autumn mites tend to leave plants which are not producing growth on which they can feed. Remove all weeds.

Strains of mites resistant to acaricides occur more frequently on infested cucumbers than on other crops. Therefore it is advisable to vary the type of chemical used and to spray HV to cover upper and lower leaf surfaces. Alternatively, use biological control (see Table 10.1, p. 258).

Suitable chemical control measures (table 10.4, p. 278) depend on the age of the plants and season. Organophosphorus compounds do not kill the eggs and further treatments may be required at 5–10-day intervals.

**Springtails (*Collembola* spp.). Adults of the species most commonly occurring in cucumber beds are about 3 mm long and dark grey: often the young stages are lighter in colour. They occasionally occur in large numbers in cucumber beds and may injure plants by feeding on roots. To control, water the beds at 2·5 litres/m run with diazinon 30 g a.i./100 litres *or* gamma-HCH 12·5 g a.i./100 litres *or* malathion 113 g a.i./100 litres *or* parathion 6·2 g a.i./100 litres (only 1 application per crop). Allow minimum intervals of 24 hours (malathion), 2 days (diazinon) and 4 weeks (parathion) before harvest. No interval is required for gamma-HCH.

*Symphylid (*Scutigerella immaculata*) (A.L. 484). This insect occasionally attacks cucumbers, checking growth and making the roots susceptible to fungal attack. If uninfested soil is used for bed construction, gamma-HCH 12·5 g a.i./100 litres applied at 1·5 litres/m² bed, is likely to prevent attack. If infestation occurs use diazinon *or* parathion as for Tomato–Symphylids, p. 306.

*Thrips (*Thrips tabaci* and *T. fuscipennis*) sometimes damage cucumbers by feeding on leaves and flowers and are detected by a small irregular white marks on the leaves. *T. tabaci* is controlled by drench methods given for aphids (p. 278), *T. fuscipennis* is controlled by spray treatments given for aphids (p. 278), but not pirimicarb.

***Whitefly (*Trialeurodes vaporariorum*) (A.L. 86). These white moth-like insects normally rest on the undersides of leaves where the conical eggs are laid erect. The larvae are scale-like and after the first two days, immobile. Sooty moulds grow on the honeydew secreted by all whitefly stages and, in the presence of large populations, disfigure the fruit. Many whitefly strains have developed resistance and either biological control Table 10.1, p. 258) or different insecticides must be used. Treatments are listed in Table 10.4, p. 278, and frequency of application must be related to the kill of the various stages of the pest. Thus, if adults only are killed it may be necessary to repeat the treatment within 5 days, but consult the product label.

**Woodlice (*Armadillidium, Oniscus, Porcellio* and *Trichoniscus* spp.) damage cucumbers by gnawing stems and lower leaves. For their control water beds with DDT *or* gamma-HCH *or* parathion as for millepedes or bait with a mixture of Paris green 1 part and either dried blood 56 parts or bran 28 parts, sprinkled on the beds at 17 g/m² and leave intervals of 2 weeks before harvest (DDT) and 4 weeks (parathion).

Diseases

*Anthracnose (*Colletotrichum lagenarium*). Pale green or reddish brown spots which enlarge until the whole leaf withers. The fungus also attacks stems and causes pinkish depressions which develop into black spots. Spray with wettable or colloidal sulphur 150 g a.i./100 litres HV or dust with sulphur 22 kg/ha.

*Basal stem rot (*Erwinia carotovora*) is a slimy soft-rot of the stem at soil level. To control, dust the stems at soil level heavily with a copper dust. Keep the base of stem dry, especially with melons.

***Black root rot (*Phomopsis sclerotioides*). This is an important root-rotting disease and most tap-root decay previously ascribed to *Fusarium, Pythium* and other causes is probably due to *Phomopsis*. Typical symptoms are black spotting on the small roots, black lesions on large

Table 10.5. **Cucumber—powdery mildew and grey mould**
(Note:) If biological control of pests is being used, see also Table 10.1, p. 258.

Fungicide	Rate		Minimum interval between treatment and harvest	Approximate Frequency of Application	Remarks
	Powdery mildew	Grey mould			
	a.i./100 litres	a.i./100 litres			
1. HV sprays					
Benomyl*	25 g	25 g	Nil	3–4 weeks	—
Carbendazim*	28–30 g	28–30 g	Nil	3–4 weeks	—
Chlorothalonil	150 g	110 g	12 hours	7–14 days	—
Copper	63–95 g metallic Cu	—	Nil	10–14 days	Copper fungicide can be added to petroleum emulsion for dilution. See Table 10.4 p. 278
Dinobuton	50 g	—	3 days	10–14 days	For established plants in summer but not more than 5 applications per crop
Dinocap	6·25 g	—	2 days	10–14 days	Young plants ⎱ Slightly incompatible with petroleum oil
	12·5 g	—	2 days	10–14 days	Established plants in summer ⎰

	a.i./100 litres	a.i./100 m³			
Pyrazophos	15 g	—	3 days	10–14 days	—
Quinomethionate	12·5 g	—	2 days	10–14 days	Only use in summer
	20 g	—	2 days	10–14 days	Use if attack is severe for first 2 applications only and use only in summer
Thiophanate-methyl*	100 g	100 g	Nil	3–4 weeks	—
Triforine	20 g	—	2 days	7–10 days	Incompatible with petroleum oil
2. Aerosols					
Dinocap	*a.i./100 m³* 7 g		2 days	7–10 days	Established plants in summer
3. Soil drenches	*a.i./100 litres*				
Benomyl*	20 g		Nil	4–5 weeks	Apply 500 ml/plant to moist soil about 3 weeks after planting
Carbendazim*	20 g		Nil	4–5 weeks	Apply 500 ml/plant to moist soil about 3 weeks after planting
Dimethirimol†	See manufacturer's instructions		Nil	4–5 weeks	—
Thiophanate-methyl*	50 g		Nil	4–5 weeks	Apply 500 ml/plant to moist soil about 3 weeks after planting

*† Where strains of fungi tolerant to systemic fungicides marked * or † occur an alternative fungicide outside the same group must be used.

roots and rotting of the tap root and hypocotyl below ground. If young plants are transplanted into infested soil, rotting develops so rapidly, especially in cold soil, below 14°C (58°F), that the plant dies before the development of the typical black lesions. Steam sterilization of the soil annually is the most effective control.

Grey mould (*Botrytis cinerea*). This disease is encouraged by high humidity and failure to remove damaged and yellowing foliage which is readily colonized by the fungus. For chemical control measures see Table 10.5, p. 282; if biological control of pests is being used, see also Table 10.1, p. 258.

Gummosis (*Cladosporium cucumerinum*) causes sunken scab-like depressions on fruit from which the sap exudes and forms an amber gum. Fruits are often distorted especially if infected when young. Infection is favoured by cool, wet conditions and inadequate ventilation. Hence raise temperature if possible, reduce humidity and remove all diseased fruits. Spraying with zineb 125 g/100 litres HV and repeating at 10-day intervals while the disease persists *or* dusting at 7-day intervals with zineb dust at 3·3 kg a.i./ha will control the disease.

Powdery mildew (*Sphaerotheca fuliginea* (*Erysiphe cichoracearum*)) is a typical white powdery mildew. Spores produced on cucumber are viable only for about 10 days. To control the disease, destroy all weeds within and near houses and, at the first sign of the disease, apply one of the treatments listed in Table 10.5, p. 282; if biological control of pests is being used, see also Table 10.1, p. 258.

Stem and fruit rot (*Didymella bryoniae*. (*Mycosphaerella melonis*)). Lesions can develop on stem, leaves and fruit. Stem lesions usually originate at pruning wounds, fruit infection from the blossom end or where fruit touches an infected stem or leaf. Leaf lesions appear as spreading, light-brown patches which later collapse and rot. Black pycnidia which develop on the lesions exude pinkish spore masses. In trials, sprays of benomyl at 25 g a.i./100 litres decreased lesion spread and prevented sporulation on infected tissue.

Root-rot (Fusarium spp.). Cucumber roots, especially where growing in unsterilized compost or wet beds, are attacked, causing the plants to wilt and die. To control the disease, steam the border and propagating soil.

Verticillium-wilt (Verticillium albo-atrum). The plants wilt and the leaves yellow and become desiccated from base upwards. Control is as for root-rot.

CYCLAMEN

Pests

*Aphids (*Aulacorthum circumflexum,* and other species). For control measures see those for aphids on Carnation (Table 10.2, p. 266), but do not use parathion, or gamma-HCH.

Tarsonemid (cyclamen) mite (*Steneotarsonemus pallidus*) (S.T.L. 133). The minute, waxy, white-brown mites feed on developing foliage causing it to become puckered, distorted, stunted and brittle. When the flower buds are infested, they are either distorted or fail to open. The best control is obtained by spraying with endosulfan 40 g a.i./100 litres HV *or* endrin 36 g a.i./100 litres HV. Dicofol 10 g a.i./100 litres HV or dicofol/tetradifon 200 ml product/100 litres HV are effective. Repeat applications once or twice at 28-day intervals.

*Thrips (*Thrips tabaci, T. fuscipennis, Heliothrips haemorrhoidalis*) damage foliage and flowers. See Carnation thrips, p. 264, but do not use parathion on cyclamen.

Vine weevil (*Otiorrhynchus sulcatus*). The whitish, wrinkled, legless grubs usually lie in the soil in a curved manner, and feed on the corms and roots. The admixture to the compost of dusts of aldrin 19·5 g a.i./m³ *or* DDT at 310 g a.i./m³ *or* dieldrin 19·5 g a.i./m³ *or* gamma-HCH at 31 g a.i./m³ prevents attack. When grubs are attacking plants, drenching with gamma-BHC 10 g a.i./100 litres destroys grubs which have not eaten their way into corms.

Diseases

*Grey mould** (Botrytis rot) (*Botrytis cinerea*). To prevent this disease avoid very high humidity; plant in soil sterilized with steam or formaldehyde; and at the first sign of infection apply one of the control measures listed in Table 10.3, p. 274, or dust with dicloran (4 per cent).

Root rots (*Thielaviopsis basicola, Cylindrocarpon destructans*). Use soil sterilized by steam, dazomet *or* metham-sodium for propagating and potting. To control *Thielaviopsis* water with captan 125 g a.i./100 litres after each repotting and at 21-day intervals after the final potting. To control *Cylindrocarpon* in unsterilized soils mix zineb 350 g a.i./m³ compost.

FREESIA

Pests

Aphids (*Aulacorthum circumflexum, Macrosiphum euphorbiae* and *Myzus persicae*) feed on freesias and cause yellow marking on the leaves and stunted growth. *M. euphorbiae* is a vector of freesia mosaic. Control with demeton-S-methyl *or* diazinon, *or* dimethoate, *or* formothion, *or* gamma-HCH (will not control *A. circumflexum*), *or* malathion, *or* naled, *or* nicotine *or* oxydemeton-methyl *or* pirimicarb, as described for Carnation-Aphids, Table 10.2, p. 266.

*Bean seed fly** (*Delia platura*) is occasionally serious in Guernsey. The larvae of the insect enter and feed on the embryo corm of seedlings, causing them to develop a bluish tinge and lack vigour. It can be controlled by drenching with gamma-HCH 25 g a.i./100 litres to wet the soil thoroughly at the first sign of damage.

Red spider mite (*Tetranychus urticae*) may be controlled on freesias with soil drenches or sprays of demeton-S-methyl, *or* dimethoate, *or* oxydemeton-methyl or sprays of dicofol *or* tetradifon, as for Carnation—Red spider mite, Table 10.2, p. 266.

Slugs—for control see Tomato—Slugs, p. 306.

Diseases

Fusarium yellows (*Fusarium moniliforme, F. oxysporum*). *F. moniliforme* is seed-borne and, in the absence of fungicidal treatment, spreads from infected to healthy seed during 'chitting', causing decreased germination or death of seedlings. Both pathogens are soil borne and invade freesia roots and subsequently the vascular tissue of the corm causing a reddish brown rot. Foliage of the infected corms becomes yellow, commencing with the outer leaves and spreading inwards until, in severely infected plants, the foliage dies.

For control, seed should be treated with a captan or thiram seed dressing and sown in sterilized soil. Glasshouse soil in which an infected freesia crop has been gown must be sterilized, preferably with steam, but where this is impossible dazomet *or* formaldehyde *or* metham-sodium may be used, see p. 253. Spread of the disease in a seed-sown crop can be decreased by drenching the soil with captan 100 g a.i./100 litres at 5·5 litres/m² *or* benomyl 12·5 g a.i./100 litres applied at 500 ml of diluted material per 250 mm pot every 4—6 weeks from the 4-leaf stage. The treatment of corms with suspected *Fusarium* infection, by soaking for 30 minutes in formaldehyde (20 ml 40 per cent formaldehyde/litre), is recommended in the USA. Trials in which corms were dipped for 15—30 minutes in benomyl 100 g a.i./100 litres *or* carbendazim 100—120 g a.i./100 litres and then dried before storage or dipped immediately before planting have shown promise.

Grey mould (*Botrytis cinerea*) attacks the stems, leaves and flowers of densely grown freesias under humid conditions. Prevent by dusting weekly with captan 40 g a.i./100 m² *or* spraying at 10-day intervals with captan *or* chlorothalonil *or* thiram *or* fortnightly with benomyl *or* carbendazim *or* thiophanate-methyl as described in Table 10.3, p. 274. Dichlofluanid 50 g a.i./100 litres HV has given effective control but insufficient information is available concerning its safety on all varieties.

FUCHSIA

Pests

Aphid (*Myzus persicae*). Control as for Chrysanthemum—Aphids, Table 10.2, p. 266; formothion may also be used.

*Red spider mite (*Tetranychus urticae*). Control as for Chrysanthemum—Red spider mite, Table 10.2, p. 266.

**Whitefly (*Trialeurodes vaporariorum*). For description see Cucumber—Whitefly p. 281. This pest may cause serious damage to fuchsias by making the leaves fall prematurely. Control as for Chrysanthemum—Whitefly, Table 10.2, p. 266.

Diseases

*Grey mould (*Botrytis cinerea*) may attack flowers under very humid conditions. For control select a suitable treatment from Table 10.3, p. 274.

GERANIUM AND PELARGONIUM

Pests

**Aphids (*Myzus persicae, Aulacorthum solani*). Control as for Carnation—Aphids, Table 10.2, p. 266; but to zonals apply demeton-S-methyl, *or* dimethoate *or* formothion, *or* oxydemeton-methyl as soil drenches only.

*Leaf hopper (*Zygina pallidifrons*)—see Calceolaria—Leaf hopper, p. 263.

*Mealy bug (*Planococcus citri*). Control with diazinon, *or* malathion, as for Grape—Mealy bug, p. 288.

*Mites. The tarsonemid (cyclamen) mite (*Steneotarsonemus pallidus*) and the broad mite (*Hemitarsonemus latus*) may infest glasshouse geraniums. Control as for Cyclamen—Tarsonemid mite, p. 285.

*Vine weevil (*Otiorrhynchus sulcatus*). The larvae of this beetle occasionally feed on the roots of geraniums. They may be introduced in unsterilized compost or hatch from eggs laid by adults within the glasshouse. For control see Cyclamen—Vine weevil, p. 285.

*Whitefly (*Trialeurodes vaporariorum*). Control as for Chrysanthemum—Whitefly, Table 10.2, p. 266, but do not use parathion.

Diseases

*Grey mould (*Botrytis cinerea*). Leaves, flowers and stems of geraniums, especially of 'soft' plants, are susceptible to rotting if the plants are kept under damp conditions. Infection usually occurs where plant debris, such as dead flowers, remains on the plants. For control see Table 10.3, p. 274.

*Leaf spot (*Alternaria tenuis*) causes small water-soaked spots with a necrotic central fleck to appear on the undersides of the leaves. The lesions increase up to about 3 mm diameter, become necrotic and may coalesce. Control is obtained by spraying with thiram 300 g a.i./100 litres *or* zineb 150 g a.i./100 litres HV.

*Root and Stem rots. Geraniums are susceptible to several soil-borne diseases. *Xanthomonas pelargonii* causes Bacterial (Black) stem rot. Fungi such as *Sclerotinia* spp., *Fusarium* spp., *Thielaviopsis basicola* and *Pythium* spp. are also responsible for the death of plants from root or stem decay. The only satisfactory preventive is the use of sterilized cutting and potting composts and propagation from healthy stock. If *Pythium* infection occurs it is possible that disease spread may be reduced by watering the compost with cupric ammonium carbonate *or* etridiazole† according to the manufacturer's instructions. However, apply to a few plants in the first instance to check that the treatment is non-injurious.

*Rust (*Puccinia pelargonii-zonalis*). This disease, which attacks zonal pelargoniums causing concentric rings of raised, reddish brown spore-producing tissue on the undersides of leaves and corresponding yellow areas on the upper sides, may occur on cuttings or plants under glass. Zineb 125 g a.i./100 litres HV will control this disease if several applications are given at 10–14-day intervals.

GRAPE (VINES)

Pests

*Brown scale (*Parthenolecanium corni*). The adult stage is protected by a hard reddish-brown scale beneath which eggs are laid. Young scale insects emerging therefrom are active and much more readily killed. Stems mainly are attacked. Their control is as for mealy-bug but using diazinon at 32 g/100 litres HV *or* malathion at 190 g a.i./100 litres HV, repeating at 21-day intervals until the bunches are thinned.

*Mealy bug (*Planococcus* spp.) Aphid-like insects which, when adult, are covered by a white, mealy protective wax. They feed on stems, berries, and leaves and secrete honeydew, on which a sooty mould grows. For control, scrape dormant rods and spray in December with tar-oil winter wash, conforming to the Ministry of Agriculture specification, 4 litres/100 litres HV. In spring, spray with diazinon 16 g a.i./100 litres HV, *or* malathion 113 g a.i./100 litres HV *or* a mixture of nicotine 50 ml (95–98 per cent) and petroleum emulsion 1 litre/100 litres HV, and repeat several times at 14-day intervals, but not after thinning. Adequate wetting agent must be used to wet thoroughly the waxy insect.

***Red spider mite (*Tetranychus urticae*). For description see Cucumber—Red spider mite, p. 280. Control may be achieved by diazinon (HV spray or aerosol) *or* petroleum emulsion (HV) *or* parathion (smoke) at rates suggested for cucumber, Table 10.4, p. 278. Repeat treatments if necessary at 7–14-day intervals but do not use sprays after thinning or smokes or atomized solutions within a week of cutting.

---

† Proposed BSI name for 5-ethoxy-3-trichloromethyl-1,2,4-thiadiazole.

Dicofol or tetradifon are probably safe on vines but there is insufficient information to permit a firm recommendation.

Diseases

*Grey mould (*Botrytis cinerea*). Under warm, very humid conditions the fungus may attack ripening berries, causing rapid rotting. To prevent this, dust the bunches with sulphur dust at the first appearance of disease and repeat at 7–10 day intervals. The disease is unlikely to occur when sulphur is regularly vaporized to control mildew. Benomyl sprays 25 g a.i./100 litres HV applied at onset of attack and repeated fortnightly have given promising control on outdoor vines on the Continent but have not yet been tested under glass.

**Powdery mildew (*Uncinula necator*). A typical whitish powdery mildew attacking leaves, young shoots, flowers and fruits, controlled by vaporizing sulphur as for Chrysanthemum-Mildew, p. 273, or by spraying with benomyl (see Grey mould) *or* colloidal or wettable sulphur 200–300 g a.i./100 litres HV *or* dinocap 12 g a.i./100 litres HV when the shoots are growing freely; repeat when flowers open and again when the berries have set, but not after thinning.

HIPPEASTRUM (AMARYLLIS)

Pests

*Bulb scale mite (*Steneotarsonemus laticeps*) A.L. 456. Infestation by this mite is shown by dark reddish spots on the leaves, flower stalks and bulb scales accompanied by distortion. Hot water treatment of dry, dormant bulbs for 1·5 hours at 43°C (110°F) followed by gradual cooling will kill the mites. Infested growing plants should be sprayed with endosulfan *or* endrin, see Cyclamen mite, p. 285. Treatment with thionazin may be found effective.

*Mealy bug (*Planococcus citri*). To control this pest spray thoroughly to wet leaf bases with diazinon *or* malathion as for Grape—Mealy bug, p. 288.

*Red spider mite (*Tetranychus urticae*). Control with demeton-S-methyl, *or* diazinon, *or* dicofol, *or* dimethoate, *or* oxydemeton-methyl, *or* tetradifon, as in Table 10.2, p. 266.

Diseases

*Leaf scorch (*Stagonospora curtisii*) causes raised, red or brownish streaky spots on the flower stalks and leaves with bending or deformation. The fungus perennates in dormant bulbs and the disease is most severe under warm, humid conditions. To control, spray the emerging foliage and buds at the first sign of disease, with Bordeaux mixture equivalent to 100 g metallic copper/100 litres HV, and repeat monthly (some cultivars are susceptible to copper damage), *or* mancozeb 125 g a.i./100 litres HV *or* tank-mixed zineb (see Tomato—Leaf mould, p. 309) and repeat at 2–3 week intervals.

Hydrangeas are very susceptible to damage by sprays applied in sunshine or if the plants are dry at their roots.

Pests

**Red spider mite (*Tetranychus urticae*). Control by using demeton-S-methyl, *or* diazinon, *or* dicofol, *or* oxydemeton-methyl, *or* tetradifon, as for Carnation, Table 10.2, p. 266.

Diseases

*Grey mould (*Botrytis cinerea*) may occur on dense flower clusters of plants growing under excessively humid conditions. For control select a suitable treatment from Table 10.3, p. 274.

*Powdery mildew (*Microsphaera polonica*). Mildew develops on the leaves as a whitish coating. For control select a suitable treatment from Table 10.3, p. 274.

LETTUCE

Where losses due to grey mould (*Botrytis*) or *Rhizoctonia solani* (=*Thanetephorus cucumeris*) occur frequently, preventive measures should be taken. With other diseases and pests, control measures should be taken immediately the disease or pest is noticed as it is almost impossible to spray or dust adequately leaves which are in contact with the ground or have folded in.

Pests

**Aphids (A.L. 392). Several species, of which *Myzus persicae, Nasonovia ribis-nigri* and *Aulacorthum solani* are most prevalent, attack lettuce. Losses are usually due to planting infested seedlings, or planting in houses where aphids are present on plants or weeds. Hence, for control, raise plants in isolation or ensure other plants are aphid-free. Clean up the houses 7–10 days before planting and smoke with gamma-HCH if aphids are present. Dip the young plants, before setting out, in malathion 1·3 g a.i./litre *or* nicotine 0·6 ml (95–98 per cent)/litre for 3 minutes. Plants raised in soil-blocks or peat-pots, etc., should be thoroughly sprayed with nicotine or malathion at these concentrations. Rubber gloves must be worn.

For attacks after planting out, use one of the treatments in Table 10.6.

*Slugs—see Tomato—Slugs, p. 306.

**Springtails For description see Cucumber—Springtails, p. 280. For control, dust the soil surface with gamma-HCH 16 g a.i./100 m² or spray soil surface with gamma-HCH 12·5 g a.i./100 l.

**Symphylids—see Tomato—Symphylids, p. 306.

*Wireworms—see Tomato—Wireworms, p. 307.

Diseases

*Damping-off (*Rhizoctonia solani*). This fungus sometimes causes the death of seedlings and may attack older leaves in contact with the soil. Propagate in sterilized soil and treat seed with captan or thiram seed dressing. Apply quintozene dust at 6·7 g a.i./m² to the seed bed before sowing. Where seedlings are being attacked, remove all seedlings seen to be affected and water with thiram 150 g a.i./100 litres.

**Downy mildew (*Bremia lactucae*) is first noticed as pale green or yellow areas on lower leaves. White or greyish fungal growth soon develops on undersides of leaves. For control, avoid over-watering and a humid atmosphere; raise temperature to 15–17°C (60–65°F) and

Table 10.6. **Lettuce—aphids**

Compound	Rate	Minimum interval between treatment and harvest
1. HV sprays *(a.i. 100 litres)*		
Demeton-S-methyl*	22 g	21 days
Derris	3·7 g	24 hours
Diazinon	16 g	14 days
Dimethoate*	30 g	7 days
Formothion*	44 g	7 days
Gamma-HCH**	12·5 g	14 days
Malathion	113 g	1 day†
Nicotine	50 ml (95–98 per cent)	48 hours
Oxydemeton-methyl*	22 g	21 days
Pirimicarb*	25 g	14 days
2. Smokes		
Gamma-HCH	} See manufacturer's instructions	48 hours
Nicotine		24 hours
Pirimicarb		14 days

* Systemic compound.
† Preferably 4 days to avoid taint.
** On established transplants only.

increase ventilation; spray with mancozeb/zineb 175 g w.p./100 litres HV, *or* thiram 300 g a.i./100 litres HV *or* zineb 125 g a.i./100 litres HV and repeat at 14-day intervals, or apply zineb dust at 3·4 kg a.i./ha and repeat at 7-day intervals. On seedlings thiram is safer than zineb and also more effective against grey mould.

***Grey mould** (*Botrytis cinerea*). To control this disease remove all debris and organic matter from soil surface and obtain an even but not a very fine tilth. Spray the superstructure of the vacant house and soil surface with formaldehyde, 2 litres *c.* 40 per cent formalin/100 litres at least 3 weeks before planting and ventilate freely after 2 days. Avoid planting too deeply, poor temperature and humidity control, and maintain adequate soil moisture to prevent plant wilting. Chemical control measures include dusting the soil surface before and at intervals after planting (according to the manufacturer's product label) with dicloran *or* tecnazene *or* thiram. Alternatively spray HV every 14 days with benomyl *or* carbendazim *or* chlorothalonil *or* thiophanate-methyl *or* thiram at rates given in Table 10.3, p. 274. Minimum interval between treatment and harvest is 7 days (thiram), 14 days (chlorothalonil), 3 weeks (dicloran), zero interval for benomyl, carbendazim and thiophanate-methyl. In all cases apply the fungicides to cover the undersides of the leaves and stem bases.

MUSHROOM

Disinfection of mushroom-houses

To destroy spores of pathogenic and competing fungi on walls, floors, woodwork, etc., the houses should be treated with a fungicide after the diseased crop has been removed, even though the old compost has been 'cooked-out' by heating to 57–71°C (135–160°F). This is most important if brown plaster-mould (*Papulaspora byssina*), olive-green mould (*Chaetomium* spp.), or truffle were present, for the spores of these fungi are heat tolerant. Suitable sprays for this purpose are formaldehyde, 5 litres *c.* 40 per cent a.i./100 litres HV *or* sodium pentachlorophenolate 1 kg/100 litres *or* DNOC-sodium 100 g/100 litres HV (for safety regulations where DNOC-sodium is used see p. 91) *or* emulsified cresylic acid 2 kg cresols/100 litres HV. Tools and equipment should be cleansed with formaldehyde *or* cresylic acid *or* sodium pentachlorophenolate.

Composts, casing media and mushroom houses can be disinfected by fumigation with methyl bromide, see Nematodes, p. 294.

Pests

Flies. Species of 3 families, Cecidomyiidae, Phoridae, and Sciaridae, are pests of mushrooms. The life histories and behaviour of these groups differ, consequently they are dealt with separately.

***Cecid midges** (*Heteropeza pygmaea, Mycophila speyeri*). The larvae are white or orange and distinguishable from other mushroom maggots by a pair of dark 'eye spots' on the body just behind the head. They are paedogenetic, giving birth to daughter larvae and, as this process may continue for a long time, the small delicate adult midges are rarely seen. With heavy infestations larvae ascend mushrooms and enter surface tissues and the mycelium may be damaged.

Adequate peak-heating of the compost at 55–60°C (130–140°F) will give control but, as the larvae are dispersed in a similar manner to eelworms, strict attention should be paid to hygiene. It is seldom possible to reduce larval populations in composts without damage to the crop, though thorough mixing of the compost with thionazin 19 g a.i./tonne should give adequate control. Treatment of the casing with gamma-HCH 56 g a.i./tonne of peat and chalk, prior to covering the beds, acts as a barrier to larval migration and reduces infestation of the mushrooms.

***Phorid flies** (*Megaselia nigra, M. halterata, M. bovista*) are small, dark, stout-bodied flies. The maggots, which are white, legless, and lacking the black head of sciarid larvae, feed on mushroom mycelium though one species tunnels in mushrooms. Their attacks are most frequent in summer and autumn and come mainly from flies entering spawn-running rooms.

To avoid attack, screen ventilators and keep spawn-running rooms closed as much as possible. Thoroughly mixing diazinon 11 g a.i./tonne with casing and/or compost *or* incorporating into compost pirimiphos-ethyl 50 g a.i./tonne *or* thionazin 19 g a.i./tonne should prevent infestation. Phorid flies are very susceptible to dichlorvos vapour and a concentration-time product of 0.7 μghr/litre (0.07 ghr/100 m^3) for the spawn run and for the first week after casing should give control. Provided the floor surface is not unduly alkaline, watering the floor with a solution of dichlorvos 100 g a.i./16 litres/100 m^3 usually gives useful control though the result depends on the rates of ventilation and hydrolysis due to high humidity.

Alternative methods are to smoke with sulfotep as for sciarids (p. 294) *or* atomize diazinon 5.3 g a.i./100 m^3 *or* dichlorvos 21 g a.i./100 m^3 *or* malathion 8.4 g a.i./100 m^3 daily (depending on the number of flies) when the spawn is running and less frequently during cropping *or* use diazinon *or* malathion dusts at the same active ingredient rates and intervals. Minimum intervals between treatment and cutting are 24 hours for dichlorvos and malathion, 48 hours for diazinon.

***Sciarid flies.** Maggots of these flies, which are small, delicate insects with a tendency to run rather than fly, are white with black heads. They tunnel into the stalks and caps of mushrooms and sever the mycelium from the stalk, causing the buttons to die, and carry bacterial pathogens and eelworms. Flies are attracted to the compost during cool-down from peak-heat but later they concentrate in the casing. For their control peak-heat the compost to 55–60°C (130–140°F).

To prevent establishment of the larvae one of the following should be admixed with the compost at spawning; chlorfenvinphos 60 g a.i./tonne *or* diazinon 50 g a.i./tonne *or* pirimiphos-ethyl 50 g a.i./tonne, *or* thionazin 19 g a.i./tonne.

Larval infestations may also be controlled before cropping by incorporating chlorfenvinphos 30 g a.i./tonne of casing *or* drenching the casing with malathion 100 g a.i./100 litres using 20 litres/100 m². Malathion drenches can also be used during cropping but apply at 10 g a.i./100 litres using 20 litres/100 m² bed allowing 24 hours before cutting.

Sulfotep smokes may be used to control sciarid flies if applied just after spawning and before and immediately after casing. If an infestation of flies develops during any stage of crop production atomize diazinon *or* dichlorvos *or* malathion *or* apply malathion dust as for phorids *or* atomize gamma-HCH 3·5 g a.i./100 m³ *or* smoke with gamma-HCH allowing 24 hours (dichlorvos and malathion) and 48 hours (diazinon and gamma-HCH) before cutting.

***Nematodes** (*Ditylenchus* spp., *Aphelenchoides* spp.) feed on mycelium causing its disappearance in patches. Where they are numerous the compost may become dark, overmoist and stinking. Nematodes, under certain dry conditions, can persist for at least two years in dry compost, woodwork, etc. They can be carried from place to place on tools, boxes, etc., or by flies.

Thionazin 20 g a.i./9–23 litres/tonne applied to compost *only,* at spawning and *at no other time,* is effective against fungal-feeding nematodes only.

No acceptable means of controlling eelworms in spawned or cropping compost is available. Peak-heating kills eelworms in the unspawned compost, and cooking-out those in spent compost and structures. Methyl bromide fumigation at a concentration-time product of 60 kghr/100 m³ gives control providing the house temperature is maintained at 21°C (70°F). Such fumigations can be made only by registered contractors or trained operators. Woodwork, structures, etc., should be cleansed thoroughly as described under Disinfection of Mushroom Houses, p. 293.

*Slugs** only damage mushrooms grown in very indifferent structures, which they can enter from outside. They may be controlled with metaldehyde bait as for Tomato—Slug, p. 306, but do not place the bait on the beds.

*Woodlice** are likely to occur when mushrooms are grown in floor beds in glasshouses. Before cropping begins, apply 5% DDT dust to the floor and crevices in which woodlice hide.

Diseases

Mushroom yields are reduced both by parasitic organisms and by fungi which grow on the compost and compete for nutrients with the mushroom mycelium. Most if not all losses from either cause can be prevented by strict attention to hygiene, correct compost fermentation, adequate pasteurization of the compost by 'peak-heating' before spawning, and 'cooking-out' before emptying or removing old trays or beds. There are no satisfactory control measures for several pathogenic and competitive fungi.

Bacterial blotch (*Pseudomonas tolaasi*) appears as slightly sunken yellowish brown blotches and spots on the pilei, which coalesce making the sporophore discoloured and sticky. The disease is favoured by moist conditions therefore, avoid watering fairly mature mushrooms, ventilate after watering and, if possible, reduce the relative humidity below 80 per cent. If the disease appears, a routine watering with sodium hypochlorite equivalent to 16 g available chlorine/100 litres or with chlorinated water containing that amount (160 ppm) of chlorine, will reduce losses.

Bacterial pit. The cause of this trouble is unknown but bacteria appear to be associated. It starts as small cavities beneath the skin of the cap, the skin eventually collapsing to leave open pits. Treatment as for bacterial blotch has reduced its incidence.

Bubble (white-mould) (*Mycogone perniciosa*). Mildly infected mushrooms usually show wavy gills covered by a white velvety fungal growth. More severely attacked mushrooms are deformed, covered by the white mould from which exudes a clear golden liquid, and rapidly become brown, wet and foul-smelling. The fungus is derived usually from infected soil and it attacks mushrooms most readily when the temperature is high. Strict hygiene must be observed. Apply benomyl, *or* carbendazim *or* thiophanate-methyl 122 g a.i./100 m^2 of bed *either* incorporated into the casing *or* as a drench in 100–200 litres water after casing. Zineb, 35 g a.i. in 50 litres water/100 m^2 should be applied three times; shortly after casing, between first and second, and between second and third flushes. Apply before watering. For spot treatment, two further applications may be made between flushes.

Cobweb (*Dactylium dendroides—Hypomyces rosellus*). The pinkish white, loose cobwebby mycelium of fungus envelopes entire mushrooms and spreads on surrounding casing. For control, treat with benomyl, *or* carbendazim, *or* thiophanate-methyl, *or* zineb as for bubble. Spray the bed-boards and woodwork with quintozene 75 g a.i./100 litres HV.

Dry bubble (Verticillium) (*V. fungicola* (*malthousei*) or *V. psalliotae*) appears first as light brown blotches on caps. When severe the mushrooms are distorted, often with swollen and bulbous stalks and very small caps. Mushrooms become covered with grey-white mould but do not decay to produce the offensive odour associated with bubble (*Mycogone*). Apply benomyl *or* cabendazim as for bubble, p. 295, *or* thiophanate-methyl 244 g a.i./100 m^2 of bed, but tolerance to these fungicides can occur with this fungus. Alternatively drench the casing with chlorothalonil w.p. at 112 g a.i./100–200 litres/100 m^2, allowing at least 24 hours between treatment and harvest. Disinfect house at end of cropping—see p. 293.

Gill mildew (*Cephalosporium lamellicola*) causes symptoms similar to a mild attack of dry bubble but the blotches are darker. The gills may become covered with a fine mycelial growth. For control see bubble (*Mycogone*).

Red geotrichum (*Sporendonema purpurascens*) is a competing fungus which develops in the compost and casing. Young mycelium is white but at the sporulating stage the fungus turns buff and then bright pink. It is favoured by poor ventilation and high temperature and may be avoided by using sterilized casing and attending to humidity and temperature during spawning. Use zineb dust, 30–60 g a.i./100 m² a.i. at weekly intervals, allowing 48 hours between treatment and cutting.

Truffle (false truffle) (*Diehliomyces microsporus—Pseudobalsamia microspora*). Spores of this fungus withstand high temperatures and may survive 'peak-heating' and 'cooking-out'. The mycelium, which runs rapidly through compost, is cream and makes the compost dark brown and soggy. Fruiting bodies are up to 25 mm in diameter with a convoluted surface. Water the compost at the last turn with copper sulphate 700 g/tonne using a suitable volume of water in relation to the dryness of the compost. If the fungus develops in shelf beds, remove the compost to form a trench extending at least 0.6 m beyond the visible limit of the fungus.

NARCISSUS (FORCED)

The major pests of forced narcissi are similar to those attacking the crop outdoors and are controlled by hot-water treatment of the dormant bulbs as described for Narcissus—Stem eelworm, p. 358. The addition of 500 ml of 40 per cent formalin to 100 litres water in the hot water treatment controls narcissus basal rot caused by *Fusarium oxysporum* f. sp. *narcissi*.

ORCHID

After the application of sprays to orchids with sheath type leaves, it is necessary to drain accumulations of spray retained in the leaf axils. Spray under quick drying conditions but not in bright sunshine.

Pests

Aphids. Several species of aphid including *Aulacorthum circumflexum, Cerataphis lataniae* and *Myzus persicae.* breed on orchids. *C. lataniae,* when apterate, is reddish brown and scale-like. Control with smokes of gamma-HCH, *or* nicotine, *or* parathion, *or* pirimicarb, *or* sulfotep, *or* fumigation with naled, *or* atomized solutions of malathion, *or* sprays of gamma-HCH *or* malathion, *or* pirimicarb as for Carnation, Table 10.2, p. 266. Some genera appear tolerant to sprays and drenches of dimethoate or oxydemeton-methyl, but further experience is necessary before systemic organophosphorus compounds can be recommended.

*Cattleya fly (*Eurytoma orchidearium*), **Cattleya midge** (*Parallelodiplosis cattleyae*) and **Cattleya weevil** (*Cholus cattleyae*). These, and several other less important dipterous and coleopterous pests, have almost been completely eliminated from orchid houses by insecticides used for other pests. Should they occur, applications of gamma-HCH sprays at 12·5 g a.i./100 litres HV will give control.

**Mealy bugs (*Planococcus* spp.). Control with HV sprays of DDT at 75 g a.i./100 litres *or* malathion at 113 g a.i./100 litres. Several applications at 14–21-day intervals may be required.

***Red spider mite (*Tetranychus urticae*). These small greenish or red mites are frequently a pest on orchids. They may be controlled by diazinon (aerosols) *or* dicofol (sprays) *or* naled (fumigation, see p. 267) *or* tetradifon (sprays, or aerosols) at the rates shown in Table 10.2, p. 266.

**Scales. Many species of scale insects occur on orchids, most having been introduced on imported plants. Recommended control measures are to spray 2–3 times at 14–21-day intervals with diazinon 32 g a.i./100 litres HV *or* malathion 190 g a.i./100 litres HV.

*Thrips. Of the species occurring on orchids, *Heliothrips haemorrhoidalis, Hercinothrips femoralis* and *Anaphothrips orchidaceous* appear the commonest. Control as for aphids, p. 296, but do not use pirimicarb.

*Vine weevil, see Cyclamen—Vine weevil, p. 285.

*Woodlice. Control with DDT, *or* parathion drenches applied sparingly to orchid composts *or* Paris green baits, as for Cucumber—Woodlice, p. 281.

Diseases

*Anthracnose. Two fungi cause diseases commonly termed anthracnose. *Glomerella cincta* causes dark, decayed areas on the leaves, usually starting on the tips. The spots caused by *Gloeosporium* spp. are yellowish, light brown at first, becoming soft and sunken later. The recommended control is to spray several times at 10–14-day intervals with a copper fungicide, e.g. copper oxychloride at 38–50 g metallic copper/100 litres HV.

*Bacterial soft rot (*Pectobacterium carotovorum*). A slimy rotting of the pseudo-bulbs and basal stems. Plants may be saved by cutting out all diseased tissue and thoroughly dusting the wounds and surrounding areas with a copper dust before rotting is extensive.

*Black rot (*Pythium* spp.). Seedlings may damp-off due to attack. The fungus also enters the roots of older plants and spreads to the pseudo-bulbs which shrivel. Leaves may be attacked if the humidity is high. The immersion of seedlings for 1 hour in captan at 3 g a.i./litre on their removal from the culture flasks is recommended as a preventive. To prevent its spread on older plants, the humidity should be reduced and the plants sprayed as for anthracnose (p. 297).

POINSETTIA

The bracts of poinsettia are very susceptible to injury by pesticides from the time they begin to colour.

Pests

*Mealy bugs (*Pseudococcus citri, P. adonidum*) and ***Scales** (several spp. including *Coccus hesperidum* and *Aspidiotus hederae*) may infest poinsettia, particularly those grown with other hosts in private glasshouses. To control them in commercial houses, spray with DDT at 75 g a.i./100 litres HV several times at 14-day intervals.
**Red spider mite (*Tetranychus urticae*). Control with HV sprays *or* atomized solutions of dicofol *or* tetradifon, as for Carnation, Table 10.2, p. 266.
***Whitefly (*Trialeurodes vaporariorum*) (A.L. 86). For description, see Cucumber—Whitefly, p. 281, and for control see the treatments listed for Chrysanthemum—Whitefly in Table 10.2, p. 266.

Diseases

*Damping off (*Pythium* sp.). Use sterilized compost but if infection occurs apply a soil drench of cupric ammonium carbonate *or* drazoxolon *or* etridiazole† according to manufacturer's instructions and check that the treatment is non-injurous by applying to a few plants in the first instance.
*Grey mould (*Botrytis cinerea*) attacks leaves and bracts under excessively moist conditions. If control is not obtained by removing the diseased parts and giving adequate ventilation apply one of the fungicides listed in Table 10.3, p. 274.
*Root rots (*Rhizoctonia solani, Fusarium* spp., *Thielaviopsis basicola*). Use benomyl 50 g a.i./100 litres applied as a post-planting drench in place of the first watering.

PRIMULA (*Primula malacoides, P. obconica*)

Pests

**Aphids (*Aulacorthum circumflexum Myzus persicae*). Control, if possible before flowering, using gamma-HCH (sprays, aerosols or smokes

---

† Proposed BSI common name for 5-ethoxy-3-trichloromethyl-1,2,4-thiadiazole.

for controlling *M. persicae* but not *A. circumflexum*) *or* malathion (sprays or aerosols) *or* naled (fumigation, see p. 267) *or* nicotine (sprays) *or* pirimicarb (sprays or smokes) at the rates given in Table 10.2, p. 266. Although some cultivars may be sensitive to systemic organophosphorus compounds, demeton-S-methyl *or* formothion *or* oxydemeton-methyl have been used successfully as sprays and drenches on several cultivars of both species.

*Leaf hopper (*Zygina pallidifrons*). An active, yellowish green insect which feeds on the undersides of the leaves, causing chlorotic spotting of the upper surfaces. Easily controlled by gamma-HCH, *or* malathion, as for Carnation—Thrips, p. 264.

**Red spider mite (*Tetranychus urticae*). To control spray at 10–14-day intervals with dicofol, *or* tetradifon, at concentrations as for Carnation—Red spider mite, Table 10.2, p. 266. The systemic compounds (but not formothion or pirimicarb) mentioned under Aphids, p. 298 may be used on most cultivars.

*Vine weevil (*Otiorrhynchus sulcatus*). For description and control with gamma-HCH see Cyclamen—Vine weevil, p. 285.

Diseases

*Grey mould (*Botrytis cinerea*). For control measures see Table 10.3, p. 274.

*Root rots (*Phytophthora* spp., *Pythium* spp., *Thielaviopsis basicola*). For control measures see Begonia-Root rot p. 263.

*Powdery mildew (*Oidium* sp.). For control measures see Table 10.3, p. 274.

ROSE

Pests

***Aphids. Several spp., of which *Macrosiphum rosae* is the most important, are troublesome. For control measures see Table 10.2, p. 266. Rose growers consider aerosols are less likely to spoil rose foliage than are HV sprays.

*Caterpillars. Several species damage leaves, shoots and buds; some tortrices spin leaves together. For control see Carnation—Tortrix, p. 264. Atomized solutions or dusts are preferable.

*Leaf hoppers. For description and control see Tomato—Leafhopper, p. 306.

*Mealy bug. For description and chemical control see Grape—Mealy bug, p. 288.

***Red spider mite (*Tetranychus urticae*). For description see Cucumber—Red spider mite, p. 280. Suitable control measures can be selected from Table 10.2, p. 266, noting that many strains of this pest are resistant to several acaricides.

*Rose (scurfy) scale (*Aulacaspis rosae*). The older branches become covered with a scurfy mass of scales and colonies of young yellowish white scales may appear on shoots and leaves. Their control is as for Scale on grapes, p. 288.

Diseases

*Black spot (*Diplocarpon rosae*) occasionally occurs on roses under glass when black or purple spots develop on upper surfaces of leaves, and leaflets drop prematurely. Spray HV at the onset of attack with captan 100 g a.i./100 litres *or* dodine 75 g a.i./100 litres *or* maneb 100 g a.i./100 litres *or* mancozeb/zineb 175 g w.p./100 litres *or* dinocap/folpet 200 g w.p./100 litres *or* triforine 25 g a.i./100 litres (some cultivars are sensitive to dinocap and triforine). The treatments should be repeated at 10–14-day intervals until the disease is eradicated. For crops attacked the previous season new foliage can be protected by captan sprays (see above) applied after pruning and repeated at 10–14-day intervals. Similar use of the other compounds listed above has not been fully tested.

*Downy (black) mildew (*Peronospora sparsa*) produces irregular yellow-grey or purplish spots on leaves and occasionally on flower stems. Whitish-grey downy fungal growth appears on undersides of leaves which fall prematurely. Spray at 14-day intervals with copper/petroleum emulsion, 600 g metallic copper (e.g. 1·2 kg copper oxychloride containing 50% Cu) and 500 ml petroleum emulsion (glasshouse grade)/100 litres HV *or* cupric ammonium carbonate *or* wettable sulphur at manufacturer's rates *or* zineb 125 g a.i./100 litres HV *or* tank-mixed zineb as for Tomato—Leaf mould, p. 309.

*Grey mould (*Botrytis cinerea*). Under excessively humid conditions, flowers and occasionally stems may be attacked. Germinating spores may produce spotting of the petals and subsequent fungal growth causes rotting of the blooms. The fungus can also invade stem wounds causing die-back of woody tissue. To check the disease, remove infected blooms and decrease the humidity. Fumigate with sulphur as for Rose powdery mildew, or select a suitable treatment from Table 10.3, p. 274.

***Powdery mildew (*Sphaerotheca pannosa*). A typical white powdery mildew attacking young leaves, stems, buds and petals, and causing new growth to be dwarfed and distorted. To control the disease, vaporize sulphur as for Chrysanthemum—Mildew, p. 273, avoiding as far as possible its use on dark red varieties when at height of flower production. The cultivar Super Star appears to be somewhat sensitive to sulphur. Spray with bupirimate at manufacturer's rates *or* dodemorph 75 g a.i./100 litres HV *or* a suitable treatment selected from Table 10.3.

SAINTPAULIA (AFRICAN VIOLET)

Do not apply sprays or dips in bright sunshine or under humid, slow-drying conditions or if the pot compost is dry. Use tepid water for sprays and preferably drain the plants on their sides.

Pests

*Aphid (*Myzus persicae*). Control with malathion *or* pirimicarb as described in Table 10.2, p. 266.

   *Mealy bug (*Planococcus citri*). For control see Orchid—Mealy bugs, p. 297.

   Mites—Broad mite (*Hemitarsonemus latus*), **Tarsomenid (Cyclamen) mite (*Steneotarsonemus pallidus*) (S.T.L. 133) and *Tarsonemus confusus*. These mites cause stunting, leaf twisting and distortion of buds and flowers. For control see Cyclamen—Tarsonemid mite, p. 285.

Diseases

*Grey mould (*Botrytis cinerea*). The conditions conducive to attack are similar to those described under Azalea—Grey mould p. 261. For control measures see Table 10.3, p. 274.

SWEET PEA

The flowers are likely to be blemished if sprayed with any pesticide.

Pests

**Aphids (*Myzus persicae*). This and occasionally other species feed on the leaves and flower buds of sweet peas under glass. Control as for Carnation—Aphids, Table 10.2, p. 266, but do not use malathion.

   **Red spider mite (*Tetranychus urticae*). This mite is frequently a serious pest on sweet pea and control measures should be taken as soon as it is noticed. Use HV sprays of demeton-S-methyl, *or* diazinon, *or* dicofol, *or* dimethoate, *or* oxydemeton-methyl, *or* tetradifon, as for Carnation, Table 10.2, p. 266.

Diseases

**Powdery mildew (*Erysiphe polygoni*). A typical powdery mildew, first appearing as a whitish powdery growth on the under-surface of the leaves. Flower buds and stems may also be attacked. Avoid wide temperature and humidity fluctuations to prevent the development of this disease, but if necessary spray dinocap at 12·5 g a.i./100 litres HV at

7–10-day intervals. Good results have also been obtained by vaporizing sulphur as for Chrysanthemum—Mildew, p. 273.

SWEET PEPPER (*Capsicum annuum* L.)
(see S.T.L. 181)

Pests

*Aphid (*Myzus persicae*). For control spray with nicotine 50 ml (95–98 per cent)/100 litres HV *or* pirimicarb 25 g a.i./100 litres HV *or* fog with pirimiphos-methyl *or* smoke with nicotine shreds). Allow the following minimum intervals between treatment and harvest: 48 hours (sprays of nicotine or pirimicarb) and 24 hours (nicotine smoke).

*Leaf miner. For description see Tomato leaf miner, p. 303. For control apply diazinon 16 g a.i./100 litres HV *or* atomize diazinon 5·3 g a.i./100 m³ *or* smoke with gamma-HCH leaving minimum intervals of 14 days (diazinon spray) 48 hours (diazinon atomized, gamma-HCH smoke) between treatment and harvest.

**Red spider mite (*Tetranychus urticae*). For description see Cucumber—Red spider mite, p. 280. For control spray HV with demeton-S-methyl, *or* dicofol, *or* dimethoate, *or* oxydemeton-methyl, *or* tetradifon at rates and observing the preharvest intervals given in Table 10.7. In trials, the predatory mite, *Phystoseiulus persimilis* has given good control of red spider mites on peppers, using the method outlined for cucumbers in Table 10.1, p. 258.

**Thrips (*Thrips tabaci*). For control atomize diazinon 5·3 g a.i./100 m³ and allow 48 hours between treatment and harvest.

**Whitefly (*Trialeurodes vaporariorum*) (A.L. 86). For description see Cucumber—Whitefly, p. 281. For control apply pirimiphos-methyl, *or* resmethrin as for Cucumber Whitefly, Table 10.4, p. 278. In trials successful biological control of whitefly has been obtained using the parasitic wasp *Encarsia formosa* as outlined for Cucumber Whitefly in Table 10.1.

**Woodlice. For description and control see Cucumber—Woodlice, p. 281.

Diseases

*Damping off and **Foot rot** (*Pythium* spp., *Phytophthora* spp. and *Rhizoctonia solani*). These soil-borne fungi can cause pre-emergence death, or post-emergence damping-off of seedlings; foot rot of young plants or root rot of mature plants. Peppers are very slow rooters and sowing or planting too deeply or in cold compost (below 15°C (58°F)), and overwatering will make plants more susceptible to attack. Crops should be grown in sterilized soil or compost but if *Pythium* or

Phytophthora infection does occur remove diseased plants and water the remainder of the crop with a copper-based fungicide *or* zineb as described for Tomato—Foot rots, p. 308. Incorporating quintozene dust 6·8 g a.i./m² into the soil or compost 2–3 weeks before planting gives some control of *Rhizoctonia*.

*Grey mould (*Botrytis cinerea*). For description see Tomato—Grey mould, p. 308. To prevent attack regulate heating and ventilation to avoid high humidity and water carefully to prevent retention of moisture on the crop. For control apply fortnightly foliar sprays of benomyl 50 g a.i./100 litres HV but do not treat with this material if benomyl-tolerant strains of the fungus occur or if biological control of pests is being used. Alternatively spray with chlorothalonil 110 g a.i./100 litres HV every 7–14 days *or* thiram 300 g a.i./100 litres HV every 10 days. Allow a minimum interval between spraying and harvest of 7 days for thiram and 12 hours for chlorothalonil.

*Sclerotinia rot (*Sclerotinia sclerotiorum*). This fungus can cause damping-off of seedlings and if air-borne spores infect petioles drooping leaves may be the first sign of attack. A fluffy white fungal growth develops on infected tissue with the formation of black sclerotia in the pith cavity. For control measures see Celery—Sclerotinia rot, p. 270.

TOMATO

Pests

*Aphid. The most serious is the glasshouse-potato aphid (*Aulacorthum solani*), which feeds on the foliage causing yellowing and increasing its susceptibility to grey mould; on immature fruits it causes raised, paler areas which remain yellow when the fruit ripens. The aphid is controlled by the treatments given in Table 10.7, p. 304, repeating if necessary after 10–14 days. If biological control of other pests is being used see Table 10.1, p. 258.

*Leaf miner (*Liriomyza bryoniae*). The larvae tunnel cotyledons and stems of seedlings early in the year, often killing them. Later generations sometimes mine the leaves of established plants sufficiently to warrant control measures. The larvae pupate and overwinter in the soil. The first indications of attack are small rounded pits on cotyledons, caused by adult flies feeding. Avoid propagating tomatoes in houses where the previous crop was infested unless the soil in the houses has been disinfected. Spray seedlings and small plants with diazinon 16 g a.i./100 litres HV *or* parathion 10 g a.i./100 litres HV and repeat at 14-day intervals if fresh pitmarks are seen on leaves *or* use dimethoate 30 g a.i./100 litres HV on young plants. On established plants, when attack is serious, use sprays or atomized solutions of gamma-HCH *or* diazinon as for aphids. If biological control of other pests is being used see Table 10.1, p. 258.

Table 10.7. **Tomato—aphid, red spider mite, and whitefly**
(Note:) If biological control of pests is being used, see also Table 10.1, p. 258

Compound	Rate	Minimum interval between treatment and harvest	Suitable for use on Young plants	Suitable for use on Established plants	Active against Aphid	Active against Red spider mite	Active against Whitefly
1. HV sprays	*a.i./100 litres*						
Cyhexatin‖	25 g	48 hours	—	Yes	—	Yes	—
DDT	100 g	14 days	Yes	Yes	—	—	Adults
Demeton-S-methyl*	22 g	21 days	Yes	Yes	Yes	Yes	Adults
Derris	3·7 g	24 hours	Yes	Yes	Yes	Yes	Adults
Diazinon	16 g	14 days	—	Yes	Yes	Yes	Adults
Dicofol	12·5 g	48 hours	Yes	Yes	—	Yes	—
Dicofol/tetradifon	40 g + 12·5 g	48 hours	Summer only	Yes	—	Yes	—
Dimethoate*	30 g	7 days	—	Yes	Yes	Yes	—
Gamma-HCH	12·5 g	14 days	—	Yes	Yes	—	—
Malathion**	113 g	24 hours	—	Yes	Yes	Yes	Adults
Nicotine	50 ml (95–98 per cent)	48 hours	Yes	Yes	Yes	—	—
Oxydemeton-methyl*	22 g	21 days	Yes	Yes	Yes	Yes	—
Parathion‡	10 g	28 days	Yes	—	Yes	Yes	Adults
Petroleum emulsion (glasshouse grade)	1 litre (product)	Nil	—	Yes	—	Yes	—
Pirimicarb*	25 g	48 hours	Yes	Yes	Yes	—	—
Pyrethrin/resmethrin	See product label	Nil	Yes	Yes	—	—	Adults, scales and eggs
Tetradifon	10 g	Nil	Yes	Yes	—	Yes	—

2. Atomization

	a.i./100 m³						Pests
Diazinon	5·3 g	48 hours	—	Yes	Yes	Yes	Adults
Dichlorvos	3·5 g	24 hours	—	Yes	Yes	Yes	Adults
Dicofol†	6 g	48 hours	—	Summer only	—	—	—
Gamma-HCH	3·5 g	48 hours	—	Yes	Yes	Yes	Adults
Malathion**	8·4 g	24 hours	—	Yes	Yes	Yes	Adults
Parathion†	1·7 g	48 hours	—	Yes	Yes	—	Adults
Resmethrin	See product label	Nil	Yes	—	—	—	—
Tetradifon†	7 g	Nil	—	Summer only	—	Yes	—

3. Smokes (S), Fogs (F), Fumigation (Fu)

	a.i./100 m³						Pests
DDT (S)	See product label	48 hours	Yes	Yes	—	—	Adults
DD1/Gamma-HCH (S)		48 hours	—	Yes	Yes	—	Adults
Gamma-HCH (S)		48 hours	—	Yes	Yes	—	—
Naled§ (Fu)		48 hours	—	Yes	Yes	Yes	Adults
Nicotine (S & Fu)		48 hours	—	Yes	Yes	—	—
Parathion† (S)		24 hours	Yes	Yes	Yes	Yes	Adults
Pirimicarb (S)		Nil	Yes	Yes	Yes	—	—
Pirimiphos-methyl (F)		Nil	Yes	Yes	Yes	—	Adults, scales
Propoxur (S)		48 hours	—	Yes	Yes	—	Adults
Pyrethrin/resmethrin (F)		Nil	Yes	Yes	—	—	Adults, young scales
Sulfotep (S)		24 hours	—	Yes	Yes	Yes	Adults

* Systemic compound.
** Allow 4 day interval to avoid taint.
† Max. 5 applications/season.
‡ Max. 3 applications/season.
§ See Carnation—Aphid. p. 264.
|| Avoid young trusses, see product label.

*Leafhopper (*Zygina pallidifrons*) may attack young tomatoes. If so, spray the seedlings with parathion as for aphids. Atomized solutions of diazinon *or* malathion may be used if attacks occur on established plants in summer—see Table 10.7.

***Potato root eelworm (*Heterodera rostochiensis*) (A.L. 284). Tomato plants attacked by the eelworm are retarded, the leaves become dark green with a purplish under-surface. The plants wilt during hot bright weather and their roots may become invaded by fungi and rot. Propagate in steamed soil and avoid introducing eelworm into glasshouses. Sterilize borders with steam *or* dazomet *or* metham-sodium *or* methyl bromide *or* a product containing chloropicrin with dichloropropane-dichloropropene and methyl isothiocyanate *or* dichloro-propane-dichloropropene mixture *or* dichloropropene, see pp. 252–254. Alternatively, oxamyl 56 g a.i./100 m² may be applied to soil surface and the soil rotavated to a depth of 100 mm immediately before planting.

**Red spider mite (*Tetranychus urticae*). For description see Cucumber—Red spider mite, p. 280, and for chemical control measures see Table 10.7, p. 304; treatments may need to be repeated to obtain complete kill. For biological control see Table 10.1, p. 258.

***Root-knot eelworm (*Meloidogyne* spp.) (A.L. 307). For control disinfect the soil by steaming or use one of the following chemical treatments at the rates given (pp. 253–254): chloropicrin with dichloro-propane-dichloropropene and methyl isothiocyanate *or* dichloropropane-dichloropropene *or* dichloropropene *or* methyl bromide.

Where tomatoes have to be planted in soil known to be infested by root-knot eelworm, watering with parathion 38 g a.i./100 litres at 5 litres/m² within 3 days of planting will protect plants from early attack and increase their yields. In view of the relatively high concentration of parathion used. the glasshouses must be well ventilated during application, rubber boots must be worn when entering the glasshouse, and rubber gloves worn when handling the soil for the following 14 days; these precautions are additional to other statutory requirements (see Chapter 4).

*Slugs (A.L. 115) occasionally enter houses from weedy land outside or in dung, and are controlled by watering the soil, in the infested area, with liquid metaldehyde preparations at 68 g a.i./100 m² or by baiting with metaldehyde bait at 17–34 g a.i./100 m² *or* methiocarb pellets 1 g a.i./100 m². Allow at least 10 days for liquid metaldehyde and 7 days for methiocarb between treatment and harvest.

*Springtails—for description see Cucumber—Springtails, p. 280, and to control, dust the soil surface with gamma-HCH 15 g a.i./100 m² *or* spray soil surface with gamma-HCH 12·5 g a.i./100 litres.

**Symphylids (*Scutigerella immaculata*) (A.L. 484) are active, white centipede-like insects which descend too deep into the soil to be destroyed by soil sterilization in autumn and winter, but return to the surface in spring and stunt gowth of plants by damaging young roots. To control, water the plants (within 3 days of planting) with diazinon 10 g a.i./100 litres *or* parathion 12·5 g a.i./100 litres. Thorough incorporation of gamma-HCH 2·7 kg a.i./ha as a dust into the top 100 mm of soil within 4 weeks prior to planting has proved satisfactory.

*Thrips (*Thrips tabaci*) seldom cause direct damage but are vectors of spotted wilt virus and thus of importance where different species of plants are grown together. Control as for aphids, Table 10.7, p. 304, but not with pirimicarb. If biological control of other pests is being used see Table 10.1, p. 258.

**Tomato moth (*Lacanobia oleracea*). The green or brown caterpillars of this moth, when young, skeletonize leaves and, when larger, eat foliage fruit and stems. For their control, use DDT either as a dust at 1.1 kg a.i./ha, *or* as an atomized solution at 8·8 g a.i./100 m³, *or* as a spray at 100 g a.i./100 litres HV, *or* as smokes. On seedlings use the dust only; allow at least 14 days between spraying or dusting and picking, 2 days in the case of aerosols or smokes.

**Whitefly (*Trialeurodes vaporariorum*) (A.L. 86). For description, see Cucumber—Whitefly, p. 281. Many whitefly strains have developed resistance and either biological control (Table 10.1, p. 258) or different insecticides must be used. Treatments are listed in Table 10.7, p. 304, and frequency of application must be related to the kill of the various stages of the pest. Thus if adults only are killed it may be necessary to repeat the treatment within 5 days but consult the product label.

*Wireworms (*Agriotes obscurus*) (A.L. 199) usually occur only in the first 3—4 years after building glasshouses on grassland but may enter under walls from heavily infested land and attack adjacent plants. Before planting in new houses apply gamma-HCH dust 830 g a.i./ha and work into top 100 mm of soil. One application is sufficient to eradicate wireworms. Where plants are being attacked, water with gamma-HCH, 9·3 g a.i./100 litres at 500 ml/plant provided that the soil is well drained and the plants set out so that no fluid remains in a pool around the stems; otherwise use parathion 9·3 g a.i./100 litres (not more than 3 applications per crop and not within 4 weeks of harvest).

*Woodlice (see Cucumber—Woodlice, p. 281) occasionally destroy seedlings and small plants during propagation. Attacks can usually be prevented by attention to hygiene, but, if troublesome, dust the soil surface and staging infested with woodlice with DDT dust at 155—310 g a.i./100 m² *or* water with DDT 100 g a.i./100 litres at 2·5 litres/m² *or* gamma-HCH 50 g a.i./100 litres *or* parathion 12·5 g a.i./100 litres (But not more than 3 applications per crop), *or* use Paris green bait as for Cucumber—Woodlice, p. 281. For DDT allow 2 weeks, and for parathion 4 weeks between treatment and harvest.

Diseases

***Brown root rot—see Root rots.

*Buck-eye rot (*Phytophthora nicotianae*). The lower fruits show grey brownish red patches with a series of concentric dark brown rings and have become infected by the fungus being splashed on to the fruit from contaminated soil or the fruit being in contact with the soil (see also Foot rot). The trusses should be tied up out of contact with soil and splashing when watering should be avoided. The disease may be prevented by propagating in soil and containers sterilized by steaming *or* formaldehyde, and by planting in sterilized borders. Where the disease occurs, spray the lower part of the plants and the surface of the soil with copper sprays at 600 g metallic copper (e.g. 1·2 kg copper oxychloride containing 50 per cent Cu)/100 litres HV *or* cupric ammonium carbonate at manufacturer's dosage, *or* zineb 140 g a.i./100 litres HV.

   *****Colletotrichum root-rot**—see Root rots.

   *****Corky root**—see Root rots.

   ****Foot rot** and **Damping-off** (*Phytophthora nicotianae, P. cryptogea, Rhizoctonia solani, Pythium* spp.). These fungi may cause damping-off of seedlings, or foot rot of small plants during propagation or soon after planting out. Infection arises from the soil or from contaminated boxes, pots or water. Prevent attack by propagating in sterilized soil, etc., as for buck-eye rot. Unsterilized soil should be watered at seed sowing, potting or planting out with cupric ammonium carbonate, according to manufacturer's instructions, *or* other copper fungicides equivalent to 80 g metallic copper/100 litres, *or* zineb 100 g a.i./100 litres. Where seedlings or young plants are being attacked, healthy plants can be protected by watering with these fungicides but great care must be taken to reject all doubtful plants when potting or planting. Etridiazole† has given promising control of *Pythium* and *Phytophthora* on seedlings and young plants and should be applied according to manufacturer's instructions.

   *****Grey mould** (*Botrytis cinerea*). This fungus invades stems from infections established on damaged or senescent leaves, on petiole stumps or deleafing wounds, producing stem lesions which may ultimately kill the plant. Masses of grey spores may be produced on the infected tissue and distinguish this disease from stem rot (*q.v.*). Under humid conditions *Botrytis* spores will germinate on green, immature fruits causing the small 'ghost' spots which persist to disfigure the ripe fruit, or may infect flowers causing premature drop. Infected flowers may produce further infections of fruit or foliage.

   To control the disease, reduce humidity by adequate ventilation. Remove all plant debris from borders, and decaying leaves and fruit from plants. Trim to allow circulation of air through the lower parts of plants and remove leaves and shoots cleanly.

   Lesions on the stems should be cut out and the wound painted with a paste of dicloran (4 per cent) dust. Painting the lesions with creosote is commonly practised and is satisfactory if care is taken to limit the treated area to the actual lesion.

---

† Proposed BSI common name for 5-ethoxy-3-trichloromethyl-1,2,4-thiadiazole.

To protect foliage and fruit from infection, spray at 14-day intervals with benomyl 50 g a.i./100 litres HV *or* carbendazim 50–60 g a.i./100 litres HV or thiophanate-methyl 100 g a.i./100 litres HV *or*, when biological control of pests is being used, apply these materials as soil drenches according to the manufacturer's label. If strains of *Botrytis* tolerant to any of these systemic materials occur apply chlorothalonil 100 g a.i./100 litres HV every 7–14 days *or* dichlofluanid 50 g a.i./110 litres HV every 14 days *or* thiram 300 g a.i./100 litres HV at 10-day intervals *or* dust thiram 6·7 kg a.i./ha every 4–7 days *or* smoke with tecnazene. Minimum intervals between treatment and harvest are: 12 hours (chlorothalonil), 2 days (tecnazene), 3 days (dichlofluanid), 7 days (thiram).

***Leaf mould** (mildew) (*Fulvia fulva, (Cladosporium fulvum)*) (A.L. 263) is first noticed as pale yellow patches on the upper surface of older leaves and the reverse surface which become covered with a pale grey mould-like growth which darkens to brownish violet. The disease occurs when the temperature and overnight humidity are high. Adequate ventilation often controls the disease but, where conditions are likely to favour leaf mould, grow varieties which are resistant or of low susceptibility, or adopt one of the following routine chemical measures. Spray HV fortnightly with freshly prepared tank-mixed zineb 125 g a.i./100 litres (add with agitation a solution of 125 g zinc sulphate to a solution of 176 g nabam (93 per cent) and make the final volume to 100 litres and add an anionic or non-ionic wetting agent) *or* chlorothalonil 110 g a.i./100 litres *or* dichlofluanid 50 g a.i./100 litres *or* maneb 125 g a.i./100 litres *or* zineb 125 g a.i./100 litres. Alternatively spray LV at 7–10-day intervals with zineb 2·2 kg a.i./20–40 litres/ha *or* tank-mixed zineb (300 g zinc sulphate and 4·7 kg nabam (93 per cent)/25–50 litres/ha, *or* dust every 7 days with zineb 3·3 kg a.i./ha.

When petroleum oil emulsions are used for controlling red spider mite add copper fungicides equivalent to 63–95 g metallic copper/100 litres diluted emulsion. Minimum intervals between treatment and harvest are: 12 hours (chlorothalonil), 2 days (maneb, zineb), 3 days (dichlofluanid), 7 days (thiram).

Systemic fungicides containing benomyl, carbendazim or thiophanate-methyl can be used fortnightly as HV sprays or monthly as soil drenches at rates as for Tomato—Grey mould, p. 308. If strains of *Fulvia (Cladosporium)* tolerant to any of these systemic materials occur use one of the non-systemic fungicides listed above.

***Root rots.** Tomato roots are attacked and rotted by several fungi, the most important being *Pyrenochaeta lycopersici*, which is responsible for brown root rot and for corky root, and *Colletotrichum cocchodes* which destroys the cortex of the roots and stem below soil level. The effect of these fungi is to reduce plant vigour and crop production; plants tend to wilt in bright, hot weather. For their control, sterilize the soil by steaming or with methyl bromide as described on pp. 252–253.

**Stem rot (*Didymella lycopersici*). Infection very occasionally originates from seed saved from diseased fruit, and then appears as spots on the leaves and proceeds to attack the main stem. However, primary infection of stem bases usually originates from infected soil, and may develop about 3 weeks after planting, later spreading to other parts of the stem. The lesions are dark brown and sunken, but a grey woolly fungal growth is not produced as with *Botrytis* (see Tomato—Grey mould, p. 308). Serious attacks may occur on steamed soils which have become re-infected. Where disease has occurred remove the old crop debris carefully and spray the superstructure and soil surface with formaldehyde 2 litres *c*. 40 per cent formalin/100 litres HV. Sterilize the soil with chemicals, or steam, and avoid recontamination. As a preventive against infection, spray the bottom 100–150 mm of stem and soil surrounding the stem to a radius of 50–75 mm with captan 250 g a.i./100 litres, using 85 ml/plant, within 2 days after planting and repeat 2–3 weeks later using 110–140 ml/plant. Do not exceed these quantities or apply to plants which are dry at their roots or to pot plants. A slight temporary check to growth may result. Treatment at the 140 ml rate may prevent further infections if applied immediately an attack occurs. Alternatively apply benomyl *or* carbendazim as HV sprays or soil drenches at rates given for Tomato—Grey mould, p. 308.

Painting stem lesions with maneb 10 g a.i./litres will eradicate the fungus if the lesions are treated in the very early stage.

***Wilt diseases (*Verticillium albo-atrum, V. dahliae, Fusarium oxysporum*) (A.L. 53). Laboratory techniques are required to establish the organism responsible for wilting. *Fusarium* wilt is more serious at higher temperatures and is widespread in the Channel Isles. First symptoms of wilt are usually yellowing or wilting of lower leaves often initially on one side of the plant. The woody stem tissue becomes brown and the discolouration can be extensive or limited to the lower stem region. The control of wilt diseases depends on soil sterilization, preferably by steaming, and the use of resistant varieties.

Where *Verticillium* wilt occurs, avoid planting in cold, wet soil. Raise the temperature of the greenhouse to 25°C (77°F), shade lightly if necessary to reduce wilting, and damp plants overhead. Soil drenches of benomyl *or* carbendazim at rates given for Tomato—Grey mould, p. 308, have given some control of these wilt diseases.

TULIPS (FORCED)

Many of the pest and disease problems of forced tulips occur as a result of 'housing' infested bulbs. It is very important to select only sound, healthy bulbs for forcing, and to apply preventive treatments before planting into sterilized soil.

Pests

*Bulb mite (*Rhizoglyphus echinopus*)—See Narcissus, p. 356.
  ***Stem eelworm (*Ditylenchus dipsaci*) (A.L. 461)—See Tulip, p. 373.
  **Tulip bulb aphid (*Dysaphis tulipae*)—See Tulip, p. 373.

Diseases

**Tulip Fire (*Botrytis tulipae*) (A.L. 536). The fungus can be seen on outer and fleshy bulb scales as small, black sclerotia (similar in size to onion seed) which may be associated with brown-margined lesions. Prior to planting, the fungus may be controlled by dipping the bulbs for 30 minutes in a suspension of benomyl 100 g a.i./100 litres *or* carbendazim 100–120 g a.i./100 litres. If primary infections occur on emerging shoots, these produce masses of olive-grey spores which can cause secondary infection or spotting on other leaves and flowers. Immediately cut off all infected shoots below soil surface, avoid splashing foliage when watering and spray HV the remainder of the crop with mancozeb 120 g a.i./100 litres *or* mancozeb/zineb 200 g product/100 litres *or* thiram 300 g a.i./100 litres *or* zineb 140 g a.i./100 litres and repeat after 14 days.

  **Fusarium Bulb Rot (*Fusarium oxysporum* f. sp. *tulipae*). Bulb infection usually occurs in the summer prior to forcing, and diseased bulbs may become soft and a pinkish-white fungal mat may develop on the shrivelling bulb. If infected bulbs are planted, the ethylene gas produced by rotting tissue can cause damage to the embryonic flowers within the bulbs. As a preventive measure dip the bulbs before planting as described for Tulip Fire, p. 311.

  *Root rot (*Pythium* spp.). This disease has become more prevalent since bulbs for forcing have been planted direct into glasshouse border soil instead of into boxes. Root rot causes patches of weak, yellowish plants with poor root systems. Bulbs should be planted in steam sterilized soil. The use of etridiazole† as a bulb dip or as a soil drench has given promising control in trials but see manufacturers' label for dosage rates.

† Proposed BSI common name for 5-ethoxy-3-trichloromethyl-1,2,4-thiadiazole.

Chapter 11
Pests and diseases
of vegetable crops

With the increased use of systemic fungicides many existing and some previously untreatable vegetable diseases, e.g. deep seated pathogens of large seeds, can now be controlled more effectively and in some cases more economically than before. The scope of this chapter has been increased to include these innovations. Not all vegetable diseases can be directly controlled by fungicides, however, and the measures necessary for their alleviation include efficient field sanitation, etc., adequate crop rotations and the use of resistant varieties, as discussed in general terms in Chapter 1. More specific information can be obtained from the MAFF Bull. 123 'Diseases of Vegetables'.

Any use of aldrin, DDT or dieldrin recommended in this Chapter applies only to commercial holdings and *not* to the home garden or allotment.

11.1 General pests and diseases

Pests

Cutworms (*Noctuidae*). Cutworms are caterpillars, chiefly of the turnip moth (*Agrotis segetum*) and the garden dart moth (*Euxoa nigricans*), which inhabit the surface layers of the soil, feeding on plants just above or below ground level. Beet, brassicae, carrots, celery, lettuce, swedes and turnips are attacked, plants being either cut off at ground level or large holes eaten into the roots. Most damage occurs in June and July and to crops just after transplanting or thinning out.

Cutworms are usually much more common in crops grown after a heavy weed cover, than in those on land that has been kept free of weeds. Dusts and sprays containing DDT are effective, especially if applied when the caterpillars are young and under moist conditions during the late afternoon or in the evening. As a dust, use DDT at 2·8 kg a.i./ha. As a spray, use DDT at 1·12 kg a.i./ha *or* carbaryl at 850 g a.i./ha, apply in 400–1000 litres water/ha along the rows; it is beneficial to hoe or harrow afterwards.

Poison bait of bran mixed with DDT broadcast during the late afternoon or evening will check cutworm attacks. To prepare the bait mix DDT (250 g a.i.) with 28 kg bran; add 15 litres water and mix thoroughly

312

and broadcast 34–45 kg/ha. Methiocarb pellets used in slug control also give some control of cutworm.

Leatherjackets (*Tipula* spp.) are the larvae of crane flies. They are normal inhabitants of the soil in damp grasslands but will feed on and damage a wide range of agricultural and horticultural crops. Most damage occurs after the ploughing-up of grassland or leys and the worst attacks usually occur in spring. Thinly sown or transplanted crops are usually those most affected. The larvae feed at or just below the soil surface. Early ploughing of grassland or leys (July or early August) will prevent egg-laying. Consolidation of the soil and the preparation of a good tilth will assist the crop to resist attack. For chemical control, see Chapter 5, p. 104.

Millepedes. Several species of millepede damage crop plants; they feed on live or dead vegetable matter in the soil and the roots, tubers, etc., of most kinds of plants are eaten. Beans, particularly the germinating seeds, carrots, peas, potatoes and strawberries are especially subject to injury. Millepedes occur in most soils but, where they are numerous and causing damage, they are difficult to control. Surface applications of pesticides are apt to be ineffective as the millepedes rarely come above the soil surface. DDT *or* gamma-HCH dust or spray controls millepedes and should be applied to the open seed drills or worked into the soil before planting. Gamma-HCH should not be applied when there is a taint hazard (see p. 46).

Swift moths (*Hepialus* spp.). The caterpillars of the garden swift moth and the ghost moth feed on the roots and rhizomes of many plants and may cause much damage to nursery and market garden crops. Strawberries, anemones, flax, Michaelmas daisies and hops are perennials particularly affected, and wide range of annuals, including lettuce, carrots, parsnips, swedes and red beet may be damaged, especially if planted on infested land. Periodic digging of the soil, control of perennial weeds and the lifting and replanting of bulbous and herbaceous plants at fairly frequent intervals are all means which reduce caterpillar populations in the soil. The chemical control of the pest in established crops is difficult for the larvae may be deep in the soil. Promising results have been obtained with an emulsion containing DDT at 50 g a.i./100 litres applied as a drench at 140 ml per plant.

Wireworms (*Agriotes* spp.) See p. 103.

Diseases

Damping-off diseases. Germinating seedlings of many vegetables can be attacked by species of *Pythium* and *Phytophthora*, either before or after emerging above the soil. The damage caused by these fungi is usually most severe under conditions when germination is slow, for example, early in the season or in cold wet soils. Considerable protection against such soil-borne pathogens can be obtained by dusting the seed with captan *or* drazoxolon *or* thiram, either alone or mixed with insecticidal seed dressings. The dressings normally contain 30–75 per cent a.i. and are used at a rate sufficient to give a good cover of the seed. Recommended rates vary, according to the vegetable, from 1·4–5·0 g/kg seed.

***Seed-borne fungal diseases.** Several fungal pathogens of vegetables and other crops are carried within the seeds and cannot be eliminated by conventional dust applications. Previously, treatment was by immersing the seeds in hot water but a new 'Thiram soak treatment' has several advantages and is being widely used on certain vegetable seeds (e.g. celery, beet, brassicas—*q.v.*). The seeds are enclosed in mesh bags and immersed for 24 hours in a suspension of thiram containing 2 g a.i. (2·5 g 80 per cent w.p. or 4 ml 50 per cent colloidal suspension)/litres maintained at 30°C (86°F). The seeds imbibe water and should be dried after treatment in a stream of cool or slightly warmed air. Up to 5 kg seed can be treated in 100 litres suspension but for small quantities (up to 1·0 kg) a 10 litre bucket can be used, the temperature being maintained by an aquarium heater and thermostat. In the larger baths the suspension is circulated and the thiram thereby deposited on the seeds also gives protection against damping-off diseases.

The thiram soak treatment does not affect germination of most vegetable seeds but some varieties or samples of brassicas may be sensitive. Before treatment of large bulks of seeds whose reaction is unknown, a pilot test should be made. Further information can be obtained from the National Vegetable Research Station, Wellesbourne, Warwick, CU35 9EF.

Recent research has shown that several seed-borne fungi can be eradicated from seeds if these are dusted with dressings or treated with slurries containing certain of the benzimidazole or related fungicides. Examples of these are given in the text.

11.2   Specific crops

BROAD BEAN

Pests

***Black bean aphid** (*Aphis fabae*). This aphid is a common and often serious pest of field and broad beans. Heavy infestations may also occur on sugar and red beet, spinach, mangolds, and the French and runner beans. On wild hosts it occurs on fat hen and thistles and it overwinters in the egg stage on the spindle tree (*Euonymus europaeus*). On beans, heavy infestations on the leaves and stems cause stunting, leading to a marked reduction in yield.

Autumn or early spring sowing of beans allows the plants to become well-grown and usually in flower before the attack starts. The removal of the plant tips when infestation has just begun usually reduces the subsequent attack. Crops may become heavily infested in June, July or August and should be treated with insecticides before this occurs. For spraying use demephion at 250 g a.i./ha *or* demeton-S-methyl at 245 g a.i./ha *or* dichlorvos at 560 g a.i./ha *or* dimethoate at 335 g a.i./ha *or* formothion at 420 g a.i./ha *or* malathion at 1·26 kg a.i./ha *or* mevinphos at 140–210 ml/ha *or* nicotine at 560 ml (95–98 per cent)/ha *or* oxydemeton-methyl at 240 g a.i./ha *or* phosphamidon at 225 g a.i./ha *or* pirimicarb at 140 g a.i./ha *or* thiometon at 280 g a.i./ha. Apply in 400–1000 litres of water as a fairly coarse spray along the rows. Nicotine is not effective at low temperature: 18–21°C (65–70°F) is preferred. Observe the minimum intervals (1 day for malathion and dichlorvos, 2 days for nicotine, 3 days for mevinphos, 7 days for dimethoate and formothion, 14 days for pirimicarb and 21 days for demephion, demeton-S-methyl, oxydemeton-methyl, phosphamidon and thiometon) which must elapse between application and harvest when the crop is for direct human consumption. Sprays containing derris and pyrethrum may be preferred to nicotine for small-scale use as they are less hazardous.

Nicotine dusts, if applied under warm dry conditions, can give satisfactory results. Depending on plant size and density, 2·2–9·0 kg a.i./ha of nicotine dust (4 per cent) are required.

Alternative treatments include: menazon at 225 g a.i./100–1000 litres/ha, at least 21 days before harvest, *or* disulfoton granules at 1·1 kg a.i./ha or at 0·85 kg a.i./ha if less persistence is required, *or* phorate granules at 1·1 kg a.i./ha before flowering are most effective. They are more persistent than sprays and are less toxic to pollinating insects. The granules are applied along the rows as a band over the plants; there must be an interval of at least 6 weeks before the crop is harvested.

FRENCH AND RUNNER BEANS

Pests

Bean seed fly (*Delia platura*). The larvae of this insect tunnel in the germinating seeds and in the young stems of French and runner beans and may cause serious crop losses by killing or stunting the plants. The damage is usually worst on crops sown early and when germination is slow. The fly is attracted to decaying plant material and is becoming increasingly common in market gardening areas.

In areas where bean seed fly is not resistant to dieldrin, seed dressings based on this insecticide are very effective and economical to use. Use dressing (containing 75 per cent dieldrin) 1·6 g a.i./kg seed. Mix dressing and seed thoroughly before sowing.

Black bean aphid (*Aphis fabae*). Heavy infestations of this aphid often develop on French and runner beans during July and August and the black smothering colonies cause stunting, flower drop and malformation of the pods. Crops should be treated with insecticides before they become heavily infested and to avoid harm to bees should not be sprayed when in flower. For chemical control see p. 315 excepting phorate which is not recommended for use on French or runner beans. Take into account crop maturity when choosing chemicals so that the necessary period between treatment and harvesting may elapse. Disulfoton granules at the rate given for broad beans must be applied at sowing time 25 mm below seed. They should not be used on light sandy soil.

Diseases

Pod rot (*Botrytis cinerea*). This disease, which is most severe in wet seasons, causes grey, water-soaked lesions on the pods and may render a sample unacceptable for quick freezing. One spray, of benomyl at 0·55 kg a.i./ha, *or* carbendazim at 0·56–0·66 kg a.i./ha in not less than 200 litres water/ha *or* thiophanate-methyl at 1·1 kg a.i./ha in 300–600 litres water/ha, applied at flowering, gives a high degree of control.

Anthracnose (*Colletotrichum lindemuthianum*). This disease causes cankers on the stems and brownish lesions associated with the leaf veins. The pods bear circular lesions on which pustules may develop. Although known in Britain for many years, the disease has been of little importance in the past. The advent of mechanical harvesting, however, associated with the needs of the quick-freeze industry, has made even small amounts of disease unacceptable and control measures are therefore necessary.

As the main source of disease is infected seed, treatment of such seed is advisable and recent research has indicated that seed dusts with benomyl at 2·2 g a.i./kg seed gives adequate control. A single application of benomyl at flowering (0·55 kg a.i./ha in not less than 200 litres water) is also reported to give control of the disease on the pods.

Halo blight (*Pseudomonas phaseolicola*). The bacterium is seed-borne and affects French and runner beans. The distinctive symptoms of the disease consist of small lesions surrounded by yellow haloes (hence halo blight) on the leaves and water-soaked lesions on the pods (i.e. grease spot). The bacterium is spread by wind-driven rain and the disease is damaging in wet seasons, sometimes reducing yield but mainly affecting the marketable quality of the pods. Halo blight can be achieved by the use of healthy seeds. If infection is detected early in the growth of a crop experiments have shown that spread of the disease can be controlled with copper oxychloride sprays (0·7 kg copper/1000 litres/ha) applied at 10–14 day intervals from emergence to pod set. Such sprays are of little use, however, unless applied from emergence onwards.

BEETROOT (RED BEET)

Pests

***Mangold fly** (*Pegomya betae*). The larvae of the mangold fly feed and cause large blisters in the leaves of sugar beet, red beet, mangolds and spinach. The damage is most serious when the plants are small but larger plants may be quite heavily attacked with little effect on the final yield of roots. On spinach, however, considerable reduction in the market value of the crop can follow even a moderate attack. Crops can be helped to grow away from an attack by providing a good tilth for germination and by top dressing with a nitrogenous fertilizer. If an attack develops in the seedling stage, singling should be delayed. Good control may be obtained by spraying with DDT at 1·05 kg a.i./ha *or* dimethoate at 84 g a.i./ha *or* formothion at 120 g a.i./ha *or* trichlorphon at 400 g a.i./ha. Apply in 200–1000 litres water and direct spray on the rows.

When both mangold fly and aphids are present use dimethoate at 400 g a.i./ha *or* formothion at 525 g a.i./ha *or* phosphamidon at 224 g a.i./ha in 200–1000 litres water. Alternatively, demeton-S-methyl at 245 g a.i./ha *or* oxydemeton-methyl at 240 g a.i./ha mixed with trichlorphon at 225 g a.i./ha and applied in 200–1000 litres water will give excellent control of both pests simultaneously.

If one of the above insecticides is used on spinach, account must be taken of the minimum time which must elapse between application and harvest.

Black bean aphid (*Aphis fabae*). This aphid is frequently common on beet causing malformation of the leaves and stunting of the crop. The aphid also spreads sugar beet virus yellows and certain viruses of beet and mangolds. With the exception of nicotine, dichlorvos and malathion (which are unlikely to be as effective as systemic materials), use any of the insecticides given on p. 000 for this pest and observe the same minimum periods between application and harvest.

Diseases

***Black leg** (*Pleospora betae*). The stems of young seedlings become blackened and shrivelled. The disease is seed-borne and may be controlled by organomercury seed treatment with dusts containing 1·0–1·5 per cent mercury at 6·7 g/kg seed (dry) or 2·2 ml/kg (wet) *or* with slurries containing 0·6–2·0 per cent mercury (generally 2·2 ml/kg seed). The thiram soak treatment (p. 314) gives complete control of seed-borne infection and is accompanied by an improvement of germination.

*Silvering** (*Corynebacterium betae*). This bacterial disease affects the seed crops, especially of Cheltenham Green Top. The leaves on seeding plants become silvery and the plants then wilt and die. The pathogen is seed-borne and may be controlled by soaking the seed for 24 hours in a solution containing 200 mg/litre streptomycin as the sulphate. Dry the seed and store until required. Less effective yet giving an appreciable control is seed treatment with an organomercury dressing at the rate of 12·5 g/kg seed.

BRASSICA CROPS

Pests

***Cabbage aphid** (*Brevicoryne brassicae*) is a mealy aphid which infests the leaves and shoots of many cruciferous crops and frequently causes severe damage to Brussels sprouts, cauliflowers, cabbage, kale and swedes. The aphid overwinters as eggs and sometimes as adults on brassica plants which, if kept for seed, may become heavily infested by early summer and serve as the main source of aphids for spread to spring and summer sown crops.

Methods of control vary according to the stage of development and the variety of the crop. Insecticidal sprays are usually more effective than dusts and should be applied at 675–1700 litres/ha, depending on crop size and density. Enough spray should be applied just to wet the foliage and an additional wetting agent may be required to ensure wetting of the waxy aphids and foliage. On Brussels sprouts, in August and September, pendant lances should be used to control aphids on the lower leaves and sprouts.

The following insecticides have proved effective: demephion at 270–330 g a.i./ha *or* demeton-S-methyl at 325 g a.i./ha *or* dichlorvos at 560 g a.i./ha *or* dimethoate at 400 g a.i./ha *or* ethoate-methyl at 410 g a.i./ha *or* formothion at 420 g a.i./ha *or* malathion at 1·26 kg a.i./ha *or* menazon at 335 g a.i./ha *or* mevinphos at 110–170 ml (99 per cent)/ha *or* nicotine at 560 ml (95–98 per cent)/ha *or* phosphamidon at 225 g a.i./ha *or* thiometon at 280 g a.i./ha. The period of protection against recolonization by the aphid varies with the insecticide used. Menazon, though slow to kill existing colonies, gives up to 3 weeks protection, demeton-S-methyl, demephion and thiometon give 10–12 days and the other insecticides give a few days or less. Edible crops must not be sprayed with demeton-S-methyl after October and a minimum period between spraying and harvesting the crop must be observed when any of the insecticides are applied. This period is 21 days for demephion, demeton-S-methyl, menazon, phosphamidon and thiometon, 7 days for dimethoate and formothion, 3 for mevinphos, 2 for nicotine and 1 day for dichlorvos and malathion.

Nicotine dusts (4 per cent nicotine) at 2·2–4·5 kg a.i./ha can give satisfactory results if applied under warm dry conditions. Disulfoton granules provide good control on Brussels sprouts, cabbages and cauliflower at 1·4 kg a.i./ha, either as a foliar dressing *or* a soil application at planting time. A second application (foliar) may follow the first (foliar or soil) at not less than 4 weeks. The minimum interval between application and harvest is 6 weeks.

An alternative treatment is a soil drench of menazon, 140 g a.i./100 litres water applied at 57 ml to the base of the plant within 4 days of transplanting. Menazon may also be used as a root dip, but not for broccoli or cauliflower; the plants are dipped, for 10 seconds before planting out, in a paste of 32 g a.i./0·9 litres water, enough for 1000 plants. For continuous protection follow up the menazon treatment with a spray of menazon at 335 g a.i./1000 litres water/ha, 8 weeks after transplanting and again, if necessary, 3 weeks later. At least 21 days must elapse between the last spray and harvest.

Cabbage moth (*Mamestra brassicae*). Caterpillars of this moth may cause damage to brassica crops over the period June to October. Cabbage suffer most for the caterpillars eat into the hearts which they spoil. The treatments recommended for the control of cabbage white caterpillars (see p. 320) give control. Mevinphos at 200 ml (99 per cent) in 500–1000 litres water/ha is the most effective treatment but the crop should not be harvested until at least 3 days after spraying.

***Cabbage root fly** (*Erioischia brassicae*) occurs throughout Great Britain and is a serious pest of cabbage, cauliflower and Brussels sprouts. The larvae feed on the roots which they may partially or completely destroy. Attacked plants are stunted or they may collapse and die. The larvae also tunnel into fleshy roots of swedes, turnips and radishes, causing loss of yield and quality. There are 2 or 3 overlapping generations in a year and plants may be attacked at any stage of growth. Methods of control vary according to the stage of development and type of crop. Plants and seedlings may be attacked at any time between mid-April and September.

The development of strains of cabbage root fly resistant to aldrin, dieldrin and HCH has limited the usefulness of these insecticides to the northern parts of the country while methods by which they may be applied have been restructured through official recommendation. In areas where this resistance occurs, organophosphorus insecticides give good control and should be used.

In areas where cabbage root fly is still susceptible to dieldrin and HCH, the following offer means of control. For control in seedbeds use HCH seed treatment: add 62 ml paraffin (kerosine) to 1 kg seed, mix well and shake thoroughly with gamma-HCH seed dressing (75 per cent) 47 g a.i. This method should not be used on radish or turnip because of risk of taint. For control in the field; treat plants by dipping the roots and lower part of the stem for about 10 seconds immediately before transplanting in a dieldrin suspension 6 g a.i./litre water. Do not let the roots dry out. This dip treatment should give protection for at least 1 month unless the attack is severe; it may cause some check, especially to cauliflowers, but the use of the wettable powder is less likely to cause a check than the use of e.c. formulations.

No seed treatment with an organophosphorus insecticide can be recommended but chlorfenvinphos e.c. at 1·12 kg a.i. in 200—500 litres water/ha sprayed on the soil surface worked into 100 mm will control the pest in nursery seedbeds. Alternatively, use chlorfenvinphos granules at 4·5 kg a.i./ha broadcast and incorporated to a depth of 20—50 mm. After mid-April and when the plants have at least 2 true leaves, if the seed bed was not treated with chlorfenvinphos, spray diazinon at 2·25 kg a.i./675—1350 litres water/ha. For direct sown crops drill through a 75—150 mm band of chlorfenvinphos granules at 28·3 g a.i./40 m row *or* fonofos granules at 28·3 g a.i./40 m row *or* spray a band of chlorfenvinphos e.c. at 8 g a.i. in 1·65 litres/100 m of row; with swedes and especially turnips these treatments carry some risk of phytotoxicity.

For transplants from seedbeds, root dips of organophosphorus insecticides cannot be safely recommended but these materials can be very effective when applied either as liquid or granules to the soil at the base of the plant within 4 days of transplanting. For liquid application, use azinphos-methyl at 550 g a.i./1000 litres *or* chlorfenvinphos at 450 g a.i./1000 litres *or* diazinon at 400 g a.i./1000 litres, giving each plant 68 ml of drench, *or* thionazin at 285 g a.i./1000 litres, giving each plant 57 ml of drench. When applying granules, use chlorfenvinphos granules at 18 g a.i./1000 plants *or* diazinon granules at 45·5 g a.i./1000 plants *or* fonofos granules at 18 g a.i./1000 plants. Alternatively, dimethoate granules at 23 g a.i./1000 plants has given good results against cabbage root fly and, for some weeks following post-transplanting application, also controlled early aphid attack. There is no recommended treatment with organophosphorus insecticides for cauliflowers or other brassicas in pots or soil blocks.

Cabbage root fly larvae sometimes occur in the buttons of Brussels sprouts and for this, trichlorphon at 1·1 kg a.i./500—1000 litres/ha gives a good control. Apply 3 sprays at 7-day intervals, the first 1 month before the anticipated harvest. This may be as late as mid-September for crops harvested up to the end of October. Obtain a good cover of the bottom half of the plant by using pendant lances.

Cabbage stem weevil (*Ceutorhynchus quadridens*). This insect infests spring sown brassica plants, especially those in seed beds. Attacks occur from April to July and the larvae mine in the petioles and stems. Plants are killed or stunted. The stems of infested plants may be spongy and snap readily on transplanting. Good control has been obtained with gamma-HCH applied either as a seed dressing (containing 75 per cent) at 47 g a.i./kg seed or by applying a gamma-HCH spray at 21 g a.i./100 litres water in a band along the row at 165 ml/m of row when the seedlings are in the second or third true leaf stages.

***Cabbage white butterflies** (*Pieris* spp.). The caterpillars feed on many kinds of brassicas, on stocks and nasturtiums and on cruciferous weeds. Most damage is caused to brassica crops from July to September, the caterpillars eating the leaves and fouling the crop with excrement.

Dusts and sprays containing one of the following insecticides applied to the leaves and hearts of the plants, have proved effective. As the foliage of brassica crops is difficult to wet, it may be necessary to use additional wetter. The dusts are DDT at 2·8 kg a.i./ha or derris (0·2–0·5 per cent rotenone) at 56 kg dust/ha. The sprays are azinphos-methyl at 310 g a.i./ha or DDT at 0·63–1·05 kg a.i./ha or mevinphos at 280 ml (99 per cent)/ha or nicotine at 560 ml (95–98 per cent)/ha or tetrachlorvinphos at 840 g a.i./ha or trichlorphon at 900 g a.i./ha. When spraying, the insecticide should be applied in 450–1000 litres water/ha, the higher volumes usually giving the better results. DDT emulsions and miscible liquids are usually more effective and persistent than DDT dusts. The nicotine spray requires the addition of a wetter and is effective only against young caterpillars. At least 3 weeks should elapse between the application of azinphos-methyl and harvest, 2 weeks in the case of DDT, 7 days with tetrachlorvinphos, 3 days with mevinphos, 2 with nicotine or trichlorphon, and 1 day in the case of derris.

Diamond-back moth (*Plutella xylostella*). The caterpillars of this moth feed on the leaves of many brassicae, including cabbages, cauliflowers, Brussels sprouts, swedes and turnips. Although they normally cause little damage, outbreaks occasionally occur during July and August especially in coastal districts. These are associated with the migration of moths from the Continent. Dusts or, preferably, sprays of DDT, applied principally, to the undersides of the leaves, give effective control. For dusting with either DDT at 2·8 kg a.i./ha or derris (0·2–0·5 per cent rotenone) at 31–62 kg dust/ha; the latter is effective but less persistent than DDT. As a spray, use DDT at 1 kg a.i. in 450–1000 litres water/ha. Azinphos-methyl at 460 g a.i./ha or carbaryl at 935 g a.i./ha or trichlorphon at 1·34 kg a.i./ha applied in 450–1000 litres water/ha also give good control. The minimum time intervals between insecticide application and harvesting are as for cabbage white butterflies and in the case of carbaryl the interval is 7 days.

***Flea beetles** (*Phyllotreta* spp.) often cause serious damage to newly-emerged seedlings of cabbage, cauliflower, kale and other brassica crops. They eat holes in the leaves and stems and may check or even destroy the young plants. Most damage is caused in late April and in May; crops sown before early April or after the end of May usually suffer only slight damage. The pests may be controlled by seed dressings and post-emergence sprays and dusts.

A seed dressing based on gamma-HCH is effective and economical to use. It protects the seedlings during emergence and in the cotyledon stage; if an attack develops later a dust or spray can be applied. For moderate or heavy attacks, use dressing (75 per cent) of gamma-HCH 47 g a.i./kg seed; for light attacks use 30 g a.i./kg. When the higher rate of dressing is used, mix 15 ml paraffin (kerosine) with each kg seed before mixing with the dressings. Seed treated with the paraffin sticker should be sown within 1 week. To avoid injury to the seedlings, gamma-HCH dressings should not be used if the seeding rate exceeds 11·2 kg/ha when drilled or 22·4 kg/ha when broadcast. Radish and turnips for early bunching should not be treated with gamma-HCH because of the risk of taint.

Post-emergence sprays and dusts should be applied when the seedlings begin to emerge through the soil; re-apply, if necessary 7–10 days later. Use gamma-HCH at 280–420 g a.i./ha. The gamma-HCH spray may be applied to the drills in a band or as an overall spray at 220–900 litres/ha. The use of gamma-HCH is restricted by risk of taint (see Chapter 2, p. 46). Carbaryl at 850 g a.i./ha *or* DDT at 630 g–875 g a.i./ha can also be used. Apply these in 200–500 litres water/ha.

Turnip gall weevil (*Ceutorhynchus pleurostigma*). The rounded galls of this weevil may occur on all cultivated brassicae, on mustard and charlock. They occur on the root just below soil level and can be distinguished from the swellings caused by clubroot as each gall contains a maggot or the cavity caused by it. Most of the damage arises in the seedbed and badly galled plants may be seriously checked. Gamma-HCH seed dressing (75 per cent a.i.) used at 47 g a.i./kg seed will reduce the attack. Well-established plants usually suffer little but swedes and turnips lose value if they are galled. Effective control on transplants can be obtained with gamma-HCH at 18–25 g a.i./100 litres water. Apply 56 ml to the base of the plant in May within 4 days of setting out. The use of dieldrin as for the control of cabbage rootfly (see p. 000) will also control the weevil. A promising control of the weevil has been obtained with dusts containing a low percentage of gamma-HCH applied as a band about 50 mm wide along the rows of seedlings when they have 2 rough leaves. Similar dusts applied to newly transplanted cabbages at the rate of 4·54 kg/1000 plants have proved effective, but both recommendations are tentative. There is a risk of root crops following this treatment being tainted.

Diseases

*Canker** (*Phoma lingam*) is a seed-borne disease which causes light brown or purple lesions on the stems of Brussels sprouts and other brassicas resulting in a stunting of the plants. Seed infection can be controlled by the thiram soak treatment (p. 314).
***Clubroot** (*Plasmodiophora brassicae*) (A.L. 276), appears as swellings on the roots which eventually decay and the plants often wilt and die. It is particularly severe on summer crops and may be confused with turnip gall weevil (round swellings containing larvae) or with damage by 2,4-D or MCPA (galls on stem base). Dipping the roots of transplants into a suspension of 100 g pure calomel in 1 litre water (preferably containing 10 g methyl cellulose) immediately before planting effects control; some insecticidal chemicals may be added for cabbage root fly control (see p. 319). In experiments preplanting dips in water containing 5 g a.i./litre of

benomyl *or* carbendazim *or* thiophanate-methyl have also given control of clubroot.

Partial sterilization of soil by use of dazomet (e.g. on areas to be used as seedbeds), at least 8–10 weeks before sowing or planting, can give considerable but not permanent control of the disease. The methods of application are similar to those recommended for its use in glasshouses (p. 253) but careful attention should also be paid to the manufacturers' instructions.

In addition to applying one of these chemical treatments, ensure that the ground is well drained and has a high lime status.

Damping-off and **wirestem** (*Rhizoctonia solani*) is a dark brown or black rot on the stem base of seedlings, especially of cauliflower, usually causing a pronounced constriction of the stem. The seedlings often die but may survive as stunted plants with a typical wirestem appearance. Control is achieved by raking into the soil surface before sowing the seed, quintozene at 7 g a.i./m^2. Do not treat soil where cucumber, marrows, melons or solanaceous plants, including tomato but not potato, are to be grown.

Downy mildew (*Peronospora parasitica*). This disease may be troublesome in the seedling stage, especially on cauliflower in Dutch lights, causing stunting or death of young plants. Spray with dichlofluanid at 0·75 kg a.i./1000 litres/ha, starting as soon as the first seedlings emerge and subsequently every 3–4 days for 10 days. Continue at weekly intervals until the seedlings have fully-grown true leaves, using higher volumes if necessary.

Leaf spot (*Alternaria brassicicola*) can cause a severe disease of brassica seed crops in wet seasons when the fungus readily invades the developing pods and causes considerable loss of seed. The disease is seed-borne and also gives rise to a damping-off of young seedlings. Seed infection can be largely controlled by the thiram soak treatment (see p. 314).

CARROT

Pests

***Carrot fly** (*Psila rosae*) is a widespread and often serious pest of carrots and, in some districts, causes damage to celery. Parsnips and parsley are also often attacked. The larvae feed on the roots and young plants may be stunted or killed. The mature roots of carrots and parsnips may be made unmarketable by the tunnelling of the larvae. There are 2 generations in a year, the first attacking crops in May and June, the second in August and September. Growing crops away from hedges or other shelter, sowing carrots in June to escape from the first generation, not growing susceptible crops frequently in the rotation, are all ways of reducing attack.

On carrots, a seed dressing containing 75 per cent gamma-HCH usually gives adequate protection on mineral soils where only light or moderate attacks are usually experienced. Use at the rate of 31 g a.i./kg of seed. On soils of high organic content such as fen soils the treatment is less effective but where carrot fly attack is light sufficient protection on early and late sown carrot may be obtained by using the dressing at 31 g a.i./kg seed. On maincrop carrots grown under similar conditions use 47 g a.i./kg seed. These dressing rates should not be exceeded or off-flavour of carrots may occur.

Aldrin and dieldrin are no longer suitable for the control of carrot fly but several organophosphorus insecticides can be used effectively, though complete control of late attacks on fen soils may not always be achieved. For carrots grown on mineral soils, chlorfenvinphos granules or spray, or diazinon granules control this pest and granules of disulfoton or phorate give combined control of carrot fly and carrot-willow aphid (see below). For fen soils use diazinon, disulfoton or phorate granules.

Chlorfenvinphos can be broadcast either as granules at 2·24 kg a.i./ha or the e.c. sprayed at 1·95 kg a.i./ha and harrowed into the soil before drilling the crop. Diazinon granules at 2·24 kg a.i./ha on mineral soils and 2·80–3·36 kg a.i./ha on fen soils can be broadcast and rotavated into the top 75 mm but are more effective when drilled as a 25 mm band about 25 mm below the soil surface directly beneath the seed. The application rate for this treatment should be at 1·68 kg a.i./ha on mineral soils and 2·24 kg a.i./ha on fen soils for 375 mm rows (26 200 m/ha) and should be correspondingly adjusted for other widths of row.

Disulfoton granules at 1·1 kg a.i./ha on mineral soils and 2·1 kg a.i./ha on fen soils or phorate granules at 1·68 kg a.i./ha on mineral soils and 2·8 kg a.i./ha on fen soils, should be applied as a 75 mm band immediately in front of the seed coulter. These rates, which are for 375 mm rows, should be adjusted proportionately for other row widths. For carrots grown in close rows, the insecticides should be broadcast and worked in, disulfoton granules at 2·1 kg a.i./ha on mineral soils only or phorate granules at 2·24 kg a.i./ha on mineral soils and 3·36 kg a.i./ha on fen soils.

Alternatively carrots sown in June, and which miss first generation attack, can be sprayed with bromophos at 1·5–2·0 kg a.i./ha over the rows, or on soil around crowns and along either side of row, in the first week of August. Repeat, if crop to remain after December, in third week of October.

For parsnips on mineral soils, use disulfoton or phorate granules at the same rates and in the manner as for carrots.

On parsley, grown on mineral soils, an effective control has resulted from the use of a seed dressing containing gamma-HCH (75 per cent) at the rate of 70 g a.i./kg seed.

Alternatively apply disulfoton granules at 1·12 kg a.i./ha as a 75 mm band immediately in front of the seed coulter.

Carrot-willow aphid (*Cavariella aegopodii*). Heavy infestations of this aphid often occur from late May to early July and cause considerable loss of yield in early and mid-season crops. The aphids infest carrots in the cotyledon stage, as well as older plants, and spread the carrot motley dwarf virus which produces a yellow mottling of the leaves and stunts the plants. Parsnips and celery are also attacked by the aphid. As soon as any aphids are seen on the plants the rows should be sprayed with demephion at 250 g a.i./ha *or* demeton-S-methyl at 245 g a.i./ha *or* dimethoate at 335 g a.i./ha *or* oxydemeton-methyl at 240 g a.i./ha *or* thiometon at 280 g a.i./ha. Apply in 200–1000 litres water/ha. If, on carrots for bunching, it is necessary to control the aphid shortly before harvest, malathion or nicotine (see cabbage aphid p. 318) should be used, as the requisite intervals (1 day and 2 days, respectively) between application and harvest is short. Alternative recommendations are granule application of phorate at 1·68 g a.i./ha *or* disulfoton at 1·1 kg a.i./ha as soon as the crop has germinated, or 0·55 kg a.i./ha of the disulfoton may be applied then followed by 0·55 kg a.i./ha 3 weeks later. Both disulfoton and phorate also give a good control of the carrot fly on mineral soils.

Diseases

Black rot (*Stemphylium radicinum*) causes black sunken lesions on the mature roots and can be troublesome in store. It can also cause loss of seed in the seed crop and a damping-off of seedlings. Seed-borne infection can be controlled by the thiram soak treatment (see p. 314).

Leaf blight (*Alternaria dauci*) can cause lesions on the leaves in wet seasons but is more important (especially in precision drilled crops) as a cause of seedling damping-off. Eradication of infection from the seed by the thiram soak treatment (see p. 314) gives a clean stand of seedlings.

CELERY

Pests

Carrot fly (*Psila rosae*), for biological details see under carrots, p. 323. Several treatments are possible against this pest. The potting soil for soil blocks or boxes can be sprayed with diazinon at 17 g a.i./1·12 litres water/100 kg soil; mix thoroughly. Good results have been obtained by drenching seedlings in soil blocks or boxes with diazinon at 720 g a.i./18 litres/1000 plants; this method also succeeds with seedlings in boxes when enough soil remains adhering to the roots at planting time. A root dip with gamma-HCH at 50 g a.i./litres water can control the pest but the roots must remain in the dip no more than 10 seconds and should not dry before planting; even so some check of growth may occur and, because of the risk of taint, avoid growing potatoes on the same land in the following year. Alternatively, the soil on the roots of seedling celery can be treated at the time of transplanting with insecticide granules from an applicator mounted (outside the operator's cab) on a mechanical planter. For this purpose, use diazinon granules at 900 g a.i. *or* disulfoton granules at 450 g a.i. to every 30 000 self-blanching plants or every 14 000 late celery plants. The last 2 treatments will also reduce damage by celery fly. Another method, which also gives some celery fly control, is to apply disulfoton granules at 1·1 kg a.i./ha for celery in wide rows and at 2·1 kg a.i./ha for self-blanching celery in narrow rows, to the furrow. The granules should be at a depth of not less than 75 mm and the celery planted into this band. The pest can also be effectively controlled by applying a liquid spot treatment within a few days of planting out. For this purpose, use diazinon at 50 g a.i./100 litres applied at 130 ml/plant. Alternatively disulfoton granules can be applied to the plant bases within a week of transplanting at the same rates as for the furrow application.

Celery fly (*Philophylla heraclei*). The larvae of this fly feed and cause large blisters in the leaves of celery and parsnips. The damage is most severe on celery, especially when the plants are small, although large plants may be quite heavily attacked. Attacks may occur from May until October. Sprays of DDT at 1·40 kg a.i./ha *or* dimethoate at 420 g a.i./ha *or* malathion at 1·26 kg a.i./ha *or* trichlorphon at 450 g a.i./ha applied in 400–1100 litres water/ha, have been found effective.

Diseases

***Leaf spot** (*Septoria apiicola*) (A.L. 241) causes brown rusty spots on leaves and stems and spreads rapidly in cool damp weather, causing much damage. The fungus is carried on the seed which should receive the thiram soak treatment (p. 314) or be immersed in water at 50°C (122°F) for 25 minutes. If the disease is seen, the plants should be sprayed with one of the following fungicides, benomyl (0·55 kg a.i.) *or* carbendazim (0·55–0·66 kg a.i.) each in not less than 200 litres water/ha repeated every 10–14 days *or* Bordeaux mixture (10 kg:12 kg:1000 litres) or a copper fungicide (w.p. or colloidal) at a rate giving 2·5 kg metallic copper/1000 litres HV or LV.

*Root rot** (*Phoma apiicola*) occasionally causes severe losses of celery seedlings in nursery beds. Soil sterilization with dazomet, applied strictly according to the manufacturer's instructions, can effectively prevent the disease. Seed-borne infection is eliminated by the thiram soak treatment (see p. 314).

LETTUCE

Pests

***Aphids.** At least 5 species of aphids live on the leaves of lettuce which they stunt, malform and contaminate by their cast skins and excretions. All are found throughout the year on lettuce grown under glass but only 3 species are serious pests of outdoor lettuce. Certain of these aphids, besides directly damaging the plants, spread viruses which cause severe stunting. Control measures should be applied early, not only so that the crop is not infested when it begins to heart, but to reduce the spread of virus when the plants are young. Sprays of demeton-S-methyl at 245 g a.i./ha *or* diazinon at 225 g a.i./ha *or* dimethoate at 335 g a.i./ha *or* ethoate-methyl at 225 g a.i./ha *or* formothion at 600 g a.i./ha *or* malathion at 1·26 kg a.i./ha *or* nicotine at 560 ml (95–98 per cent)/ha *or* oxydemeton-methyl at 240 g a.i./ha *or* pirimicarb 25 g a.i./100 litres HV have proved effective, though nicotine is less satisfactory at temperatures below 18–21°C. On summer lettuce, apply the insecticide in 800–1000 litres water/ha using a fairly coarse spray directed along the rows. Winter lettuce become infested in the seed-bed during autumn and, if treated in late October or November, will remain substantially free of aphids until cutting. The period which must elapse between application and harvest is, with malathion, 1 day; with nicotine 2 days; with dimethoate, ethoate-methyl, formothion and pirimicarb, 7 days; with diazinon, 14 days and with demeton-S-methyl and oxydemeton-methyl, 21 days.

Another method of control is to dip the seedlings, before transplanting, in a wash either of malathion, 12 g a.i. in 10 litres water, *or* nicotine 7 ml (95–98 per cent) in 10 litres water. Forty-five litres of wash is enough to treat 6000 plants, dipped in batches of 1000 or less. Each batch is dipped in the wash for about 3 minutes, then removed and planted immediately. If the lettuce is not transplanted, spray with malathion in November.

The lettuce mosaic virus is seed-borne and the risk of aphids spreading the virus can be reduced by using virus tested seed.

Lettuce root aphid (*Pemphigus bursarius*). This aphid feeds on the roots causing yellowing of the foliage, stunting and, when the soil is dry, the death of the plant. It overwinters on Lombardy poplar, and in June and July, winged migrants fly to lettuce where their progeny infest the roots, increasing rapidly in numbers especially in hot weather. After mid-August, most of the aphids return to poplar but a few remain in the soil to infest subsequent crops of lettuce planted on the same ground in the autumn and following spring. Certain varieties, such as Avoncrisp and Avondefiance, are resistant to attack. For lettuce sown between mid-April and early July, a satisfactory control can be obtained by treating the soil with diazinon at 1·12 kg a.i./ha. Spray insecticide on to the soil in 300–500 litres water/ha and cultivate to a depth of 75 mm before sowing. Alternatively use diazinon granules at 1·12 kg a.i./ha applied in 100 mm band in front of drill coulter or along rows into which mid-season lettuce crops are planted in late May, June and July. Otherwise, spray crop with diazinon at 420 g a.i./ha in 1000–1500 litres water in late June, or when the winged aphids first arrive, and again 10 days later. Allow at least 2 weeks to elapse between application and harvest. Lettuce sown at other periods usually escape attack and treatment is unnecessary.

Diseases

**Downy mildew (*Bremia lactucae*) appears as pale green or yellow angular areas on the older leaves which may bear numerous whitish spores of the fungus, especially on the lower surface. The areas later die and become brown. Under cool moist conditions, copious sporing may occur on fairly normal leaves. The disease is of most importance in the early autumn or outdoor lettuce and in late autumn on frame lettuce but also attacks overwintering lettuce, especially in frames. It may be followed by *Botrytis* or soft rot. For control of downy mildew, use zineb at 1·4 kg a.i./1000 litres/ha to check the disease on field plantings and repeat every 3–4 weeks, *or* use thiram 30 g a.i./10 litres on seedlings in frames and repeat every 14 days, a method which will also check *Botrytis*.

**Grey mould (*Botrytis cinerea*). This disease is most serious under cool damp weather conditions and is therefore most prevalent in the spring on overwintered lettuce. The first sign of infection is often a complete collapse of the plant, caused by a basal stem rot, which may occur at any stage. On young plants it is known as 'red-leg'. but sometimes the most severe attack occurs as the plants are approaching maturity. The fungus produces copious grey spores on decaying leaves and stems and also forms sclerotia which persist in the soil. The disease may start on the seedlings, sometimes following an attack of *Rhizoctonia* damping-off, and remain quiescent for many weeks before causing serious damage to the plant. Protectant treatments must therefore be applied from the seedling stage onwards and careful attention should be paid to efficient culture. Quintozene dust, applied at 7 g a.i./m², before sowing the seed or putting out transplants, checks attack by *Rhizoctonia* and has been shown to reduce *Botrytis* infection in frames. Dicloran, at 7 g a.i./m² applied before planting out in frames and followed by 1 or 2 further dustings at 2 g a.i./m² at 6-weekly intervals provides control but the last application must be made not less than 3 weeks before harvest. Seedlings grown in frames for spring planting may be sprayed with thiram suspension at 3 g a.i./litre every 3–4 weeks throughout winter; adequate wetting of the stem bases is essential. This treatment also checks downy mildew (*q.v.*). Benomyl at 250 g a.i./1000 litres/ha) applied to run off at planting out and at 2 further 14 day intervals also gives very effective *Botrytis* control. At similar spray intervals promising results have been obtained with thiophanate-methyl at 1·1 kg a.i./220–1000 litres/ha or carbendazim 0·55–0·66 kg a.i./200–1000 litres/ha applied after planting out.

*Ring spot (*Marssonina panattoniana*) forms circular spots 3–7 mm in diameter, on the outer leaves and elongated spots resembling slug injury on the leaf midribs. The disease is rarely serious although rather disfiguring but is occasionally severe. The fungus is carried on seed and in plant debris. Bordeaux mixture (1 per cent metallic copper), *or* thiram 2 g a.i./litre, *or* captan 2 g a.i./litre are reported to give some control if applied from the seedling stage onwards.

Diseases

*****Rust** (*Puccinia menthae*). The first sign of infection is the production in spring of swollen distorted shoots bearing yellow pustules of spores. Later infections appear as brown pustules on the undersides of leaves and the affected leaves eventually dry up and fall. No spray treatment will cure the disease but reductions in infection of second and third cuts from outdoor crops have been obtained by spraying the stubble of earlier cuts with sulphuric acid (BOV full strength) at the rate of 270 litres/ha.

ONION

Pests

*****Onion fly** (*Delia antiqua*). Maggots of the onion fly cause serious damage to seedling onions and plants which are bulbing-up; onion sets, leeks and shallots are sometimes infested. Treatment of the seed with a dieldrin seed dressing is an effective control measure in areas where onion fly has not developed resistance to this insecticide. First mix the seed thoroughly with 15 ml paraffin (kerosine)/kg seed; then add the seed dressing (75 per cent dust) at the rate of 29 g a.i./kg seed and mix until the seeds are uniformly covered. Starch paste at 125 g/kg seed *or* the methyl cellulose sticker described below under smut (p. 330) can be used instead of paraffin as the sticker.

Some control may be obtained by the use of gamma-HCH dust at 400–500 g a.i./ha applied as a band, about 75 mm wide, along the rows. The first application should be made when the seedlings are in the 'loop' stage and a second 2 weeks later.

Diseases

****Downy mildew** (*Perenospora destructor*). Pale oval areas appear on the leaves or the tips of the leaves may become pale and die back. The leaves often fold downwards at the infected area on which grey, later brown-purple, spores of the fungus may develop. These spread extensively in cool moist conditions. Other leaf moulds also grow on the lesion and may obscure the pathogen. The fungus grows internally and can infect the bulbs which, if kept for seed raising, produce stunted leaves. The flower stalks of seeding plants may bear a conspicuous pale oval lesion at which point the stalk often breaks. Oospores of the fungus can persist in the soil and infect new crops. The disease is also carried over in perennial onions and shallots and possibly in wild Allium species.

The chief control measures are to avoid contaminated land and to grow onions, where possible, on warm well-drained soils in sites with good air circulation. Overwintering hosts should be eradicated. Some control of leaf infection has been obtained by the application of zineb at 1·4 kg a.i. with 500 ml of a non-ionic wetter per 1000 litres/ha when the disease appears and then continued at 14-day intervals.

*Smut (*Urocystis cepulae*) (A.L. 261). This disease is soil-borne. Germinating mycelium from the spore balls in the soil infects the bases of the leaves of young seedlings causing dark lead-coloured spots or streaks beneath the skin. These leaves later become thickened and twist or curl backwards, and further leaves of bulb scales may bear dark areas which later split exposing black spores of the fungus. Soil which has carried an infected crop should not be used for onion growing for as long a time as possible. When it is used, the disease can be reduced by applying formaldehyde to the drill just after sowing but before covering the seed, using 77 ml commercial 40 per cent formaldehyde/10 litres water/100 m row. In very wet weather the formaldehyde concentration should be doubled. When soil contamination is not heavy, good control can be obtained by pelleting the seed with thiram (80 per cent w.p.), using a resin-alcohol or methyl cellulose sticker, and applying 500 g a.i./kg seed. The resin sticker is prepared by dissolving 300 g resin in 1 litre methylated spirit. Mix the sticker at 62·5 ml/kg seed thoroughly before adding the fungicide and then shake vigorously for 5 minutes. Alternatively, dissolve 50 g of methyl cellulose in 1 litre of warm water and use 185 ml/kg seed. Stringent precautions should be taken against spreading infested soil on boots, implements, etc. The disease also occurs on leeks, shallots, chives and garlic.

***White rot (*Sclerotium cepivorum*) (A.L. 62). The plants are stunted with yellow leaves, and the base of the plant is rotten and often covered with a white fungal growth in which the black resting bodies (sclerotia) of the pathogen may be embedded. The fungus is soil-borne and its sclerotia may persist for many years. The disease can be controlled to some extent in the salad onion crop by applying calomel (4 per cent dust) at 40 g a.i./100 m of drill to the seed drill *or*, more effectively, by seed treatment with calomel. A methyl cellulose sticker should be used 10–20 g of the latter being dissolved in 1 litre water and the mixture applied to onion seeds to moisten them thoroughly. The calomel is best applied at a rate of 1 kg of pure calomel/kg seed but some control can be obtained by using 0·25 kg calomel/kg seed. In experiments seed dressings of benomyl, *or* carbendazim *or* thiophanate-methyl at 0·1–0·15 kg a.i./kg seed gave effective control of white rot and these compounds are in commercial use in some areas.

Disease control by seed treatments is more effective in spring- than in autumn-sown salad onions.

Neck rot (*Botrytis allii*). Severe outbreaks of neck rot, caused by *Botrytis allii*, have occurred in stored bulb onions in recent years in Britain. The disease which only becomes evident in store, causing onions to soften and rot internally with the production of the black sclerotia of the fungus on the necks of bulbs, was thought to infect plants in the field at harvest. Recent research has shown, however, that the fungus is carried internally in onion seeds and it infects the green tissues of the emerging seedling cotyledons. Condiophores produced on necrotic parts of leaves release spores which spread infection through developing onion crops. The disease is symptomless and healthy and infected onion plants are indistinguishable. The fungus on the infected bases of older leaves invades the neck tissues of the developing bulbs and grows downwards in them causing neck rot.

The disease is effectively controlled by dry or slurry seed dressings of benomyl plus thiram applied at 1·1 g a.i./kg seed.

PEA

Pests

Pea aphid (*Acyrthosiphon pisum*). Periodically heavy infestations of this aphid occur on peas. It lives throughout the year on leguminous plants, overwintering as eggs, or occasionally as adults, on clovers, lucerne, trefoil and sainfoin. Pea crops can be attacked at any time from early May to autumn but those growing in June and July are generally the most seriously affected. Azinphos-methyl spray used against pea moth *or* fenitrothion used against any of the pea pests (see below) will also control the aphid when present, but if circumstances demand the use of separate control measures, fenitrothion at 700 g a.i./ha *or* any of the insecticides, with the exception of phosphamidon and pirimicarb, recommended for black bean aphid control (see Broad bean—Black bean aphid, p. 314) may be used for controlling pea aphid. Apply at the same strength and rate as for black bean aphid and observe the same minimum periods between spraying and harvesting.

Pea midge (*Contarinia pisi*). Outbreaks of damage by pea midge sometimes occur, and although the small whitish, jumping larvae cause more injury by feeding on the flowers and growing points of pea plants they also infest and malform the pods. Azinphos-methyl *or* fenitrothion spraying against pea moth or pea aphid usually also controls pea midge; but should midge alone be troublesome spray azinphos-methyl at 310 g a.i./ha *or* carbaryl at 1·44 kg a.i./ha *or* dimethoate at 335 g a.i./ha *or* fenitrothion at 700 g a.i./ha in 400–1000 litres water/ha on Advisory Service warnings or at first appearance of adult midges. A second application may be needed after 7–10 days.

Pea moth (*Laspeyresia nigricana*). The caterpillars of this moth feed inside the pods on developing peas greatly reducing their value as food or seed. The pest is most prevalent in the southern part of England, particularly in East Anglia and Kent. All varieties of peas are attacked but those with a long growing period suffer most. Quick maturing varieties sown early and those sown after the middle of June usually escape attack. Peas which come into flower during the flight period of the moth, between mid-June and mid-August, normally suffer most damage. This damage can be much reduced by spraying azinphos-methyl at 310 g a.i./ha or carbaryl at 1·7–1·9 kg a.i./ha or fenitrothion at 700 g a.i./ha or tetrachlorvinphos at 1·27 kg a.i./ha in at least 600 litres water at medium to high pressure. A careful timing of the spray is necessary. For crops which come into flower before the middle of June, spray in the third week of June; for those which flower between mid-June and mid-August, spray 7–10 days after the beginning of flowering; spraying is usually not necessary on crops flowering after mid-August. When using azinphos-methyl or fenitrothion or when a higher level of control is required with carbaryl, as for peas for quick-freezing or canning, a second spray may be needed about 14 days after the first. The crop should not be harvested until 14 days have elapsed since the last spraying. Two sprays, timed as above, are usually necessary on dry harvested peas because they are exposed to attack for a long period. Weather conditions may affect the time of appearance and the activity of the moths; advice on timing of sprays should be sought of the local ADAS officer.

*Pea thrips (*Kakothrips robustus*). These small insects feed on the surface tissues of the young pods and foliage of peas and beans, causing silvery mottled patches and malformation of the pods. Heavy attacks may lead to severe stunting. Peas are particularly affected and the main attacks occur in June or July. A spray of azinphos-methyl or carbaryl or fenitrothion applied to control pea moth (see above) will also give adequate protection against pea thrips. Alternatively dimethoate at 335 g a.i./ha can be used in at least 600 litres water/ha.

*Cabbage thrips (*Thrips angusticeps*). These insects resemble the pea thrips but damage young pea plants in April and May causing stunting, leaf malformation and discoloration (yellow blotching). Peas following mustard or radish seed crops are especially liable to damage. When the pest first appears use azinphos-methyl at 310 g a.i./ha or carbaryl at 1·7 kg a.i./ha or dimethoate at 335 g a.i./ha or fenitrothion at 700 g a.i./ha in 400–1000 litres water/ha.

*Pea and **bean weevils** (*Sitona* spp.). These weevils feed on a wide range of leguminous crops, the adults on the leaves, the larvae on the root nodules. The adults cause the greater damage and eat semi-circular notches in the leaf edges and, when large numbers are present, defoliation or destruction of the crop may occur. The adult weevils are active in April and May, most damage being done when conditions for plant growth are poor, as in cloddy soil or cold dry weather. The weevils can be controlled by dusting the foliage with DDT at 2·8 kg a.i./ha or gamma-HCH at 410 g a.i./ha. Alternatively excellent control can be obtained by spraying with azinphos-methyl at 310 g a.i./ha *or* DDT at 1·05 kg a.i. (in e.c. formulations) or 875 g a.i. (in w.p. formulations)/ha *or* fenitrothion at 700 g a.i./ha, using at least 200 litres water/ha. If weevil attack is anticipated, phorate granules at 1·7 kg a.i./ha can be applied at sowing time, into the furrow behind the coulter, but not in direct contact with the seed.

Diseases

Leaf and **pod spot (*Ascochyta pisi*) is the main pathogen causing leaf and pod spot of peas; it is commonly seed-borne. The disease causes sunken brown cankers on the stems of plants and tan coloured lesions on the leaves and ultimately on the pods affecting the seeds. Crop losses are due to reduction of stand and yield and to the production of stained seeds unfit for processing. Spores of the fungus are disseminated through pea crops by the action of rain and wind. The disease is thus more prevalent in wet years. Dry or slurry (with 3 ml water/kg seed) dressings of benomyl plus thiram applied at 1·1 g a.i./kg seed virtually eliminates infection. The related seed-borne fungus *Mycosphaerella pinodes* (*Ascochyta pinodes*) which causes foot rot with leaf spot of peas is reduced but not eradicated by this seed treatment.

SPINACH

Pests

**Black bean aphid (*Aphis fabae*). This aphid causes the same type of damage as that on beet (see p. 317) and can be controlled by the same insecticides. Observe particularly the minimum intervals which must elapse between application and harvest.

   ***Mangold fly (*Pegomya betae*)—see Beetroot-Mangold Fly, p. 317.

Diseases

*Downy mildew (*Peronospora effusa*) forms yellow patches on the upper surfaces of leaves accompanied by grey or violet-grey mould beneath the leaves. Badly affected leaves cease growth and may curl downwards at the edges. The disease can be very severe under moist conditions, on badly drained land and at low temperatures. The life history of the pathogen and the control of this disease have not been satisfactorily worked out and it is uncertain whether oospores carried on the seed or present in the soil serve as sources of infection of new crops. Such infections can arise early and the closeness of the affected leaves to the ground makes adequate protective spraying difficult. Protection of upper leaves can be obtained by fungicides such as zineb at 1·4 kg a.i. plus

500 ml of a non-ionic wetter/1000 litres/ha. Some control has been reported from other countries by the use of copper fungicides.

TOMATO (OUTDOOR)

Diseases

***Blight** (*Phytophthora infestans*) can be severe in cool wet seasons, causing grey-brown lesions on the leaves and russet-brown marbled areas on the fruits, which become unusable. The disease is contracted from infected potatoes and the control is the same as for potato blight (see Chapter 6, p. 125). Copper and dithiocarbamate fungicides all give good protection and one of these fungicides should be applied at 3-week intervals from the end of July, or earlier if potato blight is seen in the neighbourhood.

Didymella stem rot (*Didymella lycopersici*). A dark brown shrunken canker appears near the base of the stem and if this girdles the stem, the plant wilts suddenly. Later in the season similar cankers may appear on other parts of plants. There may also be brown spots on the leaves and black encrusted lesions on the stem ends of the fruits. Affected plants should be removed and burned as soon as possible, and at the end of the season, all wirework, canes, poles, etc., should be sterilized by immersion for 24 hours in 20 ml of commercial 40 per cent formaldehyde/litre or for 15 minutes in 50 ml commercial 40 per cent formaldehyde/litre followed by wrapping in sacks for 1 week. The remaining plants in an infected area should be sprayed at the stem base with captan at 2·5 kg a.i./1000 litres at 115–140 ml/plant. In Jersey, a reduction in infection has been obtained by spraying plants with maneb (as for blight control) using 1·0–2·0 kg a.i./ha LV. Maneb has also been used satisfactorily, on outdoor tomatoes only, for stem base applications using 115–140 ml/plant of a suspension prepared by adding 10 g of 80 per cent w.p. to 1 litre water. This treatment can be applied to plants immediately adjacent to infected ones.

WATERCRESS

Diseases

*Crook root** (*Spongospora subterranea* f. sp. *nasturtii*). The roots of the watercress are attacked by the swimming spores of the causal fungus which then grows in the cells of the roots. Affected roots are stunted, swollen and distorted and eventually decay. The plants become stunted and their leaves turn bluish or yellow. The disease is particularly severe from October to March. For control apply a solution of zinc sulphate, usually 20 per cent, to the borehole water by a constant drip-feed apparatus to give a zinc concentration of 0·1 ppm. The rate of flow of the water must be measured by means of a graduated weir and the application rate of the solution adjusted accordingly. Advice on procedure should be obtained from the local ADAS officer.

Alternatively, a powdered glass frit containing zinc may be used. Apply 180 g of the zinc frit/m^2 to the top third or quarter of the cress bed near the water entrances after draining the bed. Wash the frit off the plants on to the floor of the bed. Re-admit water slowly. This method is for use on beds fed from several water sources and in which the more preferable zinc sulphate treatment cannot be applied.

Chapter 12
Pest and disease control in outdoor ornamentals and in turf

The term 'outdoor ornamentals' is intended to mean those plants grown for their ornamental value whether trees, shrubs, herbaceous perennials or annuals. Although the number of different ornamental plants is large, not many can be considered of economic importance through selling regularly in large numbers. Many are purchased by amateur gardeners in small quantities, but they need to be raised by nurserymen on a fairly large scale and many require protection from pests and disease when in the nursery. Therefore, the following advice is meant primarily for nurserymen and large-scale growers, but it can also apply to small gardens where amateurs may meet with similar troubles. In any case, many of the smaller plants are grown from seed produced in Britain and it is important for people growing a commercial seed crop to know how to protect the seed heads from infection by various diseases. For this reason, some general advice on chemical control measures is given, based on experience with those used to combat similar pests and diseases affecting commercially-important crops.

Of the very large number of ornamental plants which are grown for sale to the public and for seed production it is only possible to mention a small proportion and to deal with the control of only the more important and destructive of the pests which attack them.

A number of important soil pests has been omitted here since they may attack almost every plant mentioned. These include cutworms, swift moth caterpillars, wireworms, chafer grubs, leatherjackets, millepedes, woodlice and slugs, all of which are dealt with in other chapters. The control of chafer grubs and leatherjackets in turf is, however, described, since the methods employed differ from those used on arable land.

Detailed information on the propagation and culture of many of the plants mentioned in this Chapter may be found in the appropriate Bulletins issued by the MAFF, e.g. Bulletin No. 190—'Outdoor Flowers for Cutting'. References to certain aspects included in these Bulletins are included under the host plant in the following pages.

ACER

Diseases

*Tar spot (*Rhytisma acerinum*). During the summer large black spots with a bright yellow edge develop on the leaves. This disease is more unsightly than harmful, but leaves may fall prematurely and these should be raked up and burnt. If necessary spray in spring with Bordeaux mixture containing 150 g copper/100 litres *or* copper oxychloride at 37 g copper/100 litres.

Pests

Aphids, e.g. peach-potato aphid (*Myzus persicae*), mottled arum aphid (*Aulacorthum circumflexum*) and other species cause direct damage by feeding on young growth and also transmit virus disease. Control as for Dahlia (p. 344).

*Caterpillars,** e.g. angleshades moth (*Phlogophora meticulosa*) and cutworms (*Agrotis* spp.) eat stems, leaves and buds. For control measures see under Dahlia (p. 345).

Diseases

Downy mildew (*Peronospora ficariae*) causes a fine whitish mould on the underside of the leaves which tend to roll upwards. Control with sprays at 10-day intervals using maneb at 2·25 kg a.i./ha *or* zineb at 140 g a.i./100 litres.

Grey mould (*Botrytis cinerea*). This fungus may cause rotting of flowers and flower buds in wet weather, especially in winter months. A benomyl spray at 25 g a.i./100 litres *or* captan at 100 g a.i./100 litres HV (2·8 kg a.i./ha LV) *or* thiophanate-methyl at 50 g a.i./100 litres *or* thiram at 160 g a.i./100 litres HV (3·4 kg a.i./ha LV) gives a worthwhile degree of protection.

Pests

*Aphids,** e.g. peach-potato aphid (*Myzus persicae*) and potato aphid (*Macrosiphum euphorbiae*), green, yellow or pinkish aphids causing stem distortion and stunting. They may be controlled by the sprays listed under Dahlia (p. 344) but do not use malathion.

Diseases

***Damping-off** (*Pythium* and *Phytophthora* spp.)—see Seedlings (p. 369).

Downy mildew (*Peronospora antirrhini*). A grey mealy growth develops on under surfaces of affected leaves which are curled and dull.

IFH—12

Growth is severely checked. Spray early with thiram at 160 g a.i./100 litres *or* zineb at 140 g a.i./100 litres.

**Grey mould (*Botrytis cinerea*). On seed crops the seed spikes may be continually destroyed by the fungus especially in damp summers and, in some cultivars, the yield of seed can be reduced to negligible proportions. It may be necessary to spray every 14 days with benomyl at 25 g a.i./100 litres *or* thiophanate-methyl at 50 g a.i./100 litres *or* every 7–14 days with captan at 100 g a.i./100 litres *or* thiram at 160 g a.i./100 litres HV (3·4 kg a.i./ha LV).

**Leaf spot and stem rot (*Phyllosticta antirrhini*). On stems dark brown spots develop on which can be seen small black fruiting bodies of the fungus. When the main stem of the young plant is attacked, dieback occurs. In older plants lesions develop on the stems causing them to split and when girdling occurs the plant is killed. Pale brown circular spots develop on the leaves, flowers and fruits. Tentative recommendations are to treat the seed with captan seed dressing at 3·4 g a.i./kg and to spray early with thiram at 160 g a.i./100 litres *or* zineb at 140 g a.i./100 litres.

***Rust (*Puccinia antirrhini*). Dark brown masses of spores develop on the leaves and stems and severely affected plants may be killed. A tentative recommendation is to spray with oxycarboxin at 47–93 g a.i./100 litres. Triforine at 25 g a.i./100 litres could also be tried but may damage certain cultivars.

AZALEA

Pests

*Azalea whitefly (*Pealius azaleae*) on evergreen azaleas are very localized. The pale green 'scales' infest the underside of leaves which become disfigured by sooty moulds. Their control is as for rhododendron whitefly (p. 366). For other pests of azaleas outdoors see under Rhododendron, p. 365. For pests of azaleas under glass see Chapter 10, p. 261.

AZALEA
(Deciduous types)

The growth of algae and lichen on branches may be corrected by a caustic wash made by dissolving sodium hydroxide (caustic soda) 1 kg/100 litres water and adding a good spreader. This spray must be applied during December or early January while the flower buds are still quite dormant. The operator should wear old clothes, goggles and gloves, but the spray is quite safe if used with care. An alternative is lime sulphur at 2·5 litres/100 litres.

AZALEA
(Evergreen types, e.g. Kurume, etc.)

Diseases

**Gall or false bloom (*Exobasidium vaccinii*). The young, developing leaves or flower buds turn reddish and swell into small galls which then turn waxy white. Pick off the galls and spray with a copper-containing fungicide, such as Bordeaux mixture, equivalent to 150 g copper/100 litres water, *or* copper oxychloride at 37 g copper/100 litres, *or* zineb at 140 g a.i./100 litres.

 ***Root rot (*Phytophthora cinnamomi* and other species)—see Heather (p. 350).

BEECH

Pests

**Beech aphid (*Phyllaphis fagi*). These are yellowish-green aphids covered with long, flocculent, white 'wool'. They feed on the underside of leaves, covering them with honeydew on which grow sooty moulds. The leaves may shrivel in severe infestations. They are controlled by the measures recommended for aphids on fruit trees (see Chapter 9, p. 172).

 **Beech scale (*Cryptococcus fagi*) is a small yellowish scale covered with a white, felted mass of 'wool', which forms conspicuous colonies on the bark. They are controlled by the measures recommended for scale insects on fruit trees (see Chapter 9, p. 187).

BEGONIA
(Outdoor bedding types)

Diseases

**Powdery mildew (*Oidium begoniae* and *Micosphaera begoniae*) causes white powdery spots or patches on leaves and stems. The damage can be extensive and control may be difficult. Several fungicides, such as copper oxychloride at 37 g a.i./100 litres *or* dinocap at 12·5 g a.i./100 litres *or* thiram at 320 g a.i./100 litres, at 3–7 day intervals, applied 3 times, are reported to be effective; although dinocap may cause slight scorch of the leaf edge of certain cultivars. Benomyl at 25 g a.i./100 litres *or* thiophanate-methyl at 50 g a.i./100 litres at 14 day intervals, can also be used, or triforine at 25 g a.i./100 litres could be tried on plants not in flower.

BIRCH

Diseases

Honey fungus (*Armillariella mellea* syn. *Armillaria mellea*)—(see Privet (p. 345)).

CAMELLIA

Pests

**Scale insects. Soft scale (*Coccus hesperidum*) and cushion scale (*Chloropulvinaria floccifera*) suck sap from the underside of the leaves and young stems. The leaves often become blackened with sooty mould. For control measures see willow scale (p. 378).

Diseases

*Leaf spot (*Pestalotiopsis guepini* syn. *Pestalotia guepini*) appears as yellowish patches on the leaves and these later turn greyish white with pinpoint size black dots on the surface. The disease is only likely to be serious in the case of cuttings and young plants in glasshouses and frames.

It is often possible to cut out the affected part of a leaf and the remainder should be sprayed with copper oxychloride at 37 g copper/100 litres.

CARNATION (BORDER)

Pests

**Aphids, e.g. the peach-potato aphid (*Myzus persicae*), green, yellow or pinkish aphids on shoots, cause stunting and distortion. They may be controlled by the methods given for aphids on Dahlia (p. 344). The summer forms of the elder aphid (*Aphis sambuci*) infest the roots and prevent normal growth. The plants wilt in dry weather. Control is difficult but the methods of controlling lettuce root aphid should be tried, i.e. water soil before planting with diazinon 1·12 kg a.i. in at least 340 litres water/ha. Established plants found infested should be drenched with diazinon 20 g a.i./100 litres. For pests of carnations under glass, see Chapter 10, p. 264.

Diseases

***Ring spot (*Didymellina dianthi* syn. *Heterosporium echinulatum*) causes round, grey spots in which the skin erupts into pustules arranged in roughly concentric rings. Infected leaves wither, affected stems snap and the plants flower poorly (see MAFF Bull. 151). Control may be difficult but repeated spraying with Bordeaux mixture, equivalent to 150 g copper/100 litres water, gives the best control.

Rust (*Uromyces dianthi*) appears as brown spore clusters on both sides of the leaves and stems, usually starting on the lower leaves. Many sprays have been recommended, such as lime sulphur at 2·5 litres/100 litres, repeated at 10–14 day intervals *or* mancozeb/zineb at 175 g product/100 litres *or* thiram at 320 g a.i./100 litres *or* zineb at 200 g a.i. + 50 ml spreader/100 litres, applied at 10–14 day intervals after the disease is first seen, to keep the new foliage and stems free from severe attack. Benodanil at 50 g a.i./100 litres *or* oxycarboxin at 63–125 g a.i./100 litres could also be tried. Sprays should not be applied on open blooms for fear of staining.

CHAMAECYPARIS

Diseases

***Root rot** (*Phytophthora cinnamomi* and other species)—see Heather (p. 350).

CHINA ASTER

Pests

Aphids, e.g. potato aphid (*Macrosiphum euphorbiae*), leaf-curling plum aphid (*Brachycaudus helichrysi*), are green, pink or greenish black aphids and cause general stunting of growth. For their control see aphids on Dahlia (p. 344).

Capsid bugs, e.g. tarnished plant bug (*Lygus rugulipennis*) cause tattering of leaves and distorted growth. Control measures are given under Dahlia, (p. 345).

Diseases

*Foot rot** (*Phytophthora cryptogea*) causes a brownish or blackish rot of roots and stem bases. The seed may be dressed with captan at 3·4 g a.i./kg seed *or* with thiram at 2·2 g a.i./kg seed. The soil can be watered with captan at 125 g a.i./100 litres *or* with Cheshunt compound at 310 g copper/100 litres *or* with zineb at 140 g a.i./100 litres.

CHRYSANTHEMUM

The pests and diseases of chrysanthemums grown as glasshouse crops are dealt with in Chapter 10 (p. 270).

Pests

**Aphids, e.g. chrysanthemum aphid (*Macrosiphoniella sanborni*), peach-potato aphid (*Myzus persicae*) and leaf-curling plum aphid (*Brachycaudus helichrysi*). These aphids feed on shoots and foliage, causing stunting and distortion of new growth and poor quality blooms and may also transmit virus diseases. They are controlled by the methods given for aphids on Dahlia (p. 344) but do not use dimethoate or formothion. Certain varieties are also damaged by demeton-S-methyl and oxydemeton-methyl (see product labels).

   *Capsid bugs—see Dahlia (p. 345).

   **Chrysanthemum leaf miner (*Phytomyza syngenesiae*)—see Chapter 10, p. 271.

   **Earwig (*Forficula auricularia*). These dark brown elongate insects with leathery wing cases and stout pincers on the hind end of the body feed at night on petals making them ragged and unsightly. During the day they hide deep among the petals or under debris on the ground. To control them, clear away all debris. If possible, shake open blooms to dislodge insects and then spray the ground and lower parts of plants with gamma-HCH 280 g a.i./ha *or* trichlorphon at 800 g a.i./ha.

   **Leaf and bud eelworm (*Aphelenchoides ritzemabosi*)—see Chapter 10, p. 271.

Diseases

*Blotch or leaf spot (*Septoria* spp.). Round brown or black spots or blotches up to 25 mm diameter develop on the lower leaves. Fruiting bodies of the fungus show on the blotches as pinpoint sized black dots. The disease is especially severe in wet weather, when the spots coalesce, the leaves wither and fall prematurely. Spray at 7–10 day intervals with copper sprays containing 75 g copper/100 litres avoiding blooms, *or* with zineb at 140 g a.i./100 litres. Some cultivars may be damaged by copper.

   **Grey mould (*Botrytis cinerea*) is favoured by warm, moist weather when this disease may cause much damage to the blooms (damping) which become covered with the greyish mass of spores. Regularly remove all infected blooms. For control apply benomyl at 25 g a.i./100 litres *or* captan at 100 g a.i./100 litres *or* thiophanate-methyl at 50 g a.i./100 litres *or* thiram at 160 g a.i./100 litres HV (3·4 kg a.i./ha LV). Alternatively try chlorothalonil at 110 g a.i./100 litres.

   **Petal blight (*Itersonilia perplexans*) appears as small oval water-soaked spots on the outer petals, rapidly spreading to spoil the bloom. Control can be achieved by spraying with zineb w.p. at 140 g a.i./100 litres *or* tank-mix zineb (for preparation see Chapter 10, p. 309). Commence when buds first show colour and repeat at 5–7 day intervals.

   *Powdery mildew (*Oidium chrysanthemi*) (see MAFF Bull. 92) shows as a superficial white powdery deposit on the leaves. It can be controlled by spraying with copper-oil sprays containing 75 g copper and 0·5 per cent white oil/100 litres HV *or* dinocap (w.p. or e.c.) at 6·25 g a.i./100 litres but care is necessary as some cultivars are copper-sensitive or may be damaged by dinocap. Other fungicides which can be used are benomyl at 25 g a.i./100 litres *or* quinomethionate at 12·5 g a.i./100 litres *or* sulphur at 300 g a.i./100 litres *or* thiophanate-methyl at 50 g a.i./100 litres *or* thiram at 300 g a.i. + 345 ml petroleum/100 litres *or* triforine at 25 g a.i./100 litres. Applications should be made at the first sign of the disease, repeating at 10–14 day intervals.

Rust (*Puccinia chrysanthemi*) causes brown erumpent pustules on the undersides of the leaves and on stems, and may be serious under moist conditions. It appears first on the lower leaves which are often removed when potting on. To control, use copper sprays at 75 g copper/100 litres at 2- or 3-weekly intervals, but keep these sprays off the blooms, and some care is necessary as certain cultivars are sensitive to copper, *or* zineb at 210 g a.i./100 litres; oxycarboxin at 70–93 g a.i./100 litres, *or* triforine at 25 g a.i./100 litres could also be tried.

CLARKIA

Diseases

*****Grey mould** (*Botrytis cinerea*) attacks the bases of the stems; for control spray with benomyl at 25 g a.i./100 litres HV *or* thiophanate-methyl at 50 g a.i./100 litres HV.

CLEMATIS

Pests

*****Clay-coloured weevil** (*Otiorrhynchus singularis*)—see Rhododendron (p. 366).

*****Earwig** (*Forficula auricularia*)—see Chrysanthemum (p. 342).

Diseases

Powdery mildew (*Erysiphe polygoni*) occasionally causes damage in southern counties. The disease can be checked by spraying with dinocap at 12·5 g a.i./100 litres but it may damage some cultivars. Benomyl at 25 g a.i./100 litres *or* thiophanate-methyl at 50 g a.i./100 litres can also be used at 14-day intervals.

Wilt (is now believed to be due to *Ascochyta clematidina*). One or more shoots die back rapidly, often to the base of the plant, but new shoots usually develop in due course. Spray the developing shoots with a copper spray such as Bordeaux mixture equivalent to 150 g copper/100 litres *or* copper oxychloride at 37 g copper/100 litres.

CONIFER SEEDLINGS

CONIFER SEEDLINGS
See Spruce, p. 371.

CORNFLOWER

Diseases

Petal blight (*Itersonilia perplexans*). Small oval water-soaked spots develop on the outer petals and spread rapidly thus spoiling the bloom. Commence spraying when the buds first show colour and repeat at weekly intervals using zineb w.p. at 140 g a.i./100 litres *or* tank-mix zineb (for preparation see Chapter 10, p. 309).

Powdery mildew (Oidium spp.) shows as a white powdery coating on leaves, stems and flowers. Spraying with dinocap at 12·5 g a.i./100 litres. Benomyl at 25 g a.i./100 litres *or* thiophanate-methyl at 50 g a.i./100 litres at 14 day intervals *or* triforine at 25 g a.i./100 litres may also control this disease.

Rust (Puccinia cyani) appears as brown spore masses on the undersides of the leaves. Spraying with thiram at 160 g a.i./100 litres as soon as the disease is first seen, should check it. Oxycarboxin at 49–94 g a.i./100 litres *or* triforine at 25 g a.i./100 litres could also be tried.

COTONEASTER

Pests

Brown scale (*Parthenolecanium corni*)—see Chapter 9, p. 209.
 Hawthorn webber (*Scythropia crataegella*)—see Hawthorn (p. 349).
 Winter moths (*Operophtera brumata, Alsophila aescularia* etc.)—see Chapter 9, p. 189.
 Woolly aphid (*Eriosoma lanigerum*)—see Chapter 9, p. 191.

Diseases

Fire blight (*Erwinia amylovora*)—see Chapter 9, p. 206.
 Honey fungus (*Armillariella mellea* syn. *Armillaria mellea*)—see Privet (p. 364).

CROCUS

Pests

Tulip bulb aphid (Dysaphis tulipae)—see Tulip (p. 373).

DAHLIA

Pests

Aphids, e.g. bean aphid (*Aphis fabae*) (A.L. 54) and leaf-curling plum aphid (*Brachycaudus helichrysi*) are black or greenish aphids infesting the foliage and young shoots, causing stunting and distortion and transmitting virus diseases. They may be controlled by spraying as necessary with demeton-S-methyl 22 g a.i./100 litres *or* diazinon 20 g a.i./100 litres *or* dimethoate 30 g a.i./100 litres *or* formothion 30 g a.i./100 litres *or* gamma-HCH 280 g a.i./ha *or* malathion 112 g a.i./100 litres *or* oxydemeton-methyl 22 g a.i./100 litres *or* pirimicarb 25 g a.i./100 litres. Demeton-S-methyl, formothion and oxydemeton-methyl may also be used as soil drenches. Aldicarb granules give good control at 5 kg a.i./ha. Further details of aphicides that may be used on dahlias and other ornamentals are given under Chrysanthemum in Chapter 10 (Table 10.2, p. 266).

Capsid bugs, e.g. tarnished plant bug (*Lygus rugulipennis*) and common green capsid (*Lygocoris pabulinus*). Capsids are active sap-sucking bugs, about 8 mm long. They feed on young leaves and shoots and on flower buds. Injured flower buds produce malformed, lop-sided blooms or the bud may be killed. Attacked shoots are scarred, stunted and twisted. On foliage, the feeding punctures appear initially as small brown spots of calloused tissue. As the leaves grow they become deformed and puckered. Damage may occur from June onward but is usually most serious later in the summer. To control this pest ensure that ground is kept clear of weeds and rubbish and spray the plants and ground underneath 2–3 times at 14-day intervals starting as early as possible with DDT at 1·4 kg a.i./ha *or* diazinon at 20 g a.i./100 litres *or* gamma-BHC at 280 g a.i./ha.

Caterpillars, e.g. angleshades moth (*Phlogophora meticulosa*). Various species of caterpillar feed on foliage and flowers. They may be controlled by spraying as necessary with carbaryl at 850 g a.i./ha *or* DDT at 1·4 kg a.i./ha *or* trichlorphon at 800 g a.i./ha *or* dusting with DDT dust *or* derris dust.

Earwig (Forficula auricularia)—see Chrysanthemum (p. 342).

Red spider mite (*Tetranychus urticae*). This mite is primarily a pest of glasshouse plants but also attacks dahlia and certain other ornamentals grown outdoors. Breeding is continuous during the summer and severe infestations build up rapidly in dry, warm weather. At first leaves are lightly speckled but bronzing and death may follow.

The chemical sprays detailed in Chapter 10, p. 266, may be used, but strains of red spider mite have developed resistance to some of these and will probably become resistant to others.

Diseases

Grey mould (Botrytis cinerea). This fungus can be a nuisance on flowers and flower buds in moist summer weather and should be checked by spraying with captan at 100 g a.i./100 litres *or* thiram at 160 g a.i./100 litres HV. Benomyl at 25 g a.i./100 litres *or* thiophanate-methyl at 50 g a.i./100 litres could also be tried. Affected flowers, leaves, etc., should be removed.

*Smut (*Entyloma calendulae* f. *dahliae*) produces circular brown spots on the lower leaves spreading upwards and, where spots coalesce, large areas may be killed and the leaf wither. The disease is only likely to be serious on the small bedding cultivars planted close together.

A copper spray, such as Bordeaux mixture at 150 g copper/100 litres *or* copper oxychloride at 37 g copper/100 litres gives control.

DAPHNE

Pests

**Aphids, e.g. peach-potato aphid (*Myzus persicae*)—for control, see aphids on Dahlia (p. 344).

Diseases

**Leaf spot (*Marssonina daphnes*) causes spotting on leaves, especially at the base and on the petiole, resulting in appreciable defoliation. Spraying with copper oxychloride at 37 g copper/100 litres is advised.

DELPHINIUM

Diseases

**Black Blotch (*Pseudomonas delphinii*). This disease causes large, black or brownish-black blotches on leaves, stems or even flower buds, but it seems to be very selective as some cultivars are badly attacked while others nearby are not affected. It is a difficult disease to check, but repeated spraying with Bordeaux mixture at 150 g copper/100 litres HV will protect young plants, if applied as soon as the shoots appear above the soil; the surface of the soil should also be sprayed.

*Powdery mildew (*Erysiphe ranunculi*) appears as the typical white, powdery coating on leaves, stems and even flowers. Spray with copper oxychloride at 37 g copper/100 litres *or* dinocap at 12·5 g a.i./100 litres *or* lime sulphur 1·25 litres/100 litres. Benomyl at 25 g a.i./100 litres *or* thiophanate-methyl at 50 g a.i./100 litres *or* triforine at 25 g a.i./100 litres could also be tried.

ELM

Diseases

***Dutch elm disease (*Ceratocystis ulmi*) (Forestry Commission Leaflets 54 and 94) causes first yellowing then browning of leaves which remain hanging in a withered condition on dead branches, the shoots of which are hooked at the tips. Affected trees usually die rapidly. It is sometimes possible to prevent infection of trees of amenity value by injecting them with benomyl *or* carbendazim hydrochloride, used according to the maker's instructions.

EUONYMUS

Pests

**Bean aphid (*Aphis fabae*) (A.L. 54) is a greenish-black aphid which overwinters on Euonymus, laying eggs on the bark of the shoots in autumn and infesting the young growth in spring causing the foliage to curl. Apply sprays as for rose (p. 367) as soon as aphids appear on new growth.

   **Spindle ermine moth (*Yponomeuta cagnagella*) (A.L. 40). The small, grey caterpillars of this moth live in colonies, binding the shoots and foliage together in a silken 'tent' and feeding inside. They are most commonly seen in May or June and are effectively controlled by spraying with the compounds listed under hawthorn webber (p. 349).

   *Scale insects, e.g. Euonymus scale (*Unaspis euonymi*) and willow scale (*Chionaspis salicis*) are small greyish-white scales which infect the stems and leaves. For control, see willow (p. 378).

Diseases

**Powdery mildew (*Oidium euonymi-japonicae*). The attack of the fungus is often severe and the foliage becomes covered in a whitish powdery deposit. The disease is said to be controlled fairly easily by spraying with dinocap at 12·5 g a.i./100 litres *or* colloidal sulphur 250 g a.i./100 litres HV *or* wettable sulphur at 200 g a.i./100 litres HV. Benomyl at 25 g a.i./100 *or* thiophanate-methyl at 50 g a.i./100 litres *or* triforine at 25 g a.i./100 litres could also be tried.

FORSYTHIA

Pests

*Common green capsid (*Lygocoris pabulinus*)—see Pelargonium (p. 361).

Diseases

**Honey fungus (*Armillariella mellea* syn. *Armillaria mellea*)—see Privet (p. 364).

FUCHSIA

The pests and diseases of fuchsias grown as pot plants are described in Chapter 10 (p. 286).

Pests

*Common green capsid (*Lygocoris pabulinus*)—see Pelargonium (p. 361).

   *Peach-potato aphid (*Myzus persicae*)—control as for aphids on Dahlia (p. 344).

GLADIOLUS

Pests

**Aphids, e.g. potato aphid (*Macrosiphum euphorbiae*), peach-potato aphid (*Myzus persicae*), shallot aphid (*Myzus ascalonicus*), etc., may infest foliage and can be controlled in the same way as aphids on Dahlia (p. 344) Grey bulb aphid (*Dysaphis tulipae*) infests dry corms, control as for Tulip (p. 373).

   *Caterpillars, e.g. angleshades moth (*Phlogophora meticulosa*) and cabbage moth (*Mamestra brassicae*) attack the foliage and are controlled as for Dahlia (p. 345).

   **Gladiolus thrips (*Taeniothrips simplex*) are tiny, elongate insects, yellow when young and dark brown when adult, which suck the sap of the stems and leaves leaving pale yellow or silvery streaks which later turn brown. They also attack the flowers, producing small white flecks on the petals.

   The thrips overwinter on corms in store, feeding under the scales, and will continue to breed if the temperature is over 10°C. Infested corms have rough greyish-brown patches on the surface. As soon as any damage is seen, apply gamma-HCH dust *or* spray with carbaryl at 85 g a.i./100 litres *or* diazinon at 20 g a.i./100 litres *or* gamma-HCH at 12·5 g a.i./100 litres *or* malathion at 112 g a.i./100 litres.

   Corms should be stored at a temperature not exceeding 10°C and dusted with gamma-HCH.

Diseases

**Core rot (*Botrytis gladiolorum*). This fungus attacks the central core of gladiolus corms and then spreads outwards to destroy the corm with a moist (but not wet) rot. In storage corms are soon destroyed and the disease can spread quickly especially if moisture is present. For control the corms should be dried thoroughly and dusted with quintozene dust at 460 g a.i./tonne corms. This treatment is also effective for rots caused by *Penicillium* spp.

Dry rot (*Sclerotinia gladioli*) causes small black spots on the flesh of the corms and these may coalesce to cause larger black areas so that badly infected corms are reduced to worthless 'mummies'.

Hard rot (*Septoria gladioli*) causes larger black spots which are somewhat sunken and these by enlarging will mummify the corms.

Scab (*Pseudomonas marginata*) shows as round, sunken craters towards the corm base and each crater is often covered by a shiny varnish-like covering. All the above corm diseases need to be controlled by a dip treatment. At planting time the corms are roughly cleaned by removal of the outermost rough scales, the badly affected ones are discarded and the others dipped in an aqueous solution of benomyl at 100 g a.i./100 litres *or* carbendazim at 100 g a.i./100 litres for 15 minutes; *or* an organomercury compound at 7·5 g mercury/100 litres for 30 minutes; *or* captan at 1 kg a.i./100 litres for 1 hour (treats 2000 corms).

GODETIA

Diseases

Grey mould (*Botrytis cinerea*)—see Clarkia, p. 343.

Stem rot (*Alternaria godetiae*). This fungus causes brown spots to appear on the stems and these extend and kill the stems. The disease is seed-borne and seed should be steeped for 30 minutes in a solution of an organomercury fungicide containing 2·5 g mercury/litre. Seed crops should also be sprayed with Bordeaux mixture at 150 g copper/100 litres HV (2·6 kg copper/ha LV).

HAWTHORN

Pests

Hawthorn webber (*Scythropia crataegella*). These small reddish or yellowish brown caterpillars feed in colonies on foliage and form a delicate silk webbing covering a considerable part of the tree. A local pest, prevalent in May and June, which may be controlled by spraying with carbaryl at 85 g a.i./100 litres *or* DDT at 1·4 kg a.i./ha *or* trichlorphon at 80 g a.i./100 litres.

Other caterpillars, e.g. *lackey moth* (*Malacosoma neustria*), **small ermine moth** (*Yponomeuta* spp.) and **vapourer moth** (*Orgyia antiqua*)—see Chapter 9, p. 189. Trichlorphon at the above rate gives good control of ermine moths.

Diseases

Fire blight (*Erwinia amylovora*)—see Chapter 9, p. 206.

Powdery mildew (*Podosphaera oxyacanthae*). The leaves and shoots become covered by the typical powdery white coating and in the case of newly planted hedges or young soft shoot growth the disease can be very injurious. To control, spray with dinocap at 12·5 g a.i./100 litres HV *or* colloidal sulphur at 250 g a.i./100 litres HV *or* wettable sulphur at 200 g a.i./100 litres HV. Benomyl at 25 g a.i./100 litres *or* thiophanate-methyl at 50 g a.i./100 litres *or* triforine at 25 g a.i./100 litres could also be tried.

HEATHER

Diseases

***Root rot** (Heather wilt) (*Phytophthora cinnamomi* and other species). Rotting of the roots results in discoloration of foliage and dieback of shoots. The disease can only be confirmed by laboratory techniques. Suppression of the symptoms on plants in nurseries can be achieved by drenching with etridiazole† at 17·5 g a.i. *or* 21·8 g a.i. *or* 35 g a.i./100 litres according to whether the soil is sandy and light *or* loamy *or* with appreciable clay content. Following removal of affected plants disinfection of the soil is recommended using formaldehyde at 2 litres formalin (38–40 per cent)/100 litres water.

HELLEBORE

Diseases

Leaf spot (*Coniothyrium hellebori*). The leaves become spotted with round or elliptical black blotches marked by concentric zones which coalesce to form large patches until eventually the leaves wither. Smaller spots disfigure the flowers. Remove and burn diseased leaves and flowers and spray regularly with Bordeaux mixture equivalent to 150 g copper/100 litres *or* copper oxychloride at 37 g copper/100 litres.

HOLLY

Pests

*Holly leaf miner** (*Phytomyza ilicis*). The small yellowish white grubs tunnel in the leaf tissue. The mines are linear at first and may run along the midrib, later broadening out to form a blister which turns yellowish brown. The mines are usually evident from early June onwards. Their control is difficult but possible by spraying 3 times at 14-day intervals with gamma-HCH 12·5 g a.i./100 litres, adding a wetting agent. Applications of diazinon at 20 g a.i./100 litres may be more effective.

---

† Proposed BSI common name for 5-ethoxy-3-trichloromethyl-1,2,4-thiadiazole.

HOLLYHOCK

Diseases

Rust (*Puccinia malvacearum*). On the undersurfaces of the leaves and on stems raised pustules develop which are at first orange and later turn brown. In severe cases shrivelling of leaves or even the whole plant may occur. Raise new plants at least every other year and spray from the seedling stage onwards at fortnightly intervals with thiram at 160 g a.i./100 litres; oxycarboxin at 47–94 g a.i./100 litres *or* triforine at 25 g a.i./100 litres could also be tried.

HONEYSUCKLE

Pests

Honeysuckle aphid (*Hyadaphis foeniculi*). Heavy infestations of this greyish-black aphid develop on the shoot tips and flowers during the summer. They can cause the flowers to turn brown and the shoots to die back. Control measures are the same as for aphids on rose (p. 367).

HYACINTH

Pests

Stem eelworm (*Ditylenchus dipsaci*). Symptoms of stem eelworm attack on hyacinth are similar to those seen on narcissus (p. 358) except that definite 'spickles' do not form on the leaves. Hot water treatment may be applied as for narcissus.

Diseases

Grey bulb rot (*Rhizoctonia tuliparum* syn. *Sclerotium tuliparum*). The fungus infects the nose portion of the bulb causing a dry greyish decay thus preventing the shoot getting far above ground. The bulb is soon destroyed by the dense fungal growth which quickly produces large resting bodies (sclerotia) which contaminate soil and infect other bulbs the next season. The result is that bare patches are seen in the beds. Rake quintozene dust into the soil at $13 \cdot 5$ g a.i./m² (135 kg a.i./ha).

HYDRANGEA

Pests

Common green capsid (*Lygocoris pabulinus*)—see Pelargonium (p. 361) but do not use HCH.
  Red spider mite (*Tetranychus urticae*)—see Chapter 10, p. 280.

Diseases

**Honey fungus (*Armillariella mellea* syn. *Armillaria mellea*)—see Privet (p. 364).

**Powdery mildew (*Microsphaera polonica*) shows on the leaves as brown spots on which the mildew shows as a whitish coating. To control spray with dinocap at 12·5 g a.i./100 litres HV *or* colloidal sulphur at 250 ml a.i./100 litres HV *or* wettable sulphur at 200 g a.i./100 litres HV. Benomyl at 25 g a.i./100 litres *or* thiophanate-methyl at 50 g a.i./100 litres *or* triforine at 25 a.i./100 litres could also be tried.

IRIS

Pests

**Aphids, e.g. potato aphid (*Macrosiphum euphorbiae*) and tulip bulb aphid (*Dysaphis tulipae*)—for control see aphids on Dahlia (p. 344) and on Tulip (p. 373).

*Iris sawfly (*Rhadinoceraea micans*). Mainly a pest of *Iris pseudacorus*, the caterpillars have smooth, lead-grey bodies with black heads. They eat strips from the leaf margins and are active from late May to July. For their control, apply gamma-HCH at 12·5 g a.i./100 litres. For small infestations or where contamination of ponds is likely, handpicking should be used.

Diseases of bulbous iris

**Ink disease (*Drechslera iridis*). Black spots develop on the leaves, sometimes affecting the whole leaf surface, and also occasionally on the flowers. Black crusty patches appear on the outer scales of the bulbs, followed by decay of the inner tissues leaving an empty tunic. To control spray with mancozeb at 160 g a.i./100 litres, and burn over with straw in late summer.

**Leaf spot (*Mycosphaerella macrospora* syn. *Didymellina macrospora*) usually appears only in second year when it causes brown oval-shaped spots on the leaves which soon wither. Control is obtained by spraying with maneb at 200 g a.i./100 litres *or* zineb at 210 g a.i./100 litres. A wetter should be added.

Diseases of rhizomatous iris

**Leaf spot (*Mycosphaerella macrospora* syn. *Didymellina macrospora*). Attacks by this fungus vary according to locality, but in some places (e.g. Western counties) the foliage is often marked in early spring by brown oval-shaped spots which coalesce so that the leaves are badly injured or killed. The control is as for the leaf spot of bulbous iris (see above).

*Rhizome rot (*Erwinia carotovora*) causes the rhizome to rot with a yellowish slimy rot so that the fan of leaves above shows browning of tips followed by collapse and death. To control the disease cut out affected parts and dust the wounds, rhizomes and soil with a copper-containing dust.

JUNIPER

Pests

*Conifer spinning mite (*Oligonychus ununguis*)—see Spruce (p. 370).

**Juniper scale (*Carulaspis* spp.). These small dirty-white scales disfigure stems and foliage. Their control is as for scale on Willow (p. 378).

*Juniper webber (*Dichomeris marginella*). These small brown caterpillars feed in colonies on the foliage, tying it up with webbing. The injured foliage dies and turns brown. It is a local pest, active in May and June, which may be controlled by the methods given for hawthorn webber (p. 349).

LABURNUM

Diseases

**Honey fungus (*Armillariella mellea* syn. *Armillaria mellea*)—see Privet (p. 364).

LARCH

Diseases

**Leaf cast (*Meria laricis*) can be a serious disease of young larch trees in nurseries. The symptoms appear in early May when the leaflets turn brown from the tip downwards. The needles fall prematurely soon after becoming completely brown. Control is by application of sulphur sprays at end of March and 2–3 week intervals until end of July or earlier in a dry season, with colloidal sulphur at 150 g sulphur/100 litres HV *or* lime sulphur 1 litre/100 litres HV.

LILAC

Pests

*Lackey moth (*Malacosoma neustria*) and **winter moths** (*Operophtera brumata, Alsophila aescularia* etc.)—see Chapter 9, p. 189.

**Lilac leaf miner (*Caloptilia syringella*). The small, whitish, legless caterpillars tunnel in the leaf tissue forming large blisters which turn brown. Older caterpillars feed externally but conceal themselves in leaves which they roll from the tip and tie with silk. They are active from May to June and from August to September and may be controlled by spraying 3 times, at 14-day intervals, with diazinon 20 g a.i./100 litres *or* gamma-HCH 12·5 g a.i./100 litres *or* trichlorphon at 80 g a.i./100 litres Treatment should begin as soon as the first signs of damage are seen.

*Privet thrips (*Dendrothrips ornatus*)—see Privet (p. 363).

*Willow scale (*Chionaspis salicis*)—see Willow (p. 378).

Diseases

*Blight (*Pseudomonas syringae*) causes small angular brown spots on leaves but the serious damage is that young shoots (both flower and foliage shoots) blacken and wither away. Cut back the affected portions to a healthy bud and spray with Bordeaux mixture equivalent to 150 g copper/100 litres *or* copper oxychloride at 37 g copper/100 litres HV.

**Honey fungus (*Armillariella mellea* syn. *Armillaria mellea*)—see Privet (p. 364).

LILY

Pests

**Aphids. Tulip bulb aphid (*Dysaphis tulipae*) can build up rapidly in store and, together with other species (e.g. mottled arum aphid, *Aulacorthum circumflexum*, shallot aphid, *Myzus ascalonicus*, etc.) will infest growing plants, causing distortion and also transmitting virus diseases. Control on bulbs in store is as for Tulip (p. 373) and, for outdoor infestations, controls may be applied as for Dahlia (p. 344).

**Lily beetle (*Lilioceris lilii*). A bright red beetle with black legs, head and underparts, 8 mm long and locally common in Surrey and Berkshire. The larvae are orange-red but often hidden under slimy black faeces. Both adults and larvae feed on the leaves, flowers and seed pods. To control, spray with gamma-HCH at 12·5 litres a.i./100 litres or apply 5 per cent gamma-HCH dust.

Diseases

*Leaf blight (*Botrytis elliptica*). The leaves develop oval-shaped water-soaked spots and these spread rapidly in humid weather conditions. Flower stems can 'topple' if the fungus spreads from the leaves and enters the stalk. Spray, as soon as the flower buds can be detected, with a copper-containing fungicide, e.g. Bordeaux mixture equivalent to 150 g copper/100 litres *or* copper oxychloride at 37 g copper/100 litres.

LOBELIA

Diseases

**Leaf spot and stem rot (*Alternaria tenuis*). Pale spots develop on leaves of seedlings often giving a scorched effect. The fungus can also cause the seedlings to damp off. Tentative recommendations are to treat the seed with captan seed dressing at 4·5 g a.i./kg seed *or* to soak seed in a suspension of thiram at 2 g a.i./litre for 24 hours at 30°C and to spray early with thiram at 160 g a.i./100 litres *or* zineb at 140 g a.i./100 litres.

MAHONIA

Diseases

*Powdery mildew (*Microsphaera berberidis*) shows as a white powdery coating on the leaves especially on young growth. Spraying with benomyl at 25 g a.i./100 litres *or* dinocap at 12·5 g a.i./100 litres *or* thiophanate-methyl at 50 g a.i./100 litres *or* triforine at 25 g a.i./100 litres should control this disease.

*Rust (*Cumminsiella mirabilissima* and *Puccinia graminis*). Small red spots appear on the upper surfaces of leaves and on the lower surfaces brownish, somewhat powdery, masses of spores are produced. Spraying with oxycarboxin at 47–94 g a.i./100 litres *or* thiram at 160 g a.i./100 litres *or* triforine at 25 g a.i./100 litres may give some control of this disease.

MALUS

Pests

These trees are afflicted by most of the pests recorded for apple, Chapter 9, p. 172.

Diseases

**Honey fungus (*Armillariella mellea* syn. *Armillaria mellea*)—see Privet (p. 364).

***Scab (*Venturia inaequalis*). On many species the apple scab fungus regularly causes brown to greenish blotches on the leaves which spoil the appearance. To control the disease, spray with captan at 100 g a.i./100 litres *or* lime sulphur at 1 litre/100 litres at green bud and after petal fall. Benomyl at 25 g a.i./100 litres *or* dodine at 65 g a.i./100 litres *or* thiophanate-methyl at 50 g a.i./100 litres may also be used.

MAPLE

See Acer, p. 336.

MECONOPSIS

Diseases

**Downy mildew (*Peronospora arborescens*) shows as furry, greyish white patches on the undersurfaces of the leaves or even on flower stalks and seed pods. Seedlings are often attacked and sterilized soil should be used for seed pans and boxes. Control is also obtained by a copper-containing spray, e.g. Bordeaux mixture at 150 g copper/100 litres *or* copper oxychloride at 37 g copper/100 litres. A good spreader is essential to wet these leaves.

MICHAELMAS DAISY

Pests

***Michaelmas daisy mite (*Tarsonemus pallidus*). These microscopic mites live and feed inside the buds and in the folds of young leaves. Heavy infestations result in stunted growth, brown scarring along the stems and the production of small green rosettes instead of flowers. Cuttings being raised under glass can be treated with endosulfan at 30 g a.i./100 litres *or* endrin at 30 g a.i./100 litres. A wetter should be added. During the summer the plants should be sprayed with dicofol at 31 g a.i./100 litres in mid July, mid August and again at the end of August.

Diseases

**Powdery mildew (*Sphaerotheca fuliginea*) gives a white, powdery deposit covering the leaves. Spray with benomyl at 25 g a.i./100 litres *or* dinocap at 12·5 g a.i./100 litres *or* thiophanate-methyl at 50 g a.i./100 litres *or* triforine at 25 g a.i./100 litres.

NARCISSUS

Pests

*Bulb mite (*Rhizoglyphus echinopus*). These are shiny, yellowish white, globular mites which infest bulbs injured by handling, by disease or by other pests. They extend the original damage and prevent recovery from slight cuts and bruises. Infested bulbs should be examined for evidence of disease or damage by other pests and, if necessary, the appropriate control measures taken. The mites are destroyed by hot-water treatment as recommended for bulb scale mite, narcissus flies or eelworm. Alternatively fumigate for 5 days in airtight container containing *p*-dichlorobenzene at rate of 4 g/litre. Cover the crystals with hessian and fill the container with bulbs.

Bulb scale mite (*Steneotarsonemus laticeps*) (A.L. 456) are microscopic mites which feed between bulb scales, leaving on their surface elongate, brown streaks which can be seen on cutting open the bulb and parting the scales. In store, infested bulbs become softer and light in weight.

On the growing bulb the effects on the vigour are best seen in forcing houses. Leaves are distorted and abnormally green with yellow streaks. Flowers may be malformed or killed in the bud. In the open, plants merely show lack of vigour, poor blooms and premature death of the foliage.

Hot-water treatment of fully dormant bulbs will reduce infestations. One hour at 44°C will give adequate control on bulbs for forcing during the season after treatment and, for stocks intended for growing on outdoors, 4 hours at 44°C or 3 hours at 45°C will eradicate the mites. Dipping bulbs for $2\frac{1}{2}$ hours in an emulsion of thionazin (230g a.i./100 litres cold water) will also give good control.

Overnight frosting followed by a drench of endosulfan (100 g a.i./100 litres) *or* endrin. (100 g a.i./100 litres, plus wetter) will check the development of mites on bulbs brought indoors for forcing.

***Large narcissus fly** (*Merodon equestris*) (A.L. 183). The larva, a yellowish white, legless grub, up to 20 mm long, tunnels in the interior of bulbs, which become filled with wet brown material. The initial entry is in May or June after the eggs hatch but damage is not easily seen until the bulbs are replanted in September or October. Infested bulbs are softer than normal, especially around the neck, and, if replanted, will produce only small leaves ('grass') or may fail to make any growth at all. Where bulbs are lifted annually, narcissus fly larvae within infested bulbs can be killed by either a 3-hour dip in gamma-HCH (50 g a.i./100 litres water, plus spreader) immediately after lifting *or* by hot water treatment (1 hour at 44°C) during the dormant period. The HCH dip may be phytotoxic and the latter method is therefore preferred, if practicable.

After hot water treatment a cold dip of 15 minutes in aldrin (200 g a.i./100 litres) will prevent reinfestation during the following growing season, and may give some protection in the second or third seasons after planting.

The application of aldrin *or* gamma-HCH to the soil after planting is biologically unsound and may cause harmful side effects.

Small narcissus flies (*Eumerus tuberculatus* and *E. strigatus*) (A.L. 183). Whereas only one larva of the previous species is normally present in a bulb, there may be several of the larvae of the small narcissus fly. They are greyish white, legless and grow to 9 mm in length.

Small narcissus flies may occasionally cause primary damage to narcissus but are generally secondary pests, following and extending damage done by eelworm, large narcissus fly or slugs. Control measures taken against these pests should therefore reduce infestations by small narcissus flies.

Root lesion eelworms (*Pratylenchus* spp.) feed on the roots, making small, brown or black, slit-like lesions which allow entry of fungi, etc. Roots decay and rot away and bulbs make poor growth, producing dwarfed plants. At present severe damage occurs only in the Isles of Scilly and in Cornwall. Chemical treatment of infested soil is possible though expensive and may be worthwhile where land is valuable.

***Stem eelworm** (*Ditylenchus dipsaci*) (A.L. 460). This eelworm infests the bulb and upper growth. The foliage of attacked plants may be paler than usual, also stunted and twisted with small yellow swellings (spickels) on the surface. Flower production is late and eventually ceases altogether as the bulb rots in the ground. Infested bulbs are soft to the touch, with dull, unhealthy-looking outer scales, and when cut across, rings of brown tissue can be seen where the scales have decayed. Rogue out infested bulbs in the field. After lifting inspect bulbs and destroy those which are badly affected.

The rest should be given hot-water treatment when dormant (3 hours at 45°C). A wetter should be added to the water for better penetration, or the bulbs may be soaked in cold water containing a wetter before hot-water treatment is given. Damage by hot-water treatment may be avoided by storing bulbs for a week at a temperature of 29–30°C, both before and after treatment.

As an alternative to hot water treatment bulbs may be dipped for $2\frac{1}{2}$ hours in cold water containing thionazin at 230 g a.i./100 litres. Heavily infested bulbs must be graded out before dipping.

On infested land the interval between bulb crops should be 3 or more years after ground keepers have been removed and no susceptible plant hosts such as bluebells, onions, scilla, snowdrops and strawberries should be grown during the interval. It may be considered worthwhile to sterilize small areas of ground (cf. Chapter 10, p. 253) with dichloropropane-dichloropropene mixture at 340–450 kg/ha *or* metham-sodium (33 per cent) at 1230 litres/ha.

Diseases

Basal rot (*Fusarium oxysporum* f. *narcissi*) affects bulbs in store. It starts at the basal plate which becomes soft and spreads upwards causing a chocolate brown discoloration of the inner scales on which can be seen pink fungal growth. The disease can be controlled by dipping the bulbs for up to 30 minutes in an aqueous solution containing benomyl at 100 g a.i./100 litres within 48 hours of lifting; *or* at any time between lifting and planting in carbendazim at 100 g a.i./100 litres for 15 minutes; *or* in formaldehyde at 500 ml (38–40 per cent formalin)/100 litres *or* in an organomercury compound at 7·5 g mercury/100 litres as a part of the routine hot water treatment prior to planting.

Fire (*Sclerotinia polyblastis*) causes a rapid leaf decay during and following flowering, especially in wet weather, starting first on the flowers as small light brown spots, which appear first about picking time and soon destroy the flowers. The leaves then show elongated reddish brown blotches and are killed very rapidly. Remove plants showing primary lesions and, when 25 mm high, spray with thiram at 3·4 kg a.i./ha. Repeat at 10-day intervals but do not spray once the flower buds have emerged from the leaf sheaf. Diseased foliage should be raked up and burnt.

Leaf scorch or tip burn (*Stagonospora curtisii*). The fungus attacks the emerging leaves so that they have a scorched or burnt appearance at the tips. Spray the foliage when it is about 80 mm high with benomyl at 560 g a.i./ha *or* zineb w.p. at 140 g a.i./100 litres *or* tank mix zineb (prepared as described in Chapter 10, p. 309). Two or 3 applications may be given before flowering and 1 afterwards, but no spraying should be carried out after bud break and during the flowering period. Remove and burn diseased tips. This disease is also controlled by dipping as described for basal rot (p. 358).

Smoulder (*Sclerotinia narcissicola*) causes a decay of stored bulbs but in cold wet seasons it can rot foliage and flowers so that plants are malformed and affected tissues are covered with grey masses of spores. In warmer drier conditions infection is usually restricted to 1 or 2 leaves and the plants grow away from the disease and flower normally. Spray as soon as the disease is seen as for leaf scorch above. Dipping as for basal rot (p. 358) will also control this disease.

White mould (Ramularia vallisumbrosae) appears as dark greenish or yellowish brown patches on the leaves, especially near the tips and these, in damp weather, soon become covered with a whitish mass of fungus spores. The foliage in these conditions is prematurely killed with a consequent check to bulb growth. Spray when the foliage is about 80 mm high as for leaf scorch above.

OAK

Pests

Caterpillars of the **green oak tortrix** (*Tortrix viridana*), the **winter moth** (*Operophtera brumata*) and many other species of moth feed on leaves and may cause severe defoliation. Control measureas are the same as those given for tortrix caterpillars and winter moths in Chapter 9, p. 188.

Diseases

Powdery mildew (*Microsphaera alphitoides*). Leaves and shoot tips become covered with a floury white coating and young shoots especially can be severely injured and crippled in growth. Control is effected by spraying with dinocap at 12·5 g a.i./100 litres *or* colloidal sulphur at 250 g a.i./100 litres *or* wettable sulphur at 200 g a.i./100 litres. Benomyl at 25 g a.i./100 litres *or* thiophanate-methyl at 50 g a.i./100 litres *or* triforine at 25 g a.i./100 litres could also be tried.

PAEONY

Pests

**Leaf and bud eelworm (*Aphelenchoides ritzemabosi*)—see Chapter 10, p. 271. No information is available on the effects of thionazin drenching.

Diseases

**Wilt (*Botrytis paeoniae*). The bases of the shoots develop a brown area and soon wither. Leaves of other shoots can also develop brown, angular patches, usually at the tips and even the flower buds may turn brown and die. Dust the crowns in early spring with a Bordeaux powder dust and remove wilting stems by cutting them off (below ground level) for burning. Spray both herbaceous and tree paeonies with dichlofluanid at 100 g a.i./100 litres *or* at 1·1 kg a.i./ha at 10–14 day intervals from emergence of foliage until flowering. Tree paeonies can also be sprayed with captan at 100 g a.i./100 litres *or* thiram at 160 g a.i./100 litres *or* zineb at 140 g a.i./100 litres.

PANSY

Diseases

*Leaf spot (*Ramularia* spp.). Small circular spots develop on the leaves and these are either greenish or yellowish, or whitish or straw-coloured with a brown margin. Spray with a copper-containing fungicide, e.g. Bordeaux mixture equivalent to 150 g copper/100 litres *or* copper oxychloride at 37 g copper/100 litres *or* maneb 200 g a.i./100 litres *or* zineb 140 g a.i./100 litres.

*Stem rot (*Myrothecium roridum*). This fungus is considered to be responsible at least in part for a brown rotting at the stem bases of violas and other plants. Apart from proper attention to soil drainage, etc., it is advised that 4 per cent calomel dust be sprinkled in the planting holes.

PEACH

See Prunus, p. 364.

PELARGONIUM

Pests

Common green capsid (*Lygocoris pabulinus*). These active insects grow up to 6 mm long and are yellowish green when young, becoming greener later. The injuries caused are similar to those of the tarnished plant bug but damage to flowers is less common. Punctured leaves of geranium pucker and tear as they grow, taking on a ragged lace-like appearance. Damage may occur from May onwards. To control this pest, keep the land clear of weeds and debris. Apply diazinon at 20 g a.i./100 litres *or* gamma-HCH at 12·5 g a.i./100 litres, giving 2–3 applications at 14-day intervals, starting as early as possible. Also spray the ground around the plants.

Diseases

Rust (*Puccinia pelargonii-zonalis*) appears on the undersurfaces of leaves as brown spore masses often arranged roughly in rings. Spray at fortnightly intervals with mancozeb/zineb at 175 g product/100 litres *or* oxycarboxin at 47–63 g a.i./100 litres *or* triforine at 25 g a.i./100 litres (but oxycarboxin and triforine should be used with care as some cultivars may be damaged) *or* with zineb at 210 g a.i./100 litres.

PETUNIA

Diseases

Foot rot (*Phytophthora cryptogea*) causes a rot of the roots and stem bases. Treat the seed with captan seed dressing at 3·4 g a.i./kg *or* with thiram at 2·2 g a.i./kg seed. The soil can also be watered with captan at 125 g a.i./100 litres *or* with Cheshunt compound at 300 g copper/100 litres *or* with zineb at 130 g a.i./100 litres.

PHLOX
(Perennial)

Pests

Stem eelworm (*Ditylenchus dipsaci*). The eelworms feed in stem and leaf tissues. On infested plants apical leaves are often narrow and strap-like while lower leaves may be rolled, swollen and puffy or crinkled. Stems may be stunted, swollen or spindly, with a tendency to crack longitudinally. Infested stools may be given hot-water treatment (1 hour at 44 °C). Clean stock may be raised from infested plants by propagating from root cuttings or seed. Eelworms on infested land must be starved out by denying them alternative host-plants, e.g. aubrieta, gypsophila, helenium, annual phlox and certain weeds, and infested sites should be kept free of such host-plants for 2 or 3 years.

Diseases

*Powdery mildew (*Sphaerotheca fuliginea*). A white, powdery coating on leaves, controlled by sprays of benomyl at 25 g a.i./100 litres *or* dinocap at 12·5 g a.i./100 litres *or* thiophanate-methyl at 50 g a.i./100 litres *or* triforine at 25 g a.i./100 litres.

PHLOX DRUMMONDII
(Annual)

Pests

*Aphids, mainly peach-potato aphid (*Myzus persicae*)—control as for aphids on Dahlia (p. 344).

Diseases

**Leaf spot (*Septoria drummondii*). Brown spots with a pale centre develop on the leaves and eventually coalesce to give a scorched effect. Treat the seed with a captan 75 per cent seed dressing at 3·4 g a.i./kg. Spray early with zineb at 140 g a.i./100 litres.

PLANE

Diseases

*Leaf scorch (*Gloeosporium nervisequum*). Brown discoloration on the leaves roughly following the main veins causes the leaves and sometimes young shoots to wither. The fallen leaves should be collected and burnt in autumn. Spray with Bordeaux mixture equivalent to 150 g copper/100 litres if possible in April and again before bud burst.

POLYANTHUS

Pests

*Angleshades moth (*Phlogophora meticulosa*)—see Dahlia (p. 345).
    **Bryobia mite (*Bryobia* spp.). These yellowish green to reddish brown mites suck the sap, causing a yellow mottling on the leaves which may spread until the whole leaf is yellowed. To control them, apply demeton-S-methyl 22 g a.i./100 litres *or* dimethoate 30 g a.i./100 litres *or* malathion

112 g a.i./100 litres *or* tetradifon 12·5 g a.i./100 litres, repeating as necessary.

**Red spider mite (*Tetranychus urticae*)—see entry under Dahlia (p. 345).

**Stem eelworm (*Ditylenchus dipsaci*). Attacked flower stems become stunted, twisted and disfigured with pale or brown blisters and longitudinal furrows. The leaves become blotched with yellow and later turn brown and die, and the crown of the plant may also decay. No specific control measure is available but valuable stock may be worth treating with hot water as for Phlox (p. 361).

Diseases

**Leaf spots (*Ramularia* spp. and *Phyllosticta* spp.). Various sized discoloured spots appear on the leaves making them unsightly and in some cases damaging or killing the leaves. These fungi are controlled by spraying with copper oxychloride at 37 g copper/100 litres outdoors, but at 18·5 g a.i./100 litres under glass. Chlorothalonil 110 g a.i./100 litres could also be tried.

POPLAR

Diseases

*Leaf spots (*Marssonina* spp.). Small irregular blackish-brown spots occur on the leaves which fall prematurely. Control may be difficult but dead shoots should be cut out where possible, and smaller trees can be sprayed not less than 3 times in the spring and at least once in summer with Bordeaux mixture, equivalent to 150 g copper/100 litres *or* copper oxychloride at 37 g copper/100 litres.

*Yellow leaf blister (*Taphrina populina*). Affected leaves are distorted and bear large blisters which are bright yellow on the lower surface but remain green on the upper side. Where the disease is troublesome and it is feasible, spray in January or February with Bordeaux mixture at 150 g copper/100 litres *or* lime sulphur 3 litres/100 litres.

PRIVET

Pests

**Lilac leaf miner (*Caloptilia syringella*)—see Lilac (p. 353).

**Privet thrips (*Dendrothrips ornatus*). These tiny elongate insects are yellow when young but the adult has brown bars across the body. They feed on the sap of the foliage, causing silvery speckling and distortion. The affected foliage dies and falls during the winter. To control, spray with diazinon at 20 g a.i./100 litres *or* gamma-HCH 12·5 g a.i./100 litres *or* malathion 112 g a.i./100 litres, as necessary.

Diseases

***Honey fungus** (*Armillariella mellea* syn. *Armillaria mellea*) (A.L. 500). This fungus attacks the roots and kills the bushes especially in hedges. Spread is mainly by means of runner-like strands called 'rhizomorphs'. The fungus only shows above ground as toadstools growing at soil level on the bark of dead trees and bushes. The removal and burning of dead bushes with as much root as possible is recommended, together with soil disinfection using formaldehyde (2 litres (38–40 per cent) formalin/100 litres water) to soak the freshly forked infected site.

PRUNUS
(Ornamental)

Pests

The pests found on Cherry, Damson or Plum, see Chapter 9, p. 210, also attack ornamental Prunus.

Diseases

****Bacterial canker** (*Pseudomonas mors-prunorum*). Lesions appear on leaves and branches in early spring with often some gumming on the branches. As these cankers develop, the shoots die back and some branches tend to be flattened. Leaves may show small brown spots. Spraying with Bordeaux mixture at 150 g copper/100 litres at end of August and end of September effects control.

****Honey fungus** (*Armillariella mellea* syn. *Armillaria mellea*)—see under Privet (p. 364).

*****Peach leaf curl** (*Taphrina deformans*) (A.L. 81) causes the leaves to develop large blisters, at first red, which swell up and turn white often involving whole leaves and large numbers of them. To control, spray the bare branches *before* the flower buds burst in January or early February, again a fortnight later and once again just before leaf fall with Bordeaux mixture, equivalent to 150 g copper/100 litres *or* lime sulphur 3 litres/100 litres.

****Silver leaf** (*Stereum purpureum*) (A.L. 246), so called from the silvering of foliage though the fungus is present inside the branch or main stem. Apart from domestic plums it has been reported on ornamental species such as *Prunus spinosa, P. triloba, P. laurocerasus, P. avium.* Fungus fruit bodies appear only on dead branches, stems or roots. As it is a wound parasite, all wounds made in Prunus trees should be painted with a wound sealing preparation containing mercury or of a bituminous type. Infected dead trees should be grubbed before July 15 and are best kept dry until burnt.

PYRACANTHA

Pests

Many of the pests of apple, including green apple aphid (*Aphis pomi*), woolly aphid (*Eriosoma lanigerum*), and various caterpillars cause trouble and may be controlled by the methods given in Chapter 9, pp. 172–191.

Diseases

Fire blight (*Erwinia amylovora*)—see Chapter 9, p. 206.

   Scab (*Fusicladium pyracanthae*) covers the berries with an olive-brown or black coating so spoiling their appearance. The leaves can also be attacked and defoliation caused. As copper fungicides are likely to cause damage, spray with captan 100 g a.i./100 litres *or* lime sulphur at 2·5 litres/100 litres 3 times in March/April and twice at 1·25 litres/100 litres in June/July. Benomyl at 25 g a.i./100 litres *or* thiophanate-methyl at 50 g a.i./100 litres may also be tried.

PYRETHRUM

Pests

Leaf and bud eelworm (*Aphelenchoides ritzemabosi*)—see Chrysanthemum, Chapter 10, p. 271.

Diseases

Grey mould (*Botrytis cinerea*) is a greyish furry mould attacking flower stalks. It can be controlled by spraying with benomyl at 25 g a.i./100 litres *or* captan at 100 g a.i./100 litres *or* thiophanate-methyl at 50 g a.i./100 litres *or* thiram at 160 g a.i./100 litres HV.

RHODODENDRON

Pests

Caterpillars, e.g. March moth (*Alsophila aescularia*), winter moth (*Operophtera brumata*), tortricid caterpillars, etc., feed on young leaves and buds and can occasionally cause considerable damage. Control measures are the same as those given for caterpillars on Roses (p. 367).

*Rhododendron bug (*Stephanitis rhododendri*). These shiny, yellowish to dark brown insects, up to 4 mm long, feed in groups on the lower surface of leaves which show characteristic rusty or chocolate-brown spotting. The upper surface becomes mottled with yellow giving the shrub a sickly appearance. This damge is first seen in May or June and may be prevented by sprays of DDT at 1·4 kg a.i./ha *or* gamma-HCH at 12·5 g a.i./100 litres *or* malathion 112 g a.i./100 litres. Two or 3 applications at intervals of 3 weeks starting in mid-June or when infestations are first seen should be made with the spray directed to the underside of the leaves.

***Rhododendron leafhopper (*Graphocephala coccinea*). This strikingly marked green and red leafhopper is one of the most important pests of rhododendron as it is mainly responsible for the infection of flower buds by bud blast disease (see p. 366). Chemical control of the leafhopper reduces bud blast infections and may be achieved by spraying with DDT at 1·4 kg a.i./ha *or* malathion at 112 g a.i./100 litres. Sprays should be applied on 2 or 3 occasions during August and September, the object being to reduce the local leafhopper population and limit egg-laying as it is the wounds made in the bud scales by the female's ovipositor that provide the infection sites for the fungus.

*Rhododendron whitefly (*Dialeurodes chittendeni*). The young stages appear as small yellowish green 'scales', difficult to distinguish, on the underside of the leaves. The adults resemble miniature, white-winged moths and cluster on the foliage. Both stages suck the sap and may severely affect the vigour of the shrub. In some cases the foliage may become disfigured with yellow mottling. Honeydew is produced in copious quantities leading to further disfigurement by sooty moulds. Spray the underside of leaves with gamma-HCH at 12·5 g a.i./100 litres *or* malathion at 112 g a.i./100 litres giving 2—3 applications at 14-day intervals.

*Weevils, e.g. clay-coloured weevil (*Otiorrhynchus singularis*) and vine weevil (*O. sulcatus*) (A.L. 57). These small wingless beetles are about 12·5 mm long and black or brown. They feed on leaves, petioles and stems, usually at night, and hide in leaf litter, etc., during daylight. Dust foliage with gamma-HCH (5 per cent) *or* apply to soil under affected plants at about 1·5 g a.i./m^2.

Diseases

**Bud blast (*Pycnostysanus azaleae*). The killed buds turn brown, black or silvery in spring and black bristle-like spore heads protrude from them. The fungus is a wound parasite and it is thought to enter wounds made by the Rhododendron Leafhopper (*Graphocephala coccinea*) as it deposits its eggs in the flower buds. Apart from picking off as many infected buds as possible, spray to control the hopper (see Insect Pests, p. 000).

***Honey fungus (*Armillariella mellea* syn. *Armillaria mellea*)—see Privet (p. 364).

***Root rot** (*Phytophthora cinnamomi* and other species)—see Heather (p. 350).

ROSE

The pests and diseases of roses grown as glasshouse crops are dealt with in Chapter 10, p. 299.

Pests

***Aphids,** e.g. rose aphid (*Macrosiphum rosae*), feed on the foliage, shoots and buds, causing stunting and distortion. For their control, spray as necessary, with demeton-S-methyl at 22 g a.i./100 litres *or* diazinon at 20 g a.i./100 litres *or* dimethoate at 30 g a.i./100 litres *or* formothion at 43 g a.i./100 litres *or* malathion at 112 g a.i./100 litres *or* oxydemeton-methyl at 22 g a.i./100 litres *or* pirimicarb at 25 g a.i./100 litres. Aldicarb granules can be applied at a rate of 11·2 kg a.i./ha to the soil and thoroughly watered in.

Caterpillars, e.g. tortrix moths (*Archips podana,* etc.), winter moths (*Operophtera brumata,* etc.), lackey moth (*Malacosoma neustria*), vapourer moth (*Orgyia antiqua*), etc. Caterpillars of very many different species feed on leaves, buds and blossoms. Many are fully exposed while others may gain protection by rolling or tying leaves together, by spinning silk coverings or by tunnelling into buds. Hand-picking can be effective on a small scale. On large-scale infestations spray with carbaryl at 0·85 kg a.i./ha *or* DDT at 1·4 kg a.i./ha *or* trichlorphon at 0·8 kg a.i./ha.

Red bud borer (*Thomasiniana oculiperda*). These small reddish maggots feed in groups on the sap under new bud grafts and prevent the graft 'taking'. To prevent their attack, smear the graft with petroleum jelly after tying or apply DDT at 2·1 kg a.i./ha before or immediately after grafting.

Red spider mite (*Tetranychus urticae*)—see Chapter 10, p. 299.

*Rose leafhopper** (*Typhlocyba rosae*). These pale yellow insects, up to 3 mm long, feed on the underside of the foliage, producing a distinctive yellow spotting on the upper surface. The leaves are also disfigured by honeydew and sooty moulds. The adults are active, jumping insects. Control measures should be taken when damage is first noticed, usually from May onwards, using the chemicals given for rhododendron leafhopper (p. 366).

Rose thrips (*Thrips fuscipennis*). These tiny, elongated, yellowish to dark brown insects feed on the sap of the petals in opening buds causing them to be distorted and streaked with brown when the blooms open. They also feed on foliage which becomes disfigured with silvery mottling, and may be controlled by the methods cited for privet thrips (see p. 363).

Sawfly larvae are common rose pests, e.g. ***Leaf-rolling rose sawfly** (*Blennocampa pusilla*), the pale green larvae of which shelter inside longitudinally rolled leaves, eating away the margin, and are found from June to August. Roses are disfigured and their vigour may be affected by severe attacks. The caterpillars can be killed by thoroughly spraying with trichlorphon at 0·8 kg a.i./ha, although by then the leaf rolling damage has already been caused. Alternatively, to prevent egg-laying, DDT 2·1 kg a.i./ha may be applied at 14-day intervals from early May to mid-June. The **rose slug sawfly (*Endelomyia aethiops*) feeds, as yellowish green larvae, on 1 surface of the leaf, leaving the other intact. The injured area turns brown giving the foliage a mottled and scorched appearance; damage occurs mainly in June and again in August and September. Spray, as necessary, with gamma-HCH 0·28 kg a.i./ha *or* malathion 112 g a.i./100 litres *or* trichlorphon at 0·8 kg a.i./ha. Other sawfly pests include the **antler sawflies** (*Cladius pectinicornis* and *C. difformis*), the **banded rose sawfly** (*Allantus cinctus*) and the **large rose sawfly** (*Arge ochropus*). These yellowish green or greyish green larvae eat irregular holes out of the leaves and may cause severe defoliation. The large rose sawfly also lays its eggs in young shoots and the stalks of the blooms causing them to become blackened and twisted. They are controlled by the methods given above for rose slug sawfly.

*Common green capsid (*Lygocoris pabulinus*)—see Pelargonium (p. 361).

Scale insects, e.g. *brown scale (*Parthenolecanium corni*) and *mussel scale (*Lepidosaphes ulmi*), see Chapter 9, p. 187; *scurfy scale (*Aulacaspis rosae*) appears as small, dirty-white scales on stems causing disfigurement and poor growth. They may be controlled by spraying 2–3 times at 14-day intervals, with diazinon 40 g a.i./100 litres *or* malathion 187 g a.i./100 litres.

Diseases

***Black spot (*Diplocarpon rosae*). Circular black spots on the leaves and in some species the fungus can be found in scabby places on the shoots. Start spraying immediately after pruning and repeat at fortnightly intervals throughout the season using captan at 100 g a.i./100 litres *or* chlorothalonil at 110 g a.i./100 litres *or* dichlofluanid at 100 g a.i./100 litres or dodine at 32·5 g a.i./100 litres HV (715 g a.i./ha LV) *or* mancozeb/zineb at 175 g product/100 litres *or* maneb at 200 g a.i./100 litres *or* thiophanate-methyl at 100 g a.i./100 litres *or* triforine at 25 g a.i./100 litres.

***Honey fungus (*Armillariella mellea* syn. *Armillaria mellea*)—see Privet (p. 364).

**Powdery mildew (*Sphaerotheca pannosa*). The fungus is the white powdery coating on shoots and young leaves which are crippled. Spray with benomyl at 25 g a.i./100 litres *or* with dichlofluanid at 375 g a.i./100 litres decreasing to 250 g a.i./100 litres *or* dinocap at 12·5 g a.i./100 litres (though some petal-spotting may result on white-flowered cultivars) *or* dodemorph at 100 g a.i./100 litres *or* thiophanate-methyl at 50 g a.i./100 litres *or* triforine at 25 g a.i./100 litres.

Rust (*Phragmidium mucronatum*) causes orange coloured pustules on the leaves in summer and, in autumn, darker brown ones often on the same spots. Spray with mancozeb/zineb at 175 g product/100 litres or maneb at 200 g a.i./100 litres *or* oxycarboxin at 70–93 a.i./100 litres *or* thiram at 320 g a.i./100 litres *or* triforine at 25 g a.i./100 litres.

SEEDLINGS
(Various)

Diseases

***Damping-off** (*Pythium* spp. and *Phytophthora* spp.). These fungi cause a shrinking and rotting of stems at soil level and the plants soon collapse. The seed should be treated with captan seed dressing at 3·4 g a.i./kg. Use enough just to give the seed a fine coating of the dressing. The soil can also be watered with captan at 125 g a.i./100 litres *or* with Cheshunt compound at 310 g copper/100 litres *or* with zineb at 140 g a.i./100 litres. On nurseries etridiazole† may be used either as a powder which is raked into the soil before seeding or planting or, using the same rate of a.i. per unit area, as a drench in 5·5 litres water/m² (or sufficient water to thoroughly saturate the soil), repeating the application every 4–8 weeks as necessary. On sandy and light loam soils use 175 g a.i./100 m²; on loamy soils 245 g a.i./100 m²; on soils with appreciable clay content 350 g a.i./100 m². Etridazole may also be incorporated into compost at 26 g a.i./m³. Drazoxolon can be used as a soil drench at 19 g a.i./100 litres.

The soil may also be sterilized with dazomet following the manufacturer's instructions carefully. Apply evenly from April–October when the soil temperature is at least 7°C at a depth of 150 mm to the soil surface at 380 kg prill/ha to very light to medium soils, 570 kg prill/ha to silts, heavy soils, peats and high organic soils. Incorporate the dazomet thoroughly into the soil to a depth of 180–200 mm and seal the surface with polythene or by rolling and leave for at least 10 days, moistening the soil to prevent the surface cracking. Then cultivate the soil to aid release of the methyl isothiocyanate formed and ensure freedom from phytotoxic residues by the cress germination test (Chapter 10, p. 253) before planting.

SNOWDROP

Pests

Large narcissus fly (*Merodon equestris*)—see Narcissus (p. 357).
  Stem eelworm (*Ditylenchus dipsaci*)—see Narcissus (p. 358).
  Tulip bulb aphid (*Dysaphis tulipae*)—see Tulip (p. 373).

Diseases

**Grey mould (*Botrytis galanthina*). The leaves and flower stalks are soon destroyed by a grey mould while the sclerotia formed on the bulbs contaminate the soil. Destroy infected clumps and disinfect the soil by mixing quintozene dust (20 per cent) in the soil at 14 g a.i./m². Spray the remaining healthy clumps with captan at 100 g a.i./100 litres. Benomyl at 25 g a.i./100 litres *or* thiophanate-methyl at 50 g a.i./100 litres *or* thiram at 160 g a.i./100 litres could also be tried.

SORBUS

Diseases

**Fire blight (*Erwinia amylovora*)—see Chapter 9, p. 206.

SPRUCE

Pests

**Conifer spinning mite (*Oligonychus ununguis*). These reddish orange mites with black markings are found among fine webbing on young shoots together with round reddish orange eggs. They suck the sap of the foliage causing it to assume an unhealthy greyish or yellowish brown appearance. Damage is seen from May onwards, and may be reduced by spraying in late May or when damage is first seen, with demeton-S-methyl 22 g a.i./100 litres *or* malathion 190 g a.i./100 litres *or* oxydemeton-methyl 22 g a.i./100 litres, repeating the application 2–3 weeks later. Other organophosphorus compounds used to control fruit tree red spider mite (see p. 185) may also be effective. If regular spraying is to be carried out, other types of chemicals should be alternated with the above to avoid a possible build-up of resistant strains of mite. For example, dicofol *or* quinomethionate *or* tetradifon, though there are no data on the efficiency of these materials against this mite.

    ***Green spruce aphid (*Elatobium abietinum*). This small red-eyed greenfly feeds on the undersides of the needles of spruce throughout the year except for a short period in summer. *Picea sitchensis* and *P. pungens* var. *glauca* are outstandingly susceptible but few spruce species are immune. Sucking of the sap causes yellow mottling and loss of all but the current year's foliage. Damage is associated with mild winters and normally occurs in early spring, and also from October onwards when autumn and winter are particularly mild. Control by spraying in March/April or when the aphid is first noticed, using malathion at 112 g a.i./100 litres.

    **Spruce Gall Adelgids (*Adelges abietis* and *A. viridis*) are tiny dark insects which spend the winter near the buds and cause distinctive 'pineapple' galls to be formed on the new shoots. Damage may seriously affect the form and appearance of the tree, and may be prevented, in most seasons, by spraying before April with gamma-HCH at 25 g a.i./100 litres, with added wetter.

Diseases

***Damping-off of seedlings** (mainly *Pythium* spp.), causes damage in seedbeds and occurs in 2 phases of seedling development. In the pre-emergence phase the germinating seed is attacked below soil level, leaving bare patches in the nursery and, in the post-emergence phase, the seedling collapses due to attack at soil level, often whilst in the cotyledon stage. To control the disease a post-emergence drench of captan *or* drazoxolon *or* etridiazole† should be applied as for seedlings (p. 369). If damping-off is a recurrent problem the nursery beds should be sterilized with formaldehyde at 6·7 litres 38–40 per cent formalin/100 litres water/ 18 m² of seedbed at least 3 weeks before sowing *or* with dazomet as for seedlings (p. 369).

STOCK

Diseases

Clubroot (*Plasmodiophora brassicae*) (A.L. 276). Irregular swellings develop on the roots and the plants remain stunted. For long-term control, lime the soil using 2·5–5·0 tonne hydrated lime/ha. For immediate control in affected beds, rake in calomel dust at 2 g a.i./m² *or* dip the roots when planting in a 'sludge' made of calomel dust at 100 g a.i. and a little clay/litre water.

Downy mildew (*Peronospora parasitica*). The leaves develop yellow blotches which on the underside show a greyish furry coating and the plants may be crippled. It is important to keep plants grown for seed free from this disease by spraying with chlorothalonil at 110 g a.i./100 litres *or* zineb at 175 g a.i./100 litres.

STRANVAESIA

Diseases

Fireblight (*Erwinia amylovora*)—see Chapter 9, p. 206.

SWEET PEA

Pests

***Aphids,** e.g. peach-potato aphid (*Myzus persicae*)—control as for aphids on Dahlia (p. 344) but do not use malathion.

SWEET WILLIAM

Diseases

**Leaf spot (*Didymellina dianthi*) appears as pale coloured circular spots on the leaves and can soon destroy them. Do not grow plants too soft. Spray, at first sign of the disease, with Bordeaux mixture equivalent to 150 g copper/100 litres *or* with copper oxychloride at 37 g copper/100 litres, repeating as required.

*Rust (*Puccinia arenariae*). Brown pustules appear arranged roughly in rings on the leaves. In autumn this disease may spread rapidly and cause great destruction. Do not encourage soft lush growth which is easily attacked and spray, at the first sign of the disease, with a copper-containing spray as given for Leaf spot above *or* with thiram at 320 g a.i./100 litres *or* zineb at 175 g a.i./100 litres. Oxycarboxin could also be tried at 62–125 g a.i./100 litres.

SYCAMORE

See Acer, p. 336.

SYRINGA

See Lilac, p. 353.

THUJA

Pests

*Conifer spinning mite (*Oligonychus ununguis*)—see Spruce, p. 370.

Diseases

***Needle blight (*Didymascella* (*Keithia*) *thujina*) (F.C.L. 43). This disease can cause serious damage to *Thuja plicata* in the nursery, particularly in transplant lines, and occasionally attacks *T. occidentalis*. Individual needles turn brown and later small dark brown spots about 1 mm across appear on the upper surface of the needle. In a heavy attack most of the needles may become infected on the lower shoots and considerable dieback occurs. (These symptoms should not be confused with winter bronzing, which is a natural occurrence during severe winters, when the whole plant turns reddish brown recovering its natural colour in the following spring). Excellent control has been achieved using cyclohex-imide with ferrous sulphate (according to manufacturer's instructions), towards the end of March and a month later in the east and south of England; and in late March, late April and mid-June in the west of England.

TULIP

Pests

*Bulb mite (*Rhizoglyphus echinopus*)—see Narcissus (p. 356).

***Stem eelworm (*Ditylenchus dipsaci*) (A.L. 461). Infested plants show fine cracks and pale or purplish streaks on the flower stem. In severe infestations the stem is distorted, split or swollen and the flowers may show pale or green streaking. On forced plants the leaves may also show pale streaks and tendency to split. Infested bulbs may have grey or brownish patches just above the basal plate. Thorough rogueing, strict attention to hygiene and care in cultural practices and in buying in new stocks will help to limit eelworm attacks. Hot water treatment of the bulbs has not proved satisfactory but, as an alternative, a cold dip during July or August in thionazin at 230 g a.i./100 litres for 2 hours will give good control.

**Tulip bulb aphid (*Dysaphis tulipae*). The yellowish or brownish aphids feed under the dry outer scales of stored bulbs and reduce the latters' vitality. When the bulbs are planted the insects attack the young shoots, severely checking their growth and causing distortion. Infested bulbs should be immersed for 15 minutes in a solution of gamma-HCH 25 g a.i./100 litres water to which a wetter has been added *or* fumigated with nicotine *or* gamma-HCH. On growing plants spray as for aphids on Dahlia (see p. 344). For other aphids, e.g. peach-potato aphid (*Myzus persicae*) and potato aphid (*Macrosiphum euphorbiae*), the control is as for aphids on Dahlia.

Diseases

***Fire disease (*Botrytis tulipae*) (A.L. 536) is a most common and serious disease of tulips and is carried over by resting bodies (sclerotia) in the soil or on the bulbs. Young leaves are scorched brown at their tips and young shoots crippled. Older leaves show small spots and the flower petals are spoiled by similar spots. In some cases the shoot is injured at a leaf axil and the flower stem falls over. To control the disease mix quintozene dust with the top soil at planting time at $13 \cdot 5$ g a.i./m² (135 kg a.i./ha) *or* the bulbs can be dusted with this powder at 460 g a.i./tonne of bulbs to reduce underground infection of the shoots. The growing plant should be sprayed in spring from the time they reach 25 mm high with dichlofluanid at 150 g a.i./100 litres *or* mancozeb/zineb at 200 g product/45–270 litres *or* maneb at $1 \cdot 8$ kg a.i./ha *or* thiram at $2 \cdot 8$ kg a.i./ha *or* zineb at 175 g a.i. + spreader/100 litres *or* zineb— poly(ethylenethiuram disulphide) complex at $1 \cdot 8$ kg a.i./ha. This disease can also be controlled by dipping the bulbs for 15 minutes in an aqueous suspension of benomyl at 100 g a.i./100 litres *or* of carbendazim at 100 g a.i./100 litres; *or* for 60 minutes in an organomercury compound at 7·5 g mercury/100 litres.

***Fusarium bulb rot** (*Fusarium oxysporum* f. *tulipae*) causes a dry rot of bulbs, commencing at the basal plate and spreading upwards rapidly. Affected tissues become brown and coated with pinkish fungal growth. Some control of the disease can be achieved by dipping the bulbs in benomyl *or* carbendazim as for Fire disease above. Dipping in thiabendazole 120 g a.i./100 litres is also said to give a good control.

Grey bulb rot (*Rhizoctonia tuliparum* syn. *Sclerotium tuliparum*)—Unlike Fire disease (see above) which is serious on the foliage, this disease attacks the growing points and usually prevents the shoot getting far above ground. The nose portion is affected by a greyish rot and the bulb is soon destroyed. Resting bodies (sclerotia) are formed on the rotting portions and released into the soil where they are scattered by cultivation and will infect other bulbs the next season. The result is that bare patches are seen in the beds. Soil treatment is with quintozene at 13·5 g a.i./m² (135 kg a.i./ha). Dipping as for Fire disease above may also help to control grey bulb rot.

TURF

Pests

Chafer grubs (*Phyllopertha horticola*, etc.) are whitish fleshy grubs with brown heads and curved bodies. *P. horticola* grows to about 12 mm long but other species up to 40 mm long may also cause damage. They feed on the grass roots causing withered patches to appear. For control, use insecticides as for leatherjackets (see p. 374).

Earthworms (*Allolobophora nocturna* and *A. longa*). These species are responsible for depositing casts on the surface of turf, particularly during the spring and autumn, making it unsightly and causing unevenness. The turf should be dressed with carbaryl at 3·65 kg a.i./ha *or* derris at 3·4 kg rotenone/ha watered well in; *or* water with chlordane at 14 kg a.i./ha *or* potassium permanganate solution at a rate of 17 g a.i./5·5 litres/m². All except chlordane bring the worms to the surface where they should be swept up. Of these materials only potassium permanganate should be used if there is risk of the contamination of ponds or water containing fish.

Leatherjackets (*Tipula paludosa, T. oleracea*, etc.). The greyish-brown or greyish-black grubs with tough wrinkled skin, feed on the roots of the grass causing it to wither. Damage is more severe if the autumn of the previous year has been wet. Water with DDT 1·05 kg a.i./ha *or* apply a dressing of DDT 0·88 kg a.i./ha *or* gamma-HCH 0·28 kg a.i./60 kg bran/ha. Treatment is most successful if given in mild, humid weather in autumn or spring.

Diseases

Dollar spot (*Sclerotinia homoeocarpa*). Small brown patches about 50 mm across, at first brown and then rather bleached, appear. The disease can be controlled by the calomel treatment recommended for Red thread (p. 375).

***Fairy rings** (*Marasmius oreades* and other fungi). In summer the ring is composed of 3 zones, an outer and an inner in both of which the grass is green and vigorous while between there is a middle zone where the grasses are brown and killed by the fairy ring fungus. The fruit bodies of the fungi appear in summer and autumn especially in wet weather and towards the outer parts of the ring. This type of disease is often difficult to control because the fungus forms so dense a mat of mycelium in the soil that it is difficult to get a fluid to penetrate the affected area; the mycelium can be present to a depth of up to 0·45 m in some soils. It is usually necessary to strip the turf from the affected zones and this should be destroyed. The soil beneath should be forked or 'spiked' and then soaked with a fungicide; it has been suggested that corrosive sublimate (56 g/100 litres) could be used but formaldehyde (410 ml 38–40 per cent formalin plus 34 ml non-ionic wetting agent/5·5 litres water/m^2) has also been recommended.

Ophiobolus patch (*Ophiobolus graminis*). This disease shows as depressed circular patches a few inches in diameter in which the grass has a light straw-coloured or bronzed appearance. Affected areas may increase in size and coalesce to form large irregularly shaped patches. As the disease is usually restricted to the bents (*Agrostis* spp.), the centres of the patches become colonized by undesirable grasses and weeds. The disease may be confused with Snow mould (see below) but it is normally found only where the soil surface has a high pH and in wet conditions. Microscopic examination of the diseased grasses is required for confirmation of the disease. It can be controlled by applying chlorothalonil at 100 g a.i./100 m^2 *or* phenylmercury acetate at 4·2 g mercury/100 m^2.

Red thread (*Corticium fuciforme*). This fungus causes pinkish patches from 75–125 mm in diameter on the turf and these are composed of dead grass with coral-red horn-like fungus growths showing amongst it. The disease can be controlled by applying benomyl at 60 g a.i./11 litres/100 m^2 *or* chlorothalonil at 100 g a.i./100 m^2 *or* phenylmercury acetate at 4·2 g mercury/100 m^2 *or* thiabendazole at 42 g a.i. 22 litres/100 m^2 *or* thiophanate-methyl at 50 g a.i./11 litres/100 m^2.

Seedling diseases (*Pythium* spp., *Fusarium* spp. and *Helminthosporium* spp.). Various fungi can cause newly-sown seed to rot or prevent the emergence of seedlings and the same fungi can also cause a collapse of seedlings at ground level once they have emerged. To prevent trouble in newly-sown turf, care should be taken in the preparation of the seed bed. In addition the seed can be treated with captan seed dressing at 1·7 g a.i./kg *or* thiram seed dressing at 1·65 g a.i./kg seed.

Snow mould or **Fusarium patch (*Fusarium nivale*). Large brown patches covered with a whitish fungal mycelium appear on lawns especially after a snowfall. Water the affected patches with calomel at 100 g a.i./160 litres/100 m² of turf, repeating at monthly intervals. Addition of wetter is advisable. A mixture of 61 g calomel and 30·5 g corrosive sublimate/100 m² may also be used. This disease can also be controlled by applying benomyl at 92 g a.i./11 litres/100 m² *or* chlorothalonil at 200 g a.i./100 m² *or* phenylmercury acetate at 2·1 g mercury/100 m² *or* quintozene dust at 120 g a.i./100 m²/month (drench w.p. at 150 g a.i./11 litres/100 m²/month) *or* thiabendazole at 42 g a.i./22 litres/100 m² *or* thiophanate-methyl at 100 g a.i./11 litres/100 m². It is really desirable to use chemical control as a preventive rather than a curative measure and the first application of fungicide should be given in autumn repeating at monthly intervals during conditions favouring fungal infection.

VIBURNUM

Pests

**Aphids, e.g. Bean aphid (*Aphis fabae*) and viburnum aphid (*A. viburni*). Both these species can occur on the young foliage during the late spring, causing severe leaf curl. *Viburnum carlesii* is particularly susceptible. Control by spraying at first sign of damage with one of the chemicals given for rose aphids (p. 367).

**Viburnum whitefly (*Aleurotrachelus jelinekii*). Only occurs on *Viburnum tinus*. Adult whiteflies are present July to early September. The pupae which are black with a white waxy fringe can be found on the underside of the leaves during most months of the year. Both adults and larvae feed by sucking sap and the leaves may become fouled with honeydew and sooty mould. For controls see rhododendron whitefly (p. 366).

Diseases

**Honey fungus (*Armillariella mellea* syn. *Armillaria mellea*)—see Privet (p. 364).

VIOLA

See Pansy, p. 360.

VIOLET

Pests

**Leaf and bud eelworm (*Aphelenchoides fragariae*) causes few definite symptoms apart from malformation of leaves and poor growth generally. Infested stock should be destroyed and the land allowed to remain fallow over the winter.

Red spider mite (*Tetranychus urticae*)—see Chapter 10, p. 264.

Violet leaf midge (*Dasyneura affinis*). The small, whitish or pale orange maggots feed on the leaves which are rolled inwards towards the mid-rib and usually greatly swollen. When badly infested the plant is dwarfed and flower production reduced or prevented altogether. The control of the midge is difficult but applications of DDT 2·1 kg a.i./ha in early May, early July, mid-August and mid-October will reduce the numbers of egg-laying adults.

WALLFLOWER

Pests

Cabbage root fly (*Erioischia brassicae*). The white maggots eat away the lateral roots and tunnel in the main stem causing wilting and death of the plant. For possible controls, see Chapter 11, p. 319.

Caterpillars, e.g. **angleshades moth** (*Phlogophora meticulosa*)—see under Dahlia (p. 345).

Diseases

Clubroot (*Plasmodiophora brassicae*) (A.L. 276). The roots develop irregular swellings and the plants remain stunted. Control using the treatments given under Stock (p. 371).

Downy mildew (*Peronospora parasitica*). Yellow blotches develop on the leaves and on the under surfaces a greyish furry coating can be seen. Affected plants may be crippled. Plants grown for seed should be sprayed with zineb at 175 g a.i./100 litres.

WILLOW

Pests

Aphids, e.g. willow stem aphid (*Tuberolachnus salignus*), large dark brown aphids which feed on shoots. Spring or summer infestations are controlled by demeton-S-methyl 22 g a.i./100 litres *or* dimethoate 30 g a.i./100 litres *or* gamma-HCH 12·5 g a.i./100 litres *or* malathion 112 g a.i./100 litres *or* oxydemeton-methyl 22 g a.i./100 litres.

Willow ermine moth (*Yponomeuta rorrella*). The small grey black-headed caterpillars of this moth feed in colonies on the foliage. The stems and leaves are enclosed in a web 'tent' which protects the caterpillars as they feed. Their control is as for small ermine moths on fruit (p. 188). For the control of other caterpillars, including lackey moth (*Malacosoma* neustria), tortrix moths (*Cacoecia* spp. and *Pandemis cerasana*) vapourer moth (*Orgyia antiqua*) and winter moths (*Operophtera brumata, Alsophila aescularia,* etc.)—see Chapter 9, p. 189.

Willow scale (*Chionaspis salicis*). The whitish scales encrust the stems and may be controlled by spraying in spring and summer with diazinon 40 g a.i./100 litres *or* malathion 187 g a.i./100 litres.

Diseases

Anthracnose (*Marssonina salicicola*) appears as small brown spots on leaves and small blackish cankers on shoots of various *Salix* spp. The leaves fall prematurely and should be raked up and burnt. On large trees control is often not possible and feeding to encourage vigour and resistance may be tried. Where possible, cutting out badly affected shoots and spraying with Bordeaux mixture at 150 g copper/100 litres *or* copper oxychloride at 37 g copper/100 litres is advised, 1 application being given as the leaves unfold and at least 1 more in summer, *or* spray with quinomethionate at 25 g a.i./100 litres applied at 2-week intervals from early March. Dodine is also said to control this disease.

Honey fungus (*Armillariella mellea* syn. *Armillaria mellea*)—see Privet, p. 364.

Pests

Yew scale (*Parthenolecanium pomeranicum*). The oval, very convex, chestnut-brown scales on the stems, cover the foliage with honeydew on which grow sooty moulds. Severe infestations may cause bronzing of the foliage and defoliation. Spring and summer infestations are controlled as for Scale on Willow (p. 378).

Diseases

Leaf spot (*Coniothyrium concentricum*). The large brown well-defined spots with grey centres caused by this disease often show small dots in concentric rings on their surfaces. These appear on the leaves and spoil the appearance. Remove severely attacked leaves and spray with Bordeaux mixture at 150 g copper/100 litres *or* copper oxychloride at 37 g copper/100 litres to protect young developing leaves.

ZINNIA

Diseases

***Grey mould** (*Botrytis cinerea*). Even outdoors, zinnias are likely in wet seasons to suffer from rotting of the flowers and even collapse through stem rot caused by grey mould. To control the disease, remove all dead and dying leaves regularly. Benomyl, applied at 25 g a.i./100 litres *or* chlorothalonil at 110 g a.i./100 litres *or* thiophanate-methyl at 50 g a.i./100 litres may prove effective in controlling this disease.

Seedling blight (*Alternaria zinniae*). This disease is seed-borne and causes reddish brown spots with greyish centres on the leaves and dark brown canker-like areas on the stems. The seed should be treated with an organomercury *or*, preferably, a captan seed dressing using 3·4 g a.i./kg seed. Even in the boxes the seedlings may need to be protected by spraying with Bordeaux mixture at 150 g copper/100 litres *or* copper oxychloride at 37 g copper/100 litres *or* zineb at 140 g a.i./100 litres.

ZONAL PELARGONIUM

See Pelargonium, p. 361.

Chapter 13
Pests and diseases
of forest crops

Commercial forest crops have, relatively speaking, a very low per acre value, and the forester tends therefore to rely on the cultural and ecological approach to his pest problems rather than the chemical one. Traditionally he has reserved insecticide and fungicide application as a means of saving his trees from death, on the basis that the cost of treatment could only be justified when his actual capital (or growing stock) is at risk. The same consideration does not apply, however, in forest nurseries where the per acre value can be very high indeed and losses correspondingly costly. There is also the special case of Norway spruce which is often planted out at high density for early thinning to supply the Christmas tree market and for this reason can economically carry comparatively high costs of treatment. Many forest trees are, of course, grown as ornamentals but, as such, will be judged by the yardstick of amenity rather than that of timber producer (see Chapter 12). Here, prescriptions for control have been strictly confined to those which apply to normal forest practice.

Routine chemical prophylactic treatments are seldom considered appropriate in forestry, the two outstanding exceptions to this rule being the measures taken against the fungus *Fomes annosus* and the weevil *Hylobius abietis*. The former organism can be regarded as the most important enemy of forest trees in Britain, not only on account of its ubiquity and for the continuing losses it causes, but for its increasing potential as a serious fungal pest in second and subsequent rotations. There are no known means of eradicating the disease once forest land has become infected and protective measures against invasion should be carried out after all thinning and felling operations.

It may be noted that the range of chemicals recommended against any one pest in this chapter may be rather more restricted than those given in others. This is often due to the choice having been limited on grounds of risks to workers; forests tend to be in isolated areas and distant from aid; they are also rough in terrain and provide conditions where men sweat and are inclined to strip off protective clothing. In general then, less toxic materials are chosen when possible.

It is seldom possible to give recommendations for rates of application, other than in nurseries, due to variations in age and therefore size of plants. The practice of spot-treating rather than applying to the whole area further confuses the issue, as does variation in planting density. A general injunction to 'visibly wet' foliage when using a manually operated sprayer with fine nozzle is a useful guide with most materials, although definitely not a rule which should be applied with motorized knapsack mist-blowers. In this case rates can only be controlled by the speed of walking down the rows of trees, a single pass from one side being made in the case of small trees and a double pass from two sides in the case of larger ones.

13.1   Forest nurseries

Pests

Aphids. A number of aphid species feed on forest nursery stock but only a few ever require measures to be taken against them.

*Beech woolly aphis (*Phyllaphis fagi*). A small yellow to green aphis which feeds on the underside of beech leaves in May to July, causing them to wilt, turn brown and fall. Control may be obtained by spraying in spring with malathion at $1\cdot26$ kg a.i./ha HV.

*Cherry blackfly (*Myzus cerasi*)—see p. 210.

*Oak Phylloxera (*Phylloxera glabra*). An orange-yellow aphis which feeds on the underside of oak leaves from May to July, causing the leaves to wilt and turn a dirty mottled grey-brown. For control, spray in spring with malathion at $1\cdot26$ kg a.i./ha.

***Green spruce aphis (*Elatobium abietinum*). This small, red-eyed, greenfly feeds on the undersides of the needles of spruce throughout the year except for a short period in summer. *Picea sitchensis* and *P. pungens* var. *glauca* are outstandingly susceptible but few spruce species entirely escape attack. Sucking of the needle sap causes yellow mottling and loss of all but current year's foliage. Damage is associated with mild conditions and may occur from autumn to spring and early summer. In districts known to be prone to attack early preventative treatment should be considered. For control, spray when the aphis is noted, or, in March/April with malathion at $1\cdot26$ kg a.i./ha.

*Pine root aphis (*Prociphilus* spp.). A small greyish white aphis which covers itself and the root system of pine transplants with waxy wool. Severe infestation can cause yellowing, wilting and even death of individual trees. Infected plants are best burnt and the ground they occupied spot-treated with gamma-HCH $0\cdot5$ per cent dust.

***Conifer Woolly aphids (*Adelges* spp.)—see p. 386.

*Beech jassid (*Typhocyba cruenta* var. *douglasi*). A yellow aphis-like insect which jumps like a flea on disturbance. It causes yellowing of the foliage between May and July and is particularly common where beech hedges are growing in the nursery area. For control spray in spring with malathion at $1\cdot26$ kg a.i./ha.

*Chafer. Chafers are soil-inhabiting beetle larvae of the May or June bug (*Melolontha melolontha*), the Garden chafer (*Phyllopertha horticola*) and of the Brown chafer (*Serica brunnea*). Damage takes place throughout the summer months. Attack first becomes obvious through the sudden wilting and browning of individual plants in the bed. When these are examined it will be found they can be removed from the ground with the slightest effort, the whole root system having been severed about 5 mm or so below ground level. Damage is most common in heavier soils, particularly where there are surrounding broadleaved hedges or woodlands; conversely it is rare in heathland nurseries.

No control is possible to the growing crop though ground known to be infested can be treated prior to use by working 0·5 per cent gamma-HCH dust into the top 100 mm of soil at 625–950 g a.i./ha.

**Conifer spinning mite (*Oligonychus* (*Paratetranychus*) *ununguis*) is tiny, being barely visible to the naked eye in both egg and adult stages. The egg is orange-red and may be found, with the aid of a hand lens, on the wood of the past few years' growth. Sitka and Norway spruce are particularly prone to attack, Scots pine less so. The young and adult stages are brown-coloured and are found on the needles and shoots amongst fine silk. Damage to transplants through their sap-sucking appears as a dirty yellowing, and continues throughout the period spring to autumn. In severe cases, treatment should be carried out when the overwintering eggs have hatched in late May, by spraying dicofol at 420 g a.i./ha HV.

**Cutworms. Cutworms are caterpillars mainly of the Turnip moth (*Agrotis segetum*), the Heart and dart moth (*A. exclamationis*) and the Yellow underwing (*Noctua pronuba*) which inhabit the soil during daylight hours. Damage to seedlings takes place above ground at night, and begins in July. It reaches a peak in early autumn and ceases as temperatures fall in September or October. A short spell of intense feeding again occurs in February to May, depending on weather conditions, until the insects are fully fed and ready to pupate. The attack on conifer seedlings is quite characteristic; the caterpillars bite through the stem, usually just above soil level, and either leave the remains lying on the ground, or pull them down into the soil a short distance, leaving a portion still above ground. As soon as damage appears, spray the seedbeds with gamma-HCH at 1·12 kg a.i./ha *or* DDT at 1·05 kg a.i./ha.

**Poplar leaf beetles. *Phyllodecta vitellinae* and *P. vulgatissima* are shiny, metallic blue or green beetles some 5 mm long. Their larvae are dirty grey-bown in colour and are black dotted. They feed, as a group, skeletonizing the underside of the leaves of most poplar species. The adults of *Chrysomela populi* are about 12 mm long and are brick-red in colour, the larvae are cream with black dots and are similar to those of the Colorado beetle in general appearance. This species completely skeletonizes leaves, particularly of *Populus tremula,* and also eats holes out of them. All species can be found from May throughout the summer, and are particularly damaging in poplar stool beds. For control, apply DDT at 1·05 kg a.i./ha HV.

*Sawflies. The main species are *Trichiocampus viminalis,* the Birch sawfly (*Croesus septentrionalis*), Large willow sawfly (*Nematus salicis*) and *N. melanaspis.* The larvae of these insects display a great variety of coloration at different stages of development and between species. Their feeding habits include skeletonization and biting of leaves. One or other may be found from May until October. For control, apply DDT at 1·05 kg a.i./ha HV.

***Springtails. The springtail, *Bourletiella hortensis,* is an occasional severe pest of conifer seedlings and particularly of *Pinus contorta.* The insect is no more than the size of a speck of dust but becomes quite noticeable when disturbed through its curious flea-like jumping habit. It is most active in the first few hours of dawn, between the end of April to mid-June. Damage is to the emerging seedlings and appears as minute lesions on hypocotyl and cotyledons. The effect in the first year of growth is to produce a dumpy little seedling with a bush of distorted, fattened, needles. In the second year the damaged seedling grows on to produce a multileadered useless plant. For control apply malathion HV at 1·26 kg a.i./ha of seedbed, on first appearance of the insect (which may be before seed germination) and at 2–3 week intervals thereafter until germination and emergence has finished.

Weevils. A number of weevil species can be most damaging both to conifer and hardwood nursery stock. This is particularly so when nurseries are surrounded by woodland in which populations may first increase and later invade the nursery. Other ground vegetation, such as heather or grass may also provide the primary breeding ground.

**Barypithes araneiformis and *B. pellucidus* are two very small weevils, about 3 mm long and dark in colour. The adult beetles feed at night between the end of April to mid-July. Death of emergent seedlings results from feeding upon, and severing of, the hypocotyl and from ringbarking of later stage seedlings. Control may be obtained by spraying gamma-HCH at 420 g a.i./ha HV, or DDT at 1·05 kg a.i./ha HV.

***Clay-coloured weevil (*Otiorrhynchus singularis*) and *Strophosomus melanogrammus,* are two speckled brown weevils which attack a great variety of tree species, particularly when the nursery is surrounded by trees. The latter weevil is very common on larch, and Western hemlock; *Tsuga heterophylla* seems particularly susceptible to both. The adults are active from April until October, and it is this stage only which is damaging through feeding on the fine shoots and needles.

Spray when damage first appears, in April-May, with DDT at 1·05 kg a.i./ha HV, or gamma-HCH at 280 g a.i./ha HV.

**Otiorrhynchus ovatus. This weevil is damaging both in the adult and larval stages. The adult is about 5 mm long, with chestnut-brown body and darker head. It is active from April-May onwards and feeds at night on the finer shoots and needles of most conifers. The larva, which is indistinguishable from that of other weevils, feeds on finer roots throughout the summer months.

Control measures can be taken against both adult and larval stages. Against adults a spray of DDT at 1·05 kg a.i./ha HV should be applied in April-May or when the insects first appear; against larvae, soil known to be infested should be treated before use by working in 0·5 per cent gamma-HCH dust into the top 100 mm of soil at the rate of 625–950 g a.i./ha.

Diseases

***Damping-off** (mainly *Pythium* spp.). Damping-off causes damage in seedbeds and occurs in two phases of seedling development. In the pre-emergence phase, the germinating seed is attacked below soil level, leaving bare patches in the nursery; in the post-emergent phase, the seedling collapses due to attack at soil level, often whilst in the cotyledon stage.

A post-emergence drench of captan at 6·8 g a.i./5·5 litres water/m² of seedbed should be applied immediately symptoms are observed. If damping-off is a recurrent problem the nursery beds should be sterilized with formaldehyde at 7·7 ml *c.* 40 per cent formalin/litres water at the rate of 5·5 litres/m² *or* dazomet at 38 g a.i./m². Label instructions must be followed to ensure that no harmful residues remain at sowing time.

***Grey mould** (*Botrytis cinerea*) (Forestry Commission Leaflet No. 50). A sparse web of grey mycelium appears over affected portions of plants often in damp weather or after they have been weakened or killed by frost, or other predisposing factors. *Sequoia* and *Cupressus* are particularly susceptible to Grey mould but many conifer species are attacked in seedbeds, or, less frequently, in transplant lines.

Spray with Bordeaux mixture, equivalent to 2·9 kg metallic copper/ha HV, immediately the disease is seen and repeat if necessary at intervals of 10–14 days. Thiram at 3·4 kg a.i./ha HV, or captan at 1·1 kg a.i./ha HV, can also be used.

Leaf cast of larch (*Meria laricis*) (Forestry Commission Leaflet No. 21). The first sign of this disease is a discoloration or browning of the middle or outer end of the needles in early May. It may be confused with frost damage which tends to kill whole needles at once, and of course takes place immediately after the frost. The needles attacked by the disease fall prematurely after becoming brown whilst those killed by frost remain attached longer. The disease is most serious in plants which have stood over either in seedbeds or transplant lines.

For control, spray with either wettable or colloidal sulphur, equivalent to 1·7 kg sulphur/ha HV, at flushing time and at 2–3 week intervals until the end of July.

Needle cast of pine (*Lophodermium pinastri*) causes the death of needles on pine seedlings and transplants particularly when a pine nursery is established close to a pine plantation. Pine should not be sown in nurseries close to mature pine plantations but, where this is unavoidable and infection appears, zineb or maneb should be applied at the rate of 2·9 kg a.i./ha HV. Three annual sprays are recommended at monthly intervals at the end of July, August and September. Spraying should begin earlier if signs of infection are present.

***Needle blight of Western red cedar** (*Didymascella* (*Keithia*) *thujina* F.C.L. No. 43). This disease can cause serious damage to Western red cedar in the nursery, particularly in transplant lines. Individual needles or fronds turn brown and, at a later stage, dark brown spots appear on the upper surface of the needle. In a heavy attack the majority of the needles may become infected on the lower fronds and considerable dieback occurs.

These symptoms should not be confused with winter bronzing, which is a natural occurrence during severe winters, when the whole plant turns reddish-brown, recovering its natural colour in the following spring.

Good control may be achieved by the application of cycloheximide at 85 g a.i./1000 litres HV. In the drier easterly and south-easterly parts of the UK, two applications are recommended, one at the end of March and the second a month later. In wetter westerly areas, 3 applications are required, 1 in late March, 1 in late April and the third in mid-June.

Oak mildew (*Microsphaera alphitoides*). During the summer the leaves and young shoots are covered with the powdery white spores of this fungus. This disease can be controlled in the nursery by the use of wettable or colloidial sulphur, equivalent to 1·7 kg sulphur/ha HV, as soon as it appears.

13.2   Forest plantations

Pests

YOUNGER PLANTATIONS

Aphids. Forest plantation trees are well endowed with aphid species, but there is seldom economic justification for their chemical control. The obvious exceptions to this rule are those species which cause degrade of Christmas trees (*Picea abies*).

Green spruce aphis (*Elatobium abietinum*) (see under Nursery above). Control of this aphis is seldom considered necessary on a forest scale since, though some loss in growth increment will be associated with defoliation, recovery is good. Such defoliation may not, however, be tolerable to the Christmas tree grower in which case the control measures given under Forest Nurseries above should be applied.

Spruce shoot aphis (*Cinara pinicola* syn. *Cinaropsis pilicornis*) is a grey or brown colonial aphis which infests the new shoots of spruce in June-August. Except for slight discoloration of the foliage even large populations appear not to harm the trees directly but their copious honeydew production and the subsequent sooty moulds which grow upon this exudation can cause Christmas trees to be unsaleable at the end of the year. For control apply malathion at 1·26 kg/ha HV.

***Conifer Woolly aphids** (*Adelges* spp.) (F.C.L. No. 7 and Bulletin No. 42). These aphids have extremely complex life cycles with up to five forms often involving alternation between a spruce host and one other conifer which may be larch, Douglas fir, a pine or a Silver fir. Most stages secrete waxy wool and can cause yellowing and needle loss. To the forester only *Adelges nordmann ianae* (or one of its close allies now recognized as separate species) can cause plant death and is of sufficient economic importance to deserve control. The control method described, however, would be effective against most of the other species. Treatment must be carried out in mild conditions between November and end of February. Spray with gamma-HCH at 560 g a.i./ha HV. DNOC and Tar oil sprays have also proved effective as dormant season treatments.

On the spruce host, those *Adelges* species which produce the characteristic pineapple gall are of importance to the Christmas tree grower only (see p. 370). The control is as above but timely spraying is essential to prevent gall formation.

Beech Scale (*Cryptococcus fagi*) occurs on the boles of infected trees (F.C.L. No. 15). Adults are 0·5−1·0 mm long and pale yellow in colour. The adults and immature stages secrete a white waxy wool which is most noticeable from June−August. Eggs are present from July until November. A fungus (*Nectria coccinea*) causing death of bark may be associated with heavy infestations, a condition known as Beach bark disease (Forestry Commission Record No. 96). Winter washes applied in the dormant season are effective as are the recommendations against Mussel Scale on apple, Chapter 9, p. 187. Such HV applications may be useful in an ornamental tree context but would seldom be feasible under forest conditions. Diazinon applied LV in the growing season against adults in May-June at 8 g a.i./pole-sized tree, shows promise of good control.

***Black pine beetles** (*Hylastes* spp.) (F.C.L. No. 58). The most common species to be found on pine are *Hylastes ater* in the south and *H. brunneus* in the north, and on spruce, *H. cunicularius*. These are bark beetles which share the breeding sites of the Pine weevil (see below), and normally commence breeding on stumps some 6−8 months after felling. They are small, black, and less than 5 mm long. The damage caused by the adults of *Hylastes* spp. is to the root collar and main roots of young transplants which are ring-barked and killed. For control, see below under Pine weevil, p. 388.

*Conifer spinning mite** (*Oligonychus (Paratetranychus) ununguis*) (see under Forest Nurseries above). Infestation in the field is more common in the south than the north, and is associated most often with Norway spruce grown as Christmas trees in sites dry for this tree species. For control, see under Forest Nurseries above.

*Pine or **Fox-coloured sawfly** (*Neodiprion sertifer*) (F.C.L. No. 35) is a grey-green colonial feeding larva often with diffuse darker strips running longitudinally, and about 25 mm long. This species is very common on pines in May and June, feeding on all needles except those of the current year. The species is under excellent natural control by a virus which will normally decimate a population some 2 or 3 years after the first attack. *Pinus contorta* is particularly susceptible and when the species is grown in remote areas, where natural virus infection may be delayed, some artificial control action may become necessary. The virus suspension is unfortunately not commercially available but excellent control may be obtained by DDT at 1 kg a.i./1000 litres HV *or* gamma-HCH at 250 g a.i./1000 litres HV.

*Pine shoot moth (*Rhyacionia buoliana*) (F.C.L. No. 40) is a small caterpillar without marked features to distinguish it from near relatives. Single insects mine into buds and new developing shoots. The shoots are not completely severed but flop over and grow on, forming a large kink in the main axis (traditionally likened to a German posthorn!) Normally in Britain the proportion of stems affected is small and the economic impact is seldom great enough to warrant control measures. For control use DDT at 2·5 kg a.i./1000 litres HV *or* fenitrothion 1·1 kg a.i./1000 litres HV applied at the end of March to early April. Correct timing is important: check for the new 'resin tents' and spray within 2 weeks of the first seen.

*Spruce bell moth or **Streaked pine bell** (*Epinotia tedella*) is a small moth whose caterpillar is barely distinguishable from many other members of its family. During August to October the caterpillars mine the needles of spruce and spin them, together with excrement, on to the shoot. These clusters of brown spun needles are later removed by wind and rain, leaving bald patches on the branches. Severe infestation can cause complete defoliation. Recovery from attack is good but the plant is unsaleable by the Christmas tree grower. For control use DDT at 1·0 kg a.i./1000 litres HV *or* fenitrothion at 1·1 kg a.i./1000 litres HV.

Weevils. This group of insects is responsible for large losses to new transplants, and prophylactic measures against them, particularly in felled-over areas, are essential if adequate final stocking is to be assured.

***Clay-coloured weevil (*Otiorrhynchus singularis*) and *Strophosomus melanogrammus*. See under Forest Nurseries above for details and control. The chemical dipping and post-planting spray treatments described below for control of Pine weevil will also give protection against these species.

*Leaf weevils (*Phyllobius argentatus*, *P. pyri*, etc.) are some 5 mm or less long, often bright green in colour or a dark speckled brown. Damage is caused by adults feeding on the leaves of a wide range of hardwoods and sometimes also conifers. They are most numerous in June and July. For control apply DDT 1·25 kg a.i./1000 litres HV *or* gamma-HCH at 250 g a.i./1000 litres HV.

***Pine weevil** (*Hylobius abietis*) (F.C.L. No. 58) is a large black weevil speckled with cream coloured scales, some 15 mm long. It breeds in conifer stumps, attacking fairly soon after felling, and takes, on average, some 15–18 months to develop in the UK. It will also breed in fresh-felled lying logs under that part of the bark in contact with the ground. After emergence the young adults feed on green-barked material, such as that provided by the main stem of young transplants, before themselves seeking a breeding site. The transplants are ring-barked and killed, and the whole crop may easily be lost in this way if prophylactic measures are not taken.

Treatment against both Pine weevil and Black pine beetle may be carried out in 1 operation: (1) Prior to planting. By complete dipping in a stable water-based suspension of finely ground gamma-HCH at 16 g a.i./litres. Control of the Pine weevil alone may also be effected by dipping tops only in DDT 50 g a.i./litres. (2) After planting. If attack develops after planting, spot treatment should be carried out with DDT 10 g a.i./litres.

***Winter moths.** Mainly *Operophtera brumata,* Northern winter moth (*O. fagata*) and Mottled umber (*Erannis defoliaria*), are all looper caterpillars and may often be found in joint attack. The first 2 are basically green larvae and both have longitudinal stripes, the last is variable in colour but is usually reddish brown above and yellowish below and on the sides. They feed early in the year, in the south from April on to June and in the north into July. These species are well known pests of hardwoods and orchard trees but can also be damaging to conifers, particularly the spruces. Conifers underplanted beneath poor hardwood are often severely defoliated by caterpillars originating in the crowns of the overstorey. There have also been a number of cases in north Scotland in which the first 2 species have occurred in outbreak densities, feeding on Sitka spruce and upon the heather natural vegetation. For control, use DDT at 1 kg a.i./1000 litres HV.

OLDER PLANTATIONS

***Bark beetles.** There are many species of bark beetle attacking dead, dying or unhealthy trees and felled timber, but few of them will damage live, established tree crops. There are 2 important British exceptions, the Pine shoot beetle, *Tomicus* (*Myelophilus*) *piniperda,* on pine (F.C.L. No. 3), and the Larch shoot beetle, *Ips cembrae,* on larch. Both these small, brown and black, beetles, some 5 mm long, bore into the shoots of their respective tree hosts, pruning the crowns and causing multiple leaders and reduced growth. After feeding in this fashion they require thick-barked stem material for breeding. In the case of *T. piniperda* this may be logs felled within the previous 6 months or trees heavily defoliated, or weakened by fire, fungal attack, etc. Successful primary attack upon healthy trees for breeding purposes is rare but can occur in conditions marginal for the tree species, such as sand dunes. *I. cembrae*, so far confined to east Scotland, is potentially a rather more dangerous pest as it seems that a primary attack can take place on apparently healthy larch. Breeding commences in spring, often in March, reaches a peak in May-June and falls off in July. A second generation is rare in Britain.

The basic control of these species lies in management of fellings so that all suitable breeding material is either removed from the forest or has the bark removed within 6 weeks, during the period from April to August in the case of *T. piniperda* and April to September for *I. cembrae*. This is not always possible when calamities, e.g. large scale windblow, occur.

Chemical control is only economic if timber is first controlled in sizeable stacks. It is best done either as a log protection treatment immediately before the attack can take place, or as early as possible after initiation of attack in order to reach the beetles before their progeny have eaten their way deep under the bark. Gamma-HCH at 5 kg a.i./1000 litres is made up in diesel oil, kerosene or gas oil and sprayed to treat not only the log surfaces but obtain maximum penetration into the stack. The dosage in litres is calculated on the superficial area (S.A.) of the stack (i.e. S.A. = Area in m^2 of 2 sides, 2 ends and top). Dosage = 0·7 litres/m^2.

Douglas fir seed wasp (*Megastigmus spermatrophus*) (F.C.L. No. 8). This small, yellow insect attacks the immature seed of Douglas fir and can cause serious losses in crop yields. The adult attacks the seed throughout the month of June and protective measures must be taken to cover this period. For this purpose, apply malathion 5 kg a.i./1000 litres HV in 3 applications at intervals of 10 days.

***Pine looper** or **Bordered white moth** (*Bupalus piniarius*) (F.C.L. No. 32) is a pine needle-feeding looper caterpillar green in colour with a number of white longitudinal stripes, and some 30 mm long. The moth is on the wing in June and July and the larvae may be found from mid-June until October or even later. The damage, at its worst, may be complete defoliation, and subsequent secondary attack by Bark beetles can cause the death of trees. Infestation typically occurs in regions of low rainfall (500–650 mm), on sandy soils, and in large areas of pure pine plantations. Control is only practicable by aerial application. Tetrachlorvinphos at 0·56 kg a.i./ha has proven effective, as has DDT in e.c. formulations at 0·56–1·12 kg a.i./ha. Fenitrothion and phosalone are also active against the insect.

FELLED PRODUCE

***Ambrosia beetle** or **Pinhole borer** (*Trypodendron lineatum*). This small black beetle, about 3 mm long, bores deep into the wood of conifers, particularly that of spruce and larch. Typically it is most harmful in the wetter western parts of Britain. The adult attacks from the end of April until mid-July. Timber felled during the period November to January of the previous year is most susceptible, and that in the period February-March rather less so. The bore tunnels become blackened with fungi and the damage is essentially of a technical nature.

Chemical control is only economic if the timber is first collected into sizeable stacks. It is carried out as a protective measure immediately prior to the start of the attack period in mid-April. Dosage and technique of spraying as for Bark beetles (see under Older Plantations, p. 388), but the gamma-HCH is diluted with water.

*Bark beetles. The felled produce of most tree species may accommodate 1 of several species of Bark beetle. The loss of bark resulting from attack may lead to accelerated drying and shake or cracking of the timber. For control see under Older Plantations, Bark beetles, p. 388.

***Elm bark beetles (*Scolytus scolytus* and *S. multistriatus*). These beetles are the vectors of Dutch elm disease and prevention of their successful entry, breeding and emergence is a key factor in disease control. The beetles can breed in trees or logs up to 2 years after death or felling. Removal and burning of the bark is the main and only sure control. Where this operation is not practicable spraying of the bark with gamma-HCH is a useful alternative.

For the protection of uninfested logs gamma-HCH at 5 g a.i./litres must be applied. In autumn to March treatments the insecticide should be diluted in diesel oil or paraffin, but for spring and summer applications water will provide adequate persistence of deposit.

To reach beetles in logs already infested single logs should be sprayed with gamma-HCH at 2·5 g a.i./litres diluted in diesel oil or paraffin.

All treatments should be applied to near run-off (approximately 0·45 litres/m² bark surface).

For fungicidal control of Dutch elm disease, see also p. 346.

Diseases

***Fomes (*Fomes annosus*) (F.C.L. No. 5). *Fomes* root and butt rot is the most serious disease of both young and mature conifers in Britain. The fungus may kill pines particularly on dry alkaline soils with pH 6 and above. In other species, young trees may also be killed, but more commonly the fungus rots the tree roots (as in Douglas fir) or proceeds from the roots into the stem, where it produces an extensive butt rot.

Fructifications of *Fomes* often appear at the base of trees killed by the disease or on stumps of previously infected ones. The upper surface of the growing fructification is reddish brown with a white margin and the under-surface is white and pierced by fine pores. Immature fructifications are commonly found on infected stumps or roots; these are small white pustules often bearing half-formed pores surfaces.

The fungus enters the crop by means of airborne spores from *Fomes* fructifications that infect exposed woody tissues of conifers, particularly cut stump surfaces immediately after thinning or clear-felling. The fungus grows into the stump tissue and down into the root system, and if the roots of healthy trees are in contact with such infected tissues, they in turn become infected.

Control is based on the protection of cut stump surfaces, thus preventing infection by airborne spores. Stump protectants should be applied immediately after all thinning and felling operations.

Urea is the standard material used for stump protection, and it is used as a solution (20 g urea/litres) with the addition of a marker dye. The solution is liberally applied to the freshly cut stump surface immediately after felling. The dye is included to assist in complete application and to ensure that the treatment has been carried out. Urea is much safer than sodium nitrite which was previously used for stump protection.

In pine plantations only, there is an alternative stump treatment. In this case, spore suspensions of a competing fungus, *Peniophora gigantea,* can be applied to the freshly cut stump surface. This method of biological control is very efficient in pine plantations but with no other species.

Advice on methods of stump protection can be obtained from the Forestry Commission.

If stump protectants can be applied throughout the life of a crop, it can be kept substantially free from *Fomes* and danger to subsequent crops is then much reduced. This is particularly important as it is very difficult to eradicate *Fomes* once it becomes established within a crop.

If an infected crop is felled and replanted with conifers, rapid infection of the new plants may take place from the stumps of the previously infection crop.

The control of *Fomes* in crops planted on infested sites can be achieved by the mechanical removal of the stumps of the previous crop, but this is an expensive operation only justified where the disease is severe. The use of *Peniophora gigantea* for stump treatments in pine areas already infected by *Fomes* may help to reduce infection in the subsequent crop.

***Dutch elm disease** (*Ceratocystis ulmi*) (F.C.L. No. 54). Although the elm is not normally a plantation tree, it seems appropriate to mention the possible use of fungicides for the control of Dutch elm disease in elms of high value. Research has shown that in young trees good control can be obtained by the injection of a solution of carbendazim hydrochloride, but so far the results have been less satisfactory on mature trees. The Pathology Branch of the Forestry Commission Research Station should be contacted for the latest information. For insecticide control of the elm bark beetles in logs see p. 390.

Appendix
Properties of insecticidal and fungicidal compounds

Common name: chemical name

Structure	Main uses	Persistence	Typical formulations	Incompatibility

Me = —CH₃   Et = —C₂H₅   Ph = —C₆H₅

Me = $-CH_3$ Et = $-C_2H_5$ Ph = $-C_6H_5$

A. Organophosphorus compounds:

1 Acephate: *OS*-dimethyl acetylphosphoramidothioate

Structure	Main uses	Persistence	Typical formulations	Incompatibility
$\begin{array}{c} MeS \\ MeO \end{array}\!\!\!\searrow\!\! PO\cdot NH\cdot CO\cdot Me$	Insecticide, briefly systemic	Moderate	s.c.	Alkaline pesticides

2 Azinphos-methyl: *S*-(3,4-dihydro-4-oxobenzo[*d*]-[1,2,3]-triazin-3-ylmethyl) *OO*-dimethyl phosphorodithioate

Structure	Main uses	Persistence	Typical formulations	Incompatibility
$(MeO)_2PS\cdot SCH_2-N$	Non-systemic broad-range insecticide and acaricide	Long	22% w/v e.c.	Alkaline pesticides

3 Bromophos: *O*-(4-bromo-2,5-dichlorophenyl) *OO*-dimethyl phosphorothioate

Structure	Main uses	Persistence	Typical formulations	Incompatibility
$(MeO)_2PS\cdot O$	Non-systemic broad-range insecticide	Moderate	e.c.; w.p.; granules	Alkaline pesticides

4 Carbophenothion: *S*-4-chlorophenylthiomethyl *OO*-diethyl phosphorodithioate

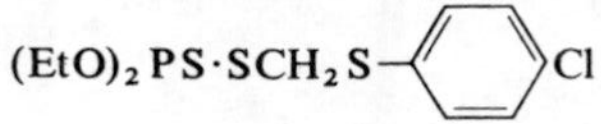

Non-systemic insecticide and acaricide — Long — Seed dressing

(EtO)$_2$PS·SCH$_2$S— —Cl

5 Chlorfenvinphos: 2-chloro-1-(2,4-dichlorophenylvinyl) diethyl phosphate

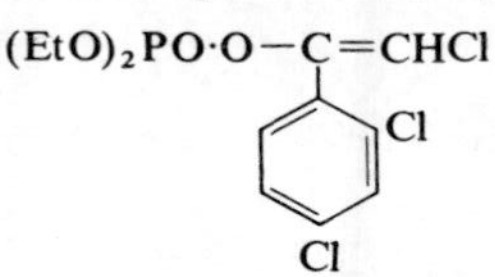

Non-systemic insecticide; soil and seed treatment — Moderate — 24% w/v e.c.; 10% w/w granules — —

(EtO)$_2$PO·O—C=CHCl

6 Chlorpyriphos: *OO*-diethyl *O*-3,5,6-trichloro-2-pyridyl phosphorothioate

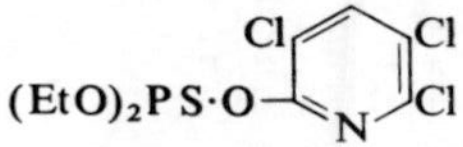

Non-systemic insecticide: rootflies — Moderate — 10% w/w granules — —

(EtO)$_2$PS·O—

7 Demephion: mixture of demephion-*O* (I) and demephion-*S* (II)

(I) (MeO)$_2$PS.OCH$_2$CH$_2$SMe
(II) (MeO)$_2$PO.SCH$_2$CH$_2$SMe

Systemic insecticide and acaricide; sap-feeding insects — Moderate — 30% w/v e.c. — Alkaline pesticides

8 Demeton-S-methyl: *S*-2-ethylthioethyl *OO*-dimethyl phosphorothioate

(MeO)$_2$PO.SCH$_2$CH$_2$SEt

Systemic insecticide and acaricide; sap-feeding insects — Moderate — 58% w/v e.c. — Alkaline pesticides

9 Demeton-S-methyl sulphone: *S*-2-ethylsulphonylethyl *OO*-dimethyl phosphorothioate

(MeO)$_2$PO.SCH$_2$CH$_2$SO$_2$Et

Systemic insecticide and acaricide; in combination with azinphos-methyl against pre-blossom complex on top fruit — — — 7·5% w/v + 25% w/w azinphos-methyl — see (2)

10 Diazinon: *OO*-diethyl *O*-2-isopropyl-6-methylpyrimidin-4-yl phosphorothioate

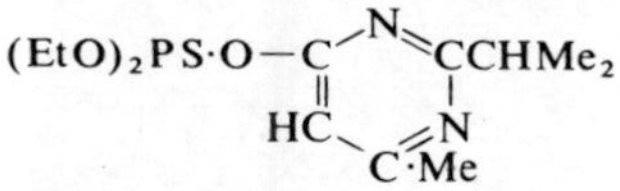

Non-systemic broad-range insecticide — Moderate — 40% w/w w.p. — Alkaline and copper pesticides

(EtO)$_2$PS·O—C

Common name: chemical name

Structure	Main uses	Persistence	Typical formulations	Incompatibility
11 Dichlorvos: 2.2-dichlorovinyl dimethyl phosphate $(MeO)_2PO.OCH:CCl_2$	Non-systemic insecticide and acaricide	Brief	50% w/v e.c.	Alkaline pesticides
12 Dimefox: tetramethylphosphorodiamidic fluoride $(Me_2N)_2POF$	Systemic insecticide; soil drench against hop aphid	Moderate	50% w/v e.c.	—
13 Dimethoate: OO-dimethyl S-methylcarbamoylmethyl phosphorodithioate $(MeO)_2PS.SCH_2CONHMe$	Systemic broad-range insecticide and acaricide	Moderate	40% w/v e.c.; 20% w/w w.p.	Alkaline pesticides
14 Dioxathion: 1,4-dioxan-2,3-diyl SS-di-(OO-diethyl phosphorodithioate) H_2C—O—CHSPS·(OEt)$_2$ H_2C—O—CHSPS·(OEt)$_2$	Non-systemic acaricide and insecticide	Moderate	50% w/v e.c.	Alkaline pesticides
15 Disulfoton: OO-diethyl S-2-ethylthioethyl phosphorodithioate $(EtO)_2PS.SCH_2CH_2SEt$	Systemic insecticide; soil pests	Moderate	7.5 and 10% w/w granules	—
16 Fenitrothion: OO-dimethyl O-4-nitro-m-tolyl phosphorothioate $(MeO)_2PS·O$—(ring with NO_2, Me)	Non-systemic broad-range insecticide	Moderate	50% w/v e.c.	Alkaline pesticides binapacryl etc.
17 Fonofos: O-ethyl S-phenyl ethylphosphonodithioate EtO—PS·SC$_6$H$_5$ Et	Non-systemic insecticide; soil pests	Moderate	10% w/w granules	—

18 Formothion: S-(N-formyl-N-methylcarbamoylmethyl) OO-dimethyl phosphorodithioate
(MeO)$_2$PS.SCH$_2$CO.N(Me)CHO
Systemic insecticide and acaricide;   Moderate   33% w/v e.c.   Alkaline pesticides
sap-feeding insects

19 Malathion: S-1,2-di(ethoxycarbonyl)ethyl OO-dimethyl phosphorodithioate
(MeO)$_2$PS.SCH.COOEt
|
CH$_2$COOEt
Non-systemic broad-range insecticide   Brief to moderate   60% w/v e.c.   Alkaline pesticides
and acaricide

20 Menazon: S-4,6-diamino-1,3,5-triazin-2-ylmethyl OO-dimethyl phosphorodithioate
Systemic aphicide   Moderate   40% w/v suspension   Alkaline pesticides

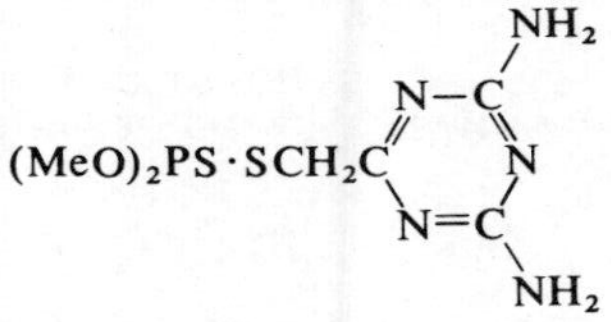

NH$_2$
|
N═C   N
| ‖
N   N═C
  \   /
   C
   |
   NH$_2$

(MeO)$_2$PS·SCH$_2$

21 Mephosfolan: diethyl 4-methyl-1,3-dithiolan-2-ylidenephosphoroamidate
Systemic insecticide; soil use   Moderate   25% w/v e.c.   Alkaline pesticides
against hop aphid

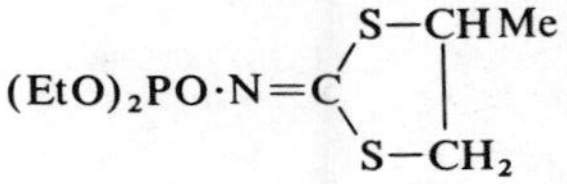

S—CHMe
/ |
(EtO)$_2$PO·N═C |
\ |
S—CH$_2$

22 Methidathion: S-2,3-dihydro-5-methoxy-2-oxo-1,3,4-thiadiazol-3-yl OO-dimethyl phosphorodithioate,
Non-systemic insecticide;   Moderate   40% w/v e.c.   Alkaline pesticides
aphid control

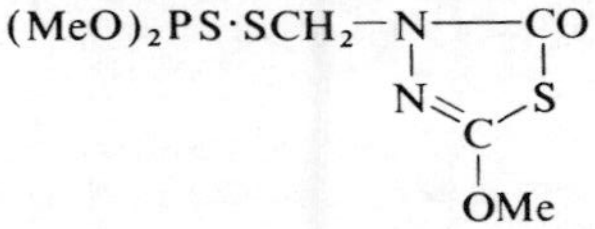

CO
/   \
N   S
| |
(MeO)$_2$PS·SCH$_2$—N   C
\   /
N
‖
C
|
OMe

23 Mevinphos: 2-methoxycarbonyl-1-methylvinyl dimethyl phosphate
(MeO)$_2$PO.OC(Me)═CH.COOMe
Systemic insecticide   Brief   Water soluble conc.   Alkaline pesticides
and acaricide

Common name: chemical name

Structure	Main uses	Persistence	Typical formulations	Incompatibility
24 Naled: 1,2-dibromo-2,2-dichloroethyl dimethyl phosphate $(MeO)_2PO.OCHBr.CBrCl_2$	Non-systemic insecticide and acaricide	Brief	50% w/v e.c.	—
25 Omethoate: OO-dimethyl S-methylcarbamoylmethyl phosphorothioate $(MeO)_2PO.SCH_2CONHMe$	Systemic insecticide and acaricide; hop aphid and wheat bulb fly	Moderate	57·5% w/v s.c.	Alkaline pesticides
26 Oxydemeton-methyl: S-2-ethylsulphinylethyl OO-dimethyl phosphorothioate $(MeO)_2PO.SCH_2CH_2SO.Et$	Systemic insecticide and acaricide; sap-feeding pests	Moderate	57% w/v e.c.	Alkaline pesticides
27 Parathion: OO-diethyl O-4-nitrophenyl phosphorothioate $(EtO)_2PS \cdot O$—⬡—NO_2	Non-systemic broad-range insecticide and acaricide	Moderate	e.c.'s of various a.i. contents	Alkaline pesticides
28 Phorate: OO-diethyl S-ethylthiomethyl phosphorodithioate $(EtO)_2PS.SCH_2SEt$	Systemic insecticide; soil use	Moderate	10% w/w granules	—
29 Phosalone: S-6-chloro-2-oxobenzoxazolin-3-ylmethyl OO-diethyl phosphorodithioate $(EtO)_2PS \cdot SCH_2$—N... Cl ...OC–O	Non-systemic insecticide and acaricide; sap-feeding pests	Moderate	33% w/v e.c.	Alkaline pesticides
30 Phosphamidon: 2-chloro-2-diethylcarbamoyl-1-methylvinyl dimethyl phosphate $(MeO)_2PO.O(MeC)\!=\!C(Cl)\text{-}CO.NEt_2$ (mixed isomers)	Systemic insecticide and acaricide	Brief	54% w/v s.c.	Alkaline pesticides

31 Pirimiphos-ethyl: *O*-2-diethylamino-6-methylpyrimidin-4-yl *OO*-diethyl phosphorothioate
Non-systemic insecticide: seed use — Moderate — 20% w/v + 7.5% w/v drazoxolon seed dressing — —

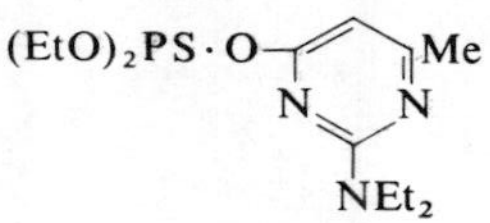

$(EtO)_2PS \cdot O$ — [pyrimidine ring: Me, N, N, NEt_2]

32 Pirimiphos-methyl: *O*-2-diethylamino-6-methylpyrimidin-4-yl *OO*-dimethyl phosphorothioate
Non-systemic insecticide and acaricide — Moderate — 50% w/v e.c. — Alkaline pesticides

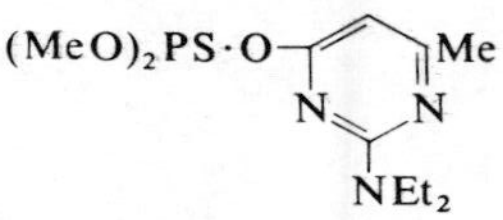

$(MeO)_2PS \cdot O$ — [pyrimidine ring: Me, N, N, NEt_2]

33 Pyrazophos: *O*-6-ethoxycarbonyl-5-methylpyrazolo[1,5-*a*]pyrimidin-2-yl *OO*-diethyl phosphorothioate
Systemic fungicide; powdery mildews — Moderate — 30% w/v e.c.; 30% w/w w.p. — Alkaline pesticides

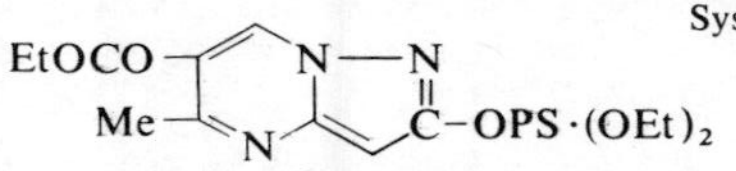

$EtOCO$ — [pyrazolopyrimidine ring] — $C - OPS \cdot (OEt)_2$, Me

34 Sulfotep: *OOO'O'*-tetraethyl dithiopyrophosphate
Non-systemic insecticidal fumigant — Brief — Smokes — —

$(EtO)_2PS.O.PS(OEt)_2$

35 Tetrachlorvinphos: (*Z*)-2-chloro-1-(2,4,5-trichlorophenyl)vinyl dimethyl phosphate
Non-systemic selective insecticide; diptera and lepidoptera — Brief — 75% w/w w.p. — —

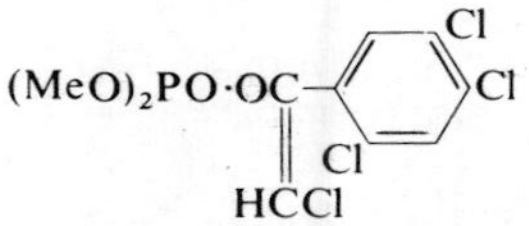

$(MeO)_2PO \cdot OC$ — [trichlorophenyl ring: Cl, Cl, Cl] — HCCl

36 Thiometon: *S*-2-ethylthioethyl *OO*-dimethyl phosphorodithioate
Systemic insecticide and acaricide: aphid control — Moderate — 25% w/v e.c. — Alkaline pesticides

$(MeO)_2PS.SCH_2CH_2SEt$

37 Thionazin: *OO*-diethyl *O*-pyrazin-2-yl phosphorothioate
Systemic insecticide and nematicide; soil and bulb treatment — Brief to moderate — 46% w/v e.c. — —

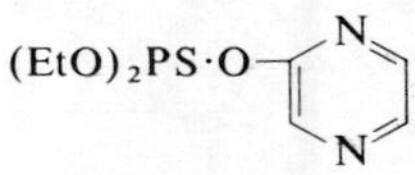

$(EtO)_2PS \cdot O$ — [pyrazine ring: N, N]

Common name: chemical name

Structure	Main uses	Persistence	Typical formulations	Incompatibility
38 Triazophos: *OO*-diethyl *O*-1-phenyl-1,2,4-triazol-3-yl phosphorothioate Ph—N⚌N N—OPS·(OEt)$_2$	Non-systemic insecticide and acaricide	Moderate	40% w/v e.c.	Alkaline pesticides
39 Trichlorphon: dimethyl 2,2,2-trichloro-1-hydroxyethylphosphonate (MeO)$_2$PO.CH(OH) CCl$_3$	Non-systemic insecticide; diptera and lepidoptera	Brief	80% w/v s.p.	Alkaline pesticides
40 Vamidothion: *OO*-dimethyl *S*-2-(1-methylcarbamoylethylthio)ethyl phosphorothioate (MeO)$_2$PO.SCH$_2$CH$_2$SC- (Me) H CO.NHMe	Systemic insecticide and acaricide; woolly aphid	Long	40% w/v e.c.	—
B. Chlorinated hydrocarbons				
41 Aldrin: 1,2,3,4,10,10-hexachloro-1,4,4a,5,8,8a-hexahydro-*exo*-1,4-*endo*-5,8-dimethanonaphthalene	Non-systemic insecticide	Long	e.c.'s of various a.i. content	—
42 Chlordane: 1,2,3,4,6,7,8,8-octachloro-3a,4,7,7a-tetrahydro-4,7-methanoindane	Non-systemic insecticide and worm killer	Long	20% w/v e.c.	—

43 DDT: 1,1,1-trichloro-2,2-di(4-chlorophenyl)ethane

Non-systemic insecticide   Long   e.c.'s of various a.i. content

44 Dichloropropene: 1,3-dichloropropene

$CH_2Cl.CH = CHCl$

Soil fumigant, nematicide   Brief   Alone or with dichloropropane

45 Dieldrin: 1,2,3,4,10,10-hexachloro-6,7-epoxy-1,4,4a,5,6,7,8,8a-octahydro-exo-1,4-endo-5,8-dimethanonaphthalene

Non-systemic insecticide   Long   15% w/v e.c.

46 Endosulfan: 6,7,8,9,10,10-hexachloro-1,5,5a,6,9,9a-hexahydro-6,9-methano-2,4,3-benzo[e]dioxathiepin-3-oxide

Non-systemic insecticide and acaricide   Long   20% and 35% w/v e.c.'s

47 Endrin: 1,2,3,4,10,10-hexachloro-6,7-epoxy-1,4,4a,5,6,7,8,8a-octahydro-exo-1,4-exo-5,8-dimethanonaphthalene

Non-systemic insecticide   Long   e.c.'s of various a.i. content

Common name: chemical name

Structure	Main uses	Persistence	Typical formulations	Incompatibility
48 Gamma-HCH: γ-1,2,3,4,5,6-hexachlorocyclohexane				
Non-systemic insecticide		Moderate to long	e.c.'s; w.p.'s; smokes; seed dressings	—
49 p-dichlorobenzene: 1,4-dichlorobenzene				
Insecticidal fumigant		Moderate	100% solid	—
C. Bridged diphenyl acaricides				
50 Dicofol: 2,2,2-trichloro-1,1-di(4-chlorophenyl)ethanol				
Non-systemic acaricide		Long	20% w/v e.c.	—
51 Tetradifon: 4-chlorophenyl 2,4,5-trichlorophenyl sulphone				
Non-systemic acaricide		Long	8% w/v e.c.	—

D. Carbamates

(a) Benzimidazole group

52 Benomyl: methyl 1-(butylcarbamoyl)benzimidazol-2-ylcarbamate — Systemic fungicide

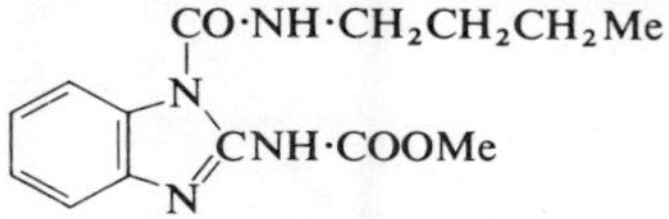

$$CO \cdot NH \cdot CH_2CH_2CH_2Me$$

Moderate   50% w/w w.p. |

53 Carbendazim: methyl benzimidazol-2-ylcarbamate — Systemic fungicide

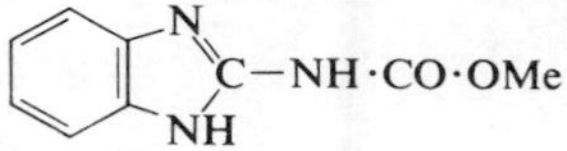

Moderate   50 and 60% w/w w.p.'s |

54 Thiabendazole: 2-(thiazol-4-yl)benzimidazole — Systemic fungicide

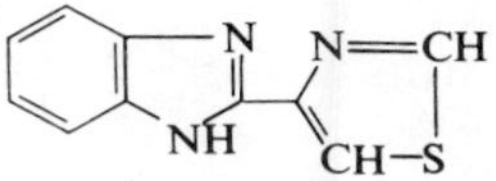

Moderate   42% w/v e.c.;   60% w/w w.p. |

55 Thiophanate-methyl: 1,2-di-(3-methoxycarbonyl-2-thioureido)benzene — Systemic fungicide

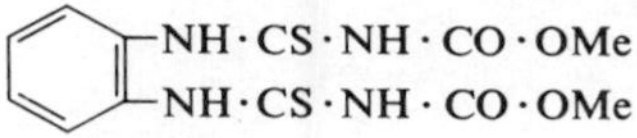

$$NH \cdot CS \cdot NH \cdot CO \cdot OMe$$
$$NH \cdot CS \cdot NH \cdot CO \cdot OMe$$

Moderate   50% w/w w.p. |

(b) Miscellaneous group

56 Aldicarb: 2-methyl-2-(methylthio)propionaldehyde O-methylcarbamoyloxime — Systemic insecticide and nematicide

$$Me$$
$$MeSCCH{:}NOCO.NHMe$$
$$Me$$

Moderate   10% w/w granules

Common name: chemical name

Structure	Main uses	Persistence	Typical formulations	Incompatibility
57 Carbaryl: 1-naphthyl methylcarbamate				
OCO·NH·Me	Non-systemic insecticide and worm killer	Moderate	85% w/w w.p.	—
58 Carboxin: 2,3-dihydro-6-methyl-5-phenylcarbamoyl-1,4-oxathiin				
H₂C—O—CMe / H₂C—S—C—CO·NHPh	Systemic fungicide: against smuts	Moderate	Seed dressing combined with organomercury compounds, gamma-HCH	—
59 Methiocarb: 4-methylthio-3,5-xylyl methylcarbamate				
Me OCO·NHMe / MeS Me	Non-systemic insecticide and molluscicide	Moderate	4% w/w pellets	—
60 Methomyl: 1-(methylthio)ethylideneamino methylcarbamate				
MeS\C=NOCO·NHMe / Me	Systemic insecticide; hop aphid control	Moderate	90% w/w s.p.	—
61 Oxamyl: NN-dimethyl-α-methylcarbamoyloxyimino-α-(methylthio)acetamide				
Me₂NCO.C = NOCO.NHMe / SMe	Systemic insecticide and nematicide	Moderate	10% w/w granules	—

62 Oxycarboxin: 2,3-dihydro-6-methyl-5-phenylcarbamoyl-1,4-oxathiin-4,4 dioxide

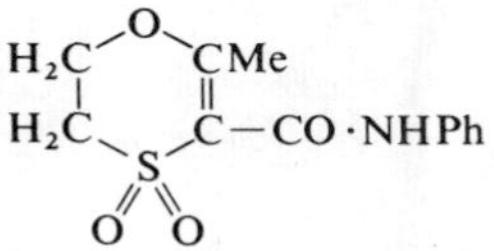

Systemic fungicide; against rusts

Moderate   75% w/w w.p.   —

H_2C—O—$\overset{CMe}{\underset{\parallel}{C}}$

H_2C—$\overset{C}{\underset{\parallel}{S}}$—CO·NHPh

O O

63 Pirimicarb: 2-dimethylamino-5,6-dimethylpyrimidin-4-yl dimethylcarbamate

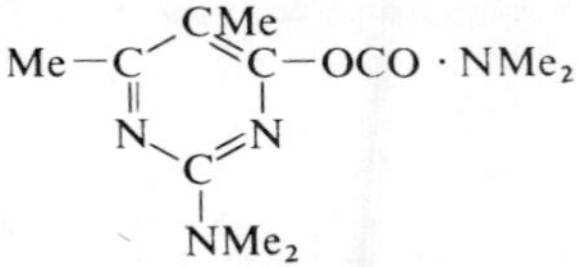

Non-systemic aphicide

Brief   50% w/w w.p.: smokes   —

Me—$\overset{CMe}{\underset{\parallel}{C}}$—C—OCO·$NMe_2$

N   N

C

NMe_2

64 Propoxur: 2-isopropoxyphenyl methylcarbamate

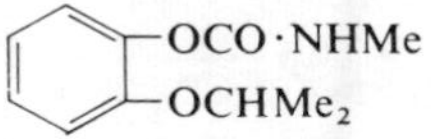

Non-systemic insecticide; hop aphid

Moderate   50% w/w w.p.; smokes   —

OCO·NHMe
OCHMe$_2$

E. Dithiocarbamates

65 Dazomet: tetrahydro-3,5-dimethyl-1,3,5-thiadiazine-2-thione

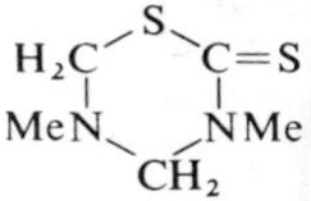

Soil fungicide and nematicide

Brief   98% w/w prill   —

H_2C—$\overset{S}{\underset{}{}}$—C=S

MeN   NMe

CH_2

66 Mancozeb: complex of zinc ion and maneb

Non-systemic fungicide; potato blight

Long   80% w/w w.p.: also with zineb   —

67 Maneb: manganese ethylenebisdithiocarbamate

|—SCS.NH.CH$_2$CH$_2$CS.SMn—|$_x$ Non-systemic fungicide

Long   80% w/w w.p.: also with zinc oxide, zineb or other metals   —

Common name: chemical name

Structure	Main uses	Persistence	Typical formulations	Incompatibility
68 Metham-sodium: sodium methyldithiocarbamate MeNH.CS.SNa	Soil fungicide and nematicide	Brief	33% w/v solution	—
69 Nabam: disodium ethylenebisdithiocarbamate CH_2.NH.CS.SNa | CH_2.NH.CS.SNa	Non-systemic fungicide	—	22% w/v solution, used with zinc sulphate	—
70 Propineb: zinc propylenebisdithiocarbamate |—SCS.NH.CH_2.CH(Me) NH.CS.SZn—$\rfloor_x$	Non-systemic fungicide; downy mildews and potato blight	Moderate	70% w/w w.p.	—
71 Thiram: tetramethylthiuram disulphide Me_2N.CS.S | Me_2N.CS.S	Non-systemic fungicide; foliage and seed use	Moderate	80% w/w w.p.	—
72 Zineb: zinc ethylenebisdithiocarbamate |—SCS.NH.CH_2.CH_2.NH.CS.SZn—$\rfloor_x$	Non-systemic fungicide; foliage use	Moderate	70% w/w w.p.	—

F. Dinitrophenyl compounds

73 Binapacryl: 2-*sec*-butyl-4,6-dinitrophenyl 3-methyl crotonate

Non-systemic acaricide and fungicide   Moderate   50% w/w suspension   —

$Me_2C{=}CH{\cdot}CO{\cdot}O$ — (dinitrophenyl ring with NO_2, NO_2, $Me{\cdot}CH{\cdot}Et$)

74 Dinobuton: 2-sec-butyl-4,6-dinitrophenyl isopropyl carbonate

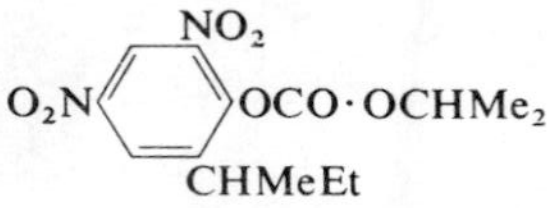

Non-systemic fungicide and acaricide — Moderate — 30% w/v e.c. — |

O_2N — NO_2 — $OCO \cdot OCHMe_2$

$CHMeEt$

75 Dinocap: mixture of 2,4-dinitro-6-octylphenyl crotonate (I) and 2,6-dinitro-4-octylphenyl crotonate (II)

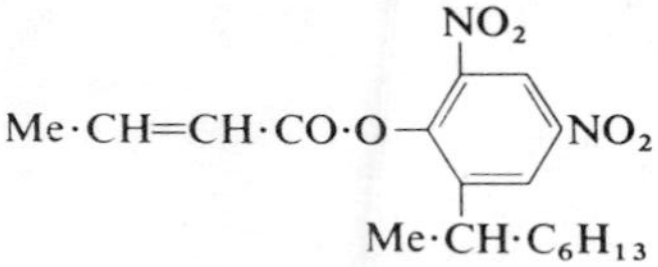

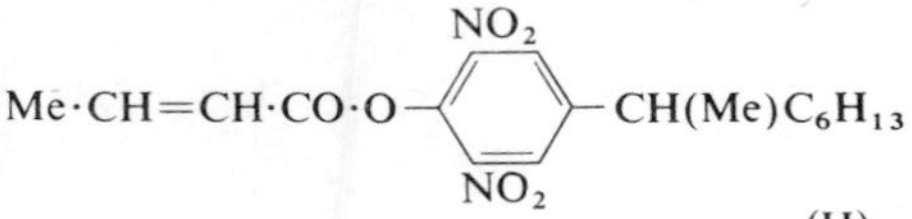

Non-systemic acaricide and fungicide — Moderate — 25% w/w w.p. — |

NO_2

NO_2 — $Me \cdot CH \cdot C_6H_{13}$

$Me \cdot CH = CH \cdot CO \cdot O$

(I)

$CH(Me)C_6H_{13}$

NO_2

$Me \cdot CH = CH \cdot CO \cdot O$ — NO_2

(II)

76 DNOC: 4,6-dinitro-o-cresol

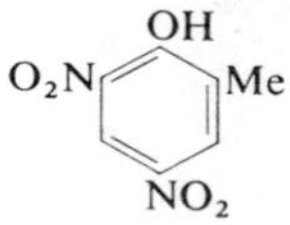

Non-systemic ovicide — Moderate — 1·35–3·5% w/v + 65–70% w/w petroleum oil s.e. — |

OH — Me

O_2N — NO_2

G. Chloronitrobenzenes

77 Quintozene: pentachloronitrobenzene

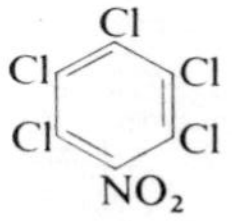

Non-systemic fungicide; seed and soil use — Long — 20% w/w dust — |

Cl — Cl — Cl — NO_2 — Cl — Cl

Common name: chemical name

Structure	Main uses	Persistence	Typical formulations	Incompatibility
78 Tecnazene: 1,2,4,5-tetrachloro-3-nitrobenzene				
(structure) Non-systemic fungicide		Long	3–6% w/w dust; smokes	—
H. Organotin compounds				
79 Cyhexatin: tricyclohexyltin hydroxide				
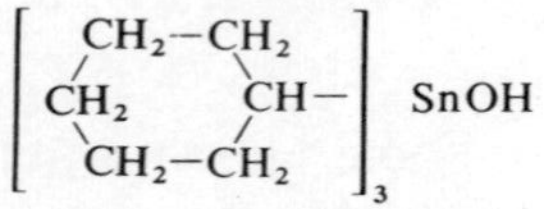 (structure)	Non-systemic acaricide	Long	25% w/w w.p.	—
80 Fentin acetate: triphenyltin acetate $Ph_3SnOCOMe$	Non-systemic fungicide; potato blight	Long	Mixture with maneb	—
81 Fentin hydroxide: triphenyltin hydroxide Ph_3SnOH	Non-systemic fungicide; potato blight	Long	20% w/w w.p.	—
I. Pyrethroids				
82 Pyrethrum: mixed esters of pyrethrolone and cinerolone with chrysathemic and pyrethric acids	Non-systemic contact insecticides	Moderate	Dusts and extracts	—

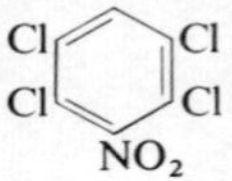

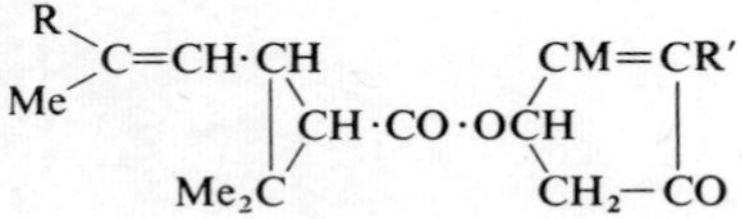

(structure for 82)

When R = —Me or —CO·OMe and
R' = —CH₂·CH=CH·CH=CH₂ or
—CH₂·CH=CH·Me

83 Resmethrin: 5-benzyl-3-furylmethyl chrysanthemate

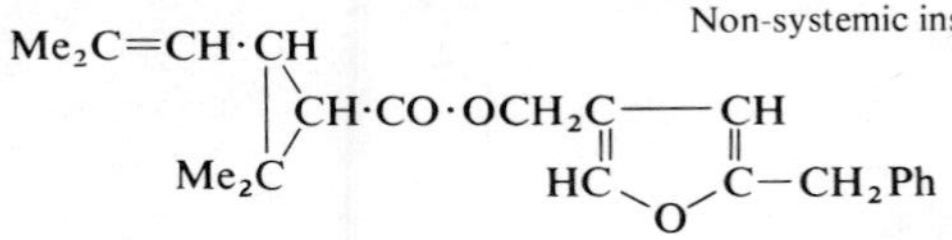

Non-systemic insecticide — Moderate — Aerosols, with or without other pyrethroids; e.c.'s with other pyrethroids — —

$$Me_2C=CH\cdot CH$$
$$CH\cdot CO\cdot OCH_2C=CH$$
$$Me_2C \qquad HC—O \qquad C—CH_2Ph$$

J. Miscellaneous compounds

84 2-aminobutane

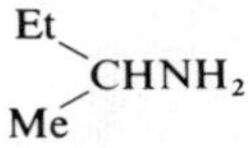

Fungicidal fumigant — Brief — — — —

$$Et—CHNH_2$$
$$Me$$

85 Benodanil: 2-iodobenzanilide

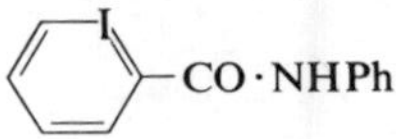

Systemic fungicide — Long — 50% w/w w.p. — —

I — CO·NHPh

86 Calomel: mercurous chloride Hg_2Cl_2

Non-systemic fungicide; soil use — Long — 4% w/w dust — —

87 Captafol: 3a,4,7,7a-tetrahydro-N-(1,1,2,2-tetrachloroethanesulphenyl)phthalimide

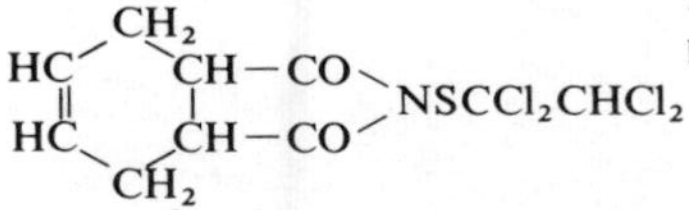

Non-systemic fungicide: potato blight — Long — 80% w/w w.p. — —

$$CH_2 \quad CH—CO$$
$$HC \qquad \qquad NSCCl_2CHCl_2$$
$$HC \qquad \qquad$$
$$CH_2 \quad CH—CO$$

Common name: chemical name

Structure	Main uses	Persistence	Typical formulations	Incompatibility
88 Captan: 3a,4,7,7a-tetrahydro-*N*-(trichloromethane phenyl)phthalimide				
HC—CH₂—CH—CO—NSCCl₃ CH—CO CH₂ (structure)	Non-systemic fungicide; foliage use	Long	50% w/w w.p.	—
89 Chloropicrin: trichloronitromethane CCl₃.NO₂	Insecticidal and nematicidal fumigant	Brief	For use only by trained personnel	—
90 Chlorothalonil: tetrachloroisophthalonitrile (structure with CN, Cl)	Non-systemic fungicide	Moderate	75% w/w w.p.; smokes	—
91 Copper sulphate: CuSO₄.5H₂O	As component of Bordeaux mixture	—	98% pure	—
92 Corrosive sublimate: mercuric chloride HgCl₂	Non-systemic fungicide and insecticide; soil use	Long	—	—
93 Cycloheximide: 4-{(2S)-[(1S,3S,5S)-(3,5-dimethyl-2-oxocyclohexyl)]-2-hydroxyethyl}piperidine-2,6-dione (structure)	Antibiotic	Moderate	s.c.	Alkaline pesticides; chlordane

h]benzopyran-6-(6a*H*)-one

| 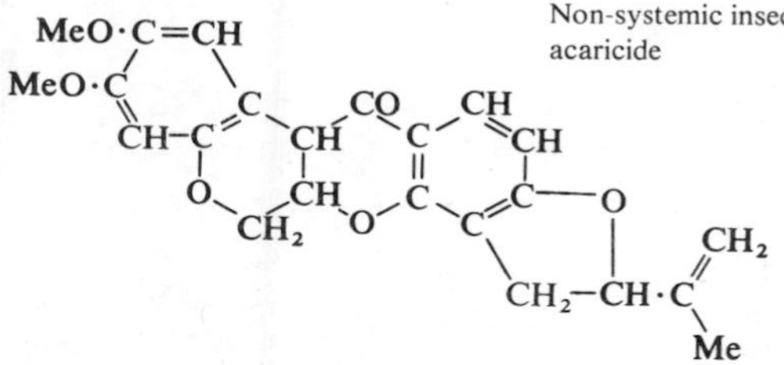 | Non-systemic insecticide and acaricide | Moderate | Ground root; extracts | — |

95 Dichlofluanid: *N*-dichlorofluoromethanesulphenyl-*N'N'*-dimethyl-*N*-phenylsulphamide

$Me_2NSO_2-N-SCCl_2F$
            |
            Ph

| 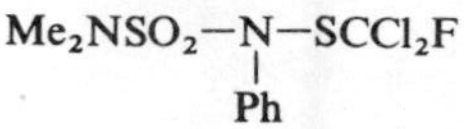 | Non-systemic fungicide; Botrytis | Moderate | 50% w/w w.p. | Alkaline pesticides |

96 Dicloran: 2,6-dichloro-4-nitroaniline

| 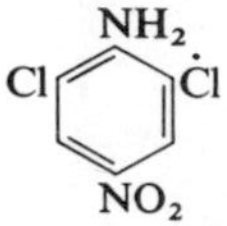 | Non-systemic fungicide | Moderate | 4% w/w dust | — |

97 Dimethirimol: 5-butyl-2-dimethylamino-6-methylpyrimidin-4-ol

| 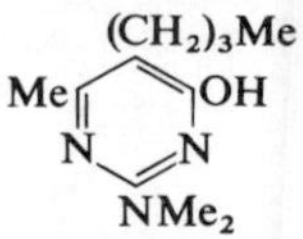 | Systemic fungicide; soil use | Moderate | 12·5% w/v e.c., as hydrochloride | — |

98 Dithianon: 2,3-dicyano-1,4-dithia-anthraquinone

| 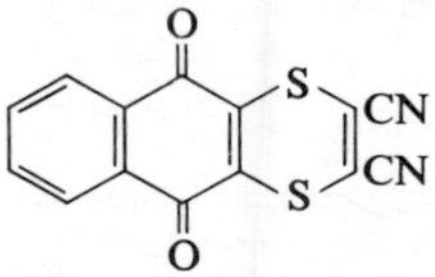 | Non-systemic fungicide; apple scab | Moderate | 75% w/w w.p. | Alkaline pesticides; oil sprays |

Common name: chemical name

Structure	Main uses	Persistence	Typical formulations	Incompatibility
99 Dodemorph: 4-cyclododecyl-2·6-dimethylmorpholine	Systemic fungicide; powdery mildews	Moderate	40% w/w e.c.; as acetate	Anionic wetters
100 Dodine: dodecylguanidine $C_{12}H_{25}NH.C.NH_2$ ‖ NH	Non-systemic fungicide; apple scab	Moderate	65% w/w w.p.; as acetate	Anionic wetters
101 Drazoxolon: 4-(2-chlorophenylhydrazono)-3-methyl-5-isoxazolone	Non-systemic fungicide; powdery mildews; seed treatment	Moderate	30% w/v aqueous suspension; seed dressing with pirimiphos-ethyl	Lime sulphur
102 Ethirimol: 5-butyl-2-ethylamino-6-methylpyrimidin-4-ol	Systemic fungicide; powdery mildews	Moderate	58% w/v aqueous suspension; seed dressing	—
103 Etridiazole: 5-ethoxy-3-trichloromethyl-1,2,4-thiadiazole	Non-systemic fungicide; turf	Moderate	35% w/w w.p.	—

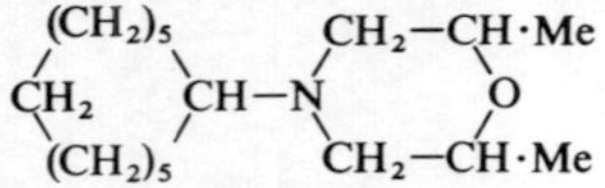

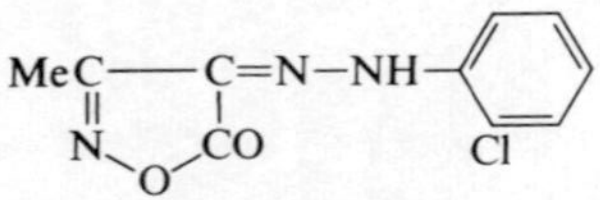

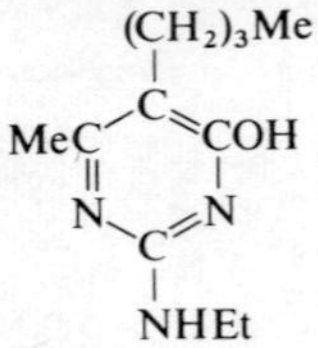

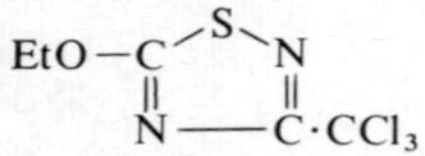

Structure 99:
$(CH_2)_5$ CH_2—CH·Me
           CH—N   O
$(CH_2)_5$ CH_2—CH·Me

Structure 101:
MeC——C=N—NH
   N   CO
    O
(phenyl ring with Cl)

Structure 102:
$(CH_2)_3$Me
MeC—C·COH
   N   N
MeC   C
   N
   NHEt

Structure 103:
EtO—C—S—N
        ‖
   N—C·CCl$_3$

No.	Name / formula	Type	Persistence	Notes	Incompatibility
104	Folpet: N-(trichloromethanesulphenyl)phthalimide	Non-systemic fungicide	Long	50% w/w w.p.	Alkaline pesticides
105	Lead arsenate: diplumbic hydrogen arsenate $PbHAsO_4$	Non-systemic stomach poison	Long	Paste or powder of specified content	Alkaline pesticides
106	Lime sulphur: calcium polysulphides $CaS.S_x$	Non-systemic fungicide and scalecide	Moderate	Not less than d 21°C 1·28	—
107	Mercuric oxide: HgO	Fungicide as paint for wounds on fruit trees	Long	—	—
108	Metaldehyde: 2,4,6,8-tetramethyl-1,3,5,7-tetroxocane	Molluscicide	Brief	2·5–6% w/w baits	
	MeCH—O—CH·Me / MeCH—O—CH·Me				
109	Methyl bromide: bromomethane MeBr	Insecticidal and nematicidal fumigant	Brief	For use only by trained personnel	—
110	Nicotine: 3-(1-methylpyrrolidin-2-yl)pyridine	Non-systemic contact insecticide	Brief	95–98% w/w concentrate	
111	Paris green: copper acetoarsenite $(MeCOO)_2Cu.3Cu(AsO_3)_2$	Insecticidal baits	Moderate	—	—

Common name: chemical name

Structure	Main uses	Persistence	Typical formulations	Incompatibility
112 Phenylmercury acetate PhHgOOCMe	Non-systemic fungicide	Moderate	—	—
113 2-phenylphenol (OH on biphenyl structure)	Fungicidal disinfectant; canker paint	Moderate	—	—
114 Quinomethionate: 6-methyl-2-oxo-1,3-dithiolo[4,5-b]quinoxaline (ring structure with Me)	Non-systemic acaricide and fungicide; powdery mildews	Moderate	25% w/w w.p., smokes	
115 Sulphur: S_x	Non-systemic fungicide and acaricide	Moderate	Dusts; w.p.; suspensions	—
116 Tridemorph: 2,6-dimethyl-4-tridecylmorpholine $C_{12}H_{25} \cdot CH_2 \cdot CH_2 - N$ (morpholine ring: CH$_2$—CHMe, O, CH$_2$—CHMe)	Systemic fungicide; powdery mildews	Moderate	75% w/v e.c.	—
117 Triforine: 1,4-di-(2,2,2-trichloro-1-formamidoethyl)-piperazine $Cl_3CCH \cdot NH \cdot CHO$ (piperazine ring) $Cl_3CCH \cdot NH \cdot CHO$	Systemic fungicide; powdery mildews	Moderate	20% w/v e.c.	

Index